SPRINGER
LAB MANUALS

Springer-Verlag Berlin Heidelberg GmbH

Bernd W. Rautenstrauß Thomas Liehr (Eds.)

FISH Technology

With 78 Figures, 44 in colour

Springer

BERND W. RAUTENSTRAUSS
Institut für Humangenetik
Universität Erlangen-Nürnberg
Schwabachanlage 10
91054 Erlangen
Germany

e-mail: berndwr@humgenet.uni-erlangen.de

THOMAS LIEHR
Institut für Humangenetik
Universität Jena
Kollegiengasse 10
077743 Jena
Germany

e-mail: j8lith@mti-n.mti.ini-jena.de

Library of Congress Cataloging-in-Publication Data
FISH technology / Bernd W. Rautenstrauss, Thomas Liehr (eds.)
p. cm. – (Springer lab manuals)
Includes bibliographical references.
ISBN 978-3-642-47739-3 ISBN 978-3-642-56404-8 (eBook)
DOI 10.1007/978-3-642-56404-8
1. Fluorescence in situ hybridization – Laboratory manuals. I. Rautenstrauss, Bernd W., 1959 - II. Liehr, Thomas, 1965- III. Springer lab manual.
QP519.9.F56.F574 2002
572.8'028'4–dc21

Originally published by Springer-Verlag Berlin Heidelberg New York in 2002
Softcover reprint of the hardcover 1st edition 2002
http://www.springer.de

Production: PRO EDIT GmbH, 69126 Heidelberg, Germany
Cover design: design & production GmbH, 69121 Heidelberg, Germany
Typesetting: Mitterweger + Partner, 68723 Plankstadt, Germany
SPIN 10674102 39/3130SO 5 4 3 2 1 0 – Printed on acid free paper

Preface

The fluorescence in situ hybridization (FISH) technique has nowadays become a powerful and versatile tool in clinical genetics, tumor cytogenetics and science. Invented in the 1970's there were many problems to overcome in working out the methodology. However, the technique is now established in many cytogenetic and other clinically orientated routine laboratories as well as in many research facilities all over the world.

This book is part of a series of labmanuals providing practical aspects of technologies used in biology, medicine and genetics. As in many cases published methods are incomplete or refer back to series of previously published papers, one goal of this labmanual for FISH is to provide a laboratory handbook that will facilitate research in connection with fluorescence in situ hybridization approaches.

The labmanual for FISH aims to provide a resource for both the novice scientist as well as for those more experienced investigators seeking to extend their technological base. Thus, apart from the basic technology and routine applications of FISH, also more sophisticated approaches have been included into this booklet, like e.g. Fiber FISH, Yeast FISH, FISH in combination with immunohistochemistry or stylish multicolor-FISH assays. Also future aspects of further developments of the field of molecular cytogenetics are covered. The chapters dealing with the recently developed FISH techniques centromere-specific multicolor FISH or the microdissection based multicolor banding method demonstrate, that the possibilities of this field still did not draw to a close.

We would like to express our gratitude to all those who have contributed to this booklet! Moreover, we are grateful to Springer-Verlag for asking us to put together these protocols of the state of the art, and latest advances in this exciting and inspiring field of molecular cytogenetics.

Jena, July 2001 — Thomas Liehr, PhD
Erlangen, July 2001 — Bernd Rautenstrauss, PhD

List of Contents

The Standard FISH Procedure – an Introduction

FISH on Native, Human Tissues

FISH on Archival, Human Tissues

Special FISH Techniques (Fiber-FISH, PNA-FISH, Co-FISH, NU-FISH)

FISH in Zoology and Virology

Multicolor FISH Applications

Not FISH but Almost

Part A

The Standard FISH Procedure – an Introduction

FISH Techniques, FISH Probes and Their Applications in Medicine and Biology – An Overview

EIGIL KJELDSEN and STEEN KØLVRAA

Introduction

FISH, as an acronym for fluorescence in situ hybridization, has been coined to describe this very powerful technique which allows direct visualization of genetic alterations on a cell by cell basis. Generally, the aim of the technique is to demonstrate either imbalances, in the form of gains or losses of segments of chromosome materials, or to demonstrate specific breakpoints with or without imbalance. Since fluorescence hybridization initially was introduced for chromosome classification (Pinkel et al. 1986) the technique has been adopted in an ever-increasing range of applications in both medicine and biology and almost 10,000 articles have now been published utilizing FISH technology. The present review will give an outline of the principles of this technique and its use in different areas of human disease and in research, with special emphasis on FISH as a diagnostic tool in chromosome and genome analysis within clinical genetics and oncology.

Historic perspectives

Although routine chromosome diagnostics utilizing conventional banding methods have been used during the last 3 decades, the first identification of chromosomes was much earlier (Table 1). Chromosomes were originally seen late in the last century and were described as heavily stained

---

✉ Eigil Kjeldsen, Aarhus University Hospital, Department of Clinical Genetics, Aarhus C, 8000, Denmark (*phone* +45 8949 4354; *fax* +45 8949 4370; *e-mail* eikjeld@nf.au.dk)
Steen Kølvraa, Aarhus University, Institute of Human Genetics, The Bartholin Building, Aarhus C, 8000, Denmark (*phone* +45-8942-1677; *fax* +45-8612-3173; *e-mail* steen@humgen.au.dk)

structures in cell preparations, hence the name (chromos for stain and soma for body). It took several decades, however, before the function of these cellular components was identified even though the chemical composition of the structures was known. It was not until Avery and co-workers in 1944 showed that DNA was the transforming factor (i.e. the factor carrying genetic information from one cell to another), that its role as genetic material was firmly established. Detailed characterization of the whole chromosome complement, including the exact number of chromosomes in man had, however, to await the development of culturing systems together with colcimide and hypotonic treatment, before nicely spread chromosomes were obtained as late as in 1956. It was not until this relatively late time that it was finally established that man has 46 chromosomes and the field of cytogenetics was born.

From then on the application of chromosome analysis in medicine was swift and the first numeric and structural chromosomal abnormalities could be described as early as 1960. Another very important step in the evolution of cytogenetics came with the discovery by Caspersson et al. (1968, 1971) that certain dyes, e.g. quinacrine mustard, had the ability to create discrete, fluorescent bands across the longitudinal axes of the chromosomes giving each chromosome its unique identity. This prompted fluorescence microscopy to be introduced into the field of human cytogenetics. The ability to identify banding patterns within each chromosome was an important step in increasing the resolution power of the analysis, enabling the characterization of more discrete structural aberrations, and then to correlate these with disease.

An increasing need for physical mapping of genes resulted in 1981 in the first gene to be localized by in situ hybridization on metaphases using radioactive labeling of a cloned gene as probe. This development initiated a number of gene mapping studies by in situ hybridization using radioactive labeling. At that time, radioactive labeling was the first choice of labeling due to its higher sensitivity than fluorescence-based labeling. However, the isotopic methods require long exposure times (weeks to months) and the precision is low. Developments of the fluorescence based in situ hybridization method resulted in 1986 in an improved method that offered higher speed and spatial resolution together with sufficient sensitivity as compared to isotopic methods. Soon after, fluorescent labeling was adopted as the label of choice, representing the birth of FISH. Various other refinements of this technique soon followed such as whole chromosome painting (1986), reverse chromosome painting (1989), multicolor labeling (1990) and 24-color karyotyping (1996; see Table 1 for more details). It should be remembered that these refinements of FISH have very much been based on achievements obtained in recombinant DNA technology, microscope hardware (epifluorescence and filter technology) as well as in computer and software technology for the analysis and storage of data.

The developments of the present day FISH-technology has mainly been driven by geneticists and pathologists involved in human disease, because these specialists have had an ever-increasing need for characterizing genotype-phenotype correlations, a need which is also illustrated by the development of large worldwide accessible databases harboring clinical as well as genetic information (e.g. OMIM – On Line Medelian Inheritance in Man). These developments were initiated by worldwide biotechnology programs such as the human genome program (HUGO) and the methods developed within these programs have subsequently been adopted by other fields in human and veterinary medicine and biology. The draft sequence of the human genome which was completed in February 2001 (Lander et al. 2001, Venter JC et al. 2001 and references in these publications) is an important step to further our understanding on how the human genome is organized, how it functions, how malfunction relates to disease processes and is a fundamental advance in self-knowledge.

Basic principles of FISH

In situ hybridization (ISH) can be defined as the morphological localization of genetic sequences. The objective of ISH is thus to determine the presence or absence of specific DNA or RNA species and to localize these species to particular cellular or chromosomal sites (Fig. 1). The identification of specific sequences within cells is achieved by exploiting a fundamental property of nucleic acids, i.e. their ability to anneal to one another in a specific manner to form hybrids. This is true not only for two complementary strands of DNA but also for RNA-DNA and RNA-RNA combinations as well as hybrids between natural and artificial nucleic acids. By labeling one of these two strands the annealed hybrids can be detected by a variety of means, including isotopic and non-isotopic (fluorescent and non-fluorescent) methods.

The basic requirements of the fluorescence in situ hybridization (FISH) technique are therefore: 1) a probe that is specific for the sequence of interest; 2) fluorescent labeling of this probe to allow appropriate detection; and 3) a biological specimen with preservation of sufficient morphological detail to determine the localization of the labeled probe after hybridization.

Table 1. Milestones in karyotyping and molecular cytogenetics

Early findings		Several authors in the 19th century	– Cell theory – Mendel's laws of inheritance – First identification of miotic figures
	1882 1883	Flemming Roux	Behavior of chromosomes in a dividing cell
	1944	Avery et al.	DNA - transforming factor
	1952	Hsu	Hypotonic treatment of cells
	1953	Watson & Crick	DNA double helix and genetic implications
Pre-banding era	1956	Tjio & Levan	46 chromosomes
	1958	Ford, Jacobs & Lajtha	First chromosome analysis in leukemias
	1959	Lejeune et al.; Ford et al.	First chromosomal diseases: – (Down's syndrome) trisomy of one of the smallest chromosomes. – Turner's syndrome (monosomy X)
	1960		Denver Conference 23 human chromosome pairs divided by morphology into 7 groups (A - G)
	1960	Edwards et al.; Patau et al.	– Trisomy for a D-group chromosome – Trisomy for an E-group chromosome
	1960	Nowell & Hungerford	First identification of chromosomal disorder related to leukemia
Banding era	1968	Caspersson et al.	Quinacrine mustard banding
	1970	Chaudhuri et al. Caspersson et al.	G- and Q-banding refined 450-550 bands/metaphase
	1970	Smith & Wilcox Kelly & Smith	Birth of molecular genetics: specific restriction endonucleases
	1971		Paris Conference - First cytogenetic nomenclature of all 23 chr. pairs
	1973	Cohen et al.	First report on cloning DNA using plasmids as vectors
	1976 1981	Yunis	High resolution chromosome banding (up to 2000 bands/prophase)

Table 1. Continuous

FISH era	1981	Harper & Saunders; Malcolm et al.; Gerhard et al.	Localization of single copy genes by in situ hybridization
	1981	Langer et al.	Biotin-labeled nucleotides
	1985		International System for Human Cytogenetic Nomenclature (ISCN 1985)
	1986	Saiki et al.	PCR technology
	1986	Pinkel et al.	Fluorescence hybridization
	1986	Deaven et al.	Chromosome specific DNA libraries
	1988	Pinkel et al.; Cremer et al.	Fluorescence in situ hybridization with human chromosome specific libraries.
	1988	Lichter et al.	CISS hybridization
	1990	Nederlof et al.	Multi-color FISH
	1991		Supplement to ISCN now including cancercytogenetics
	1992	Telenius et al.	DOP-PCR
	1992	Kallioniemi et al.	CGH-analysis
	1995	Mitelman	Supplement to ISCN now including FISH nomenclature
	1996	Schröck et al.; Speicher et al.	24-color karyotyping spectral karyotyping multicolor karyotyping
	1998	Choudoba et al.	High-resolution bar-coding

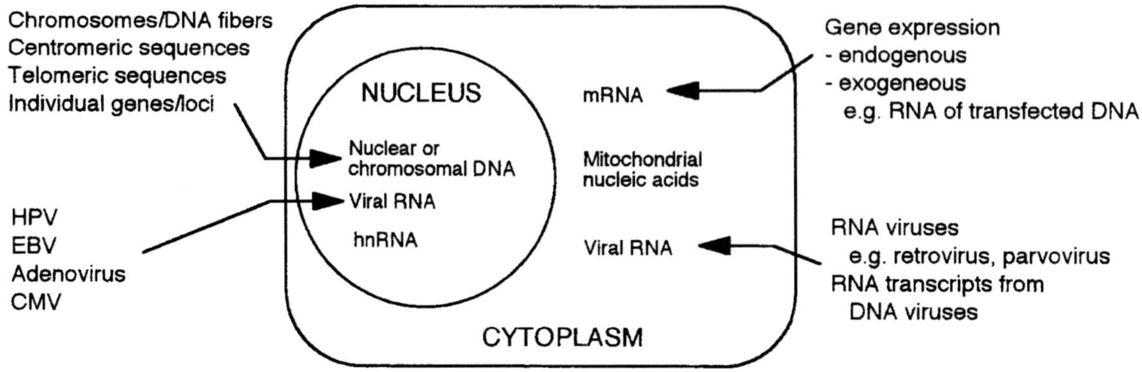

Fig. 1. Targets for in situ techniques

Technical elements

The specific annealing of probe to chromosomal or nuclear DNA is schematically illustrated in Fig. 2. The target DNA, e.g. chromosomes or nuclei, is first bound to the surface of glass slides. The DNA is then denatured by heating either before or after applying a special mixture that allows for annealing or hybridization of the single stranded probe to complementary sequences in the target DNA. Hybridization is usually performed overnight at a specific temperature which, in combination with the hybridization mixture determines the specificity of the annealing. Surplus probe is finally washed off in a stringent wash and the specifically annealed probe-fragments can subsequently be demonstrated by fluorescence microscopy.

The standard FISH protocol can be described as having six steps, each of which is crucial for obtaining a successful FISH-result (practical details can be found in Levy 1995, Choo 1994, Wilkinson 1992):

1. Sample preparation (source, fixation and unmasking)
2. Probe preparation and labeling
3. Denaturation of probe and sample
4. Hybridization of probe to sample (annealing)
5. Post-hybridization washing
6. Detection

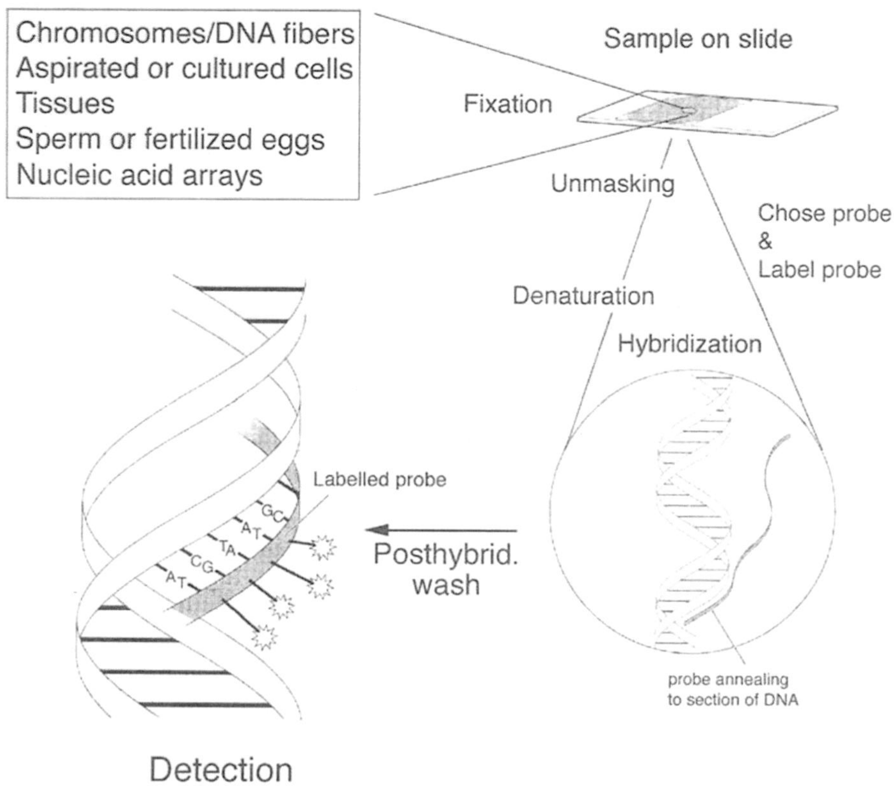

Fig. 2. Schematic illustration of the fluorescence in situ hybridization principle

Ad 1: Sample preparation (source, fixation and unmasking)

FISH can be used to analyze a wide variety of cellular materials and each sample type requires specific pretreatment measures. To preserve morphological detail, when using individual cells or chromosomes obtained from cell cultures or cells from aspirated clinical samples, a fixation step is required. Precipitating fixatives such as 70% ethanol and the conventional fixative used in cytogenetics (methanol:acetic acid; 3:1) are very useful as they allow cell suspensions to be stored indefinitely at -20°C. Another advantage is that in situ hybridization can often be performed directly on such fixed cells or with relatively mild proteolytic digestion. Cross linking fixatives are sometimes used to improve morphology but it is at the expense of having to use more rigorous proteolysis. Another important source of material is archival wax embedded tissues which has the advantage that morphology as well as nucleic acids are optimally preserved.

Once cells or tissues have been fixed, the nucleic acids contained within them have to be made available to the probe (unmasking). This process often involves a proteolytic step most frequently using proteinase K or pepsin HCl although other enzymes have been used. This step removes components of the cell nucleus and cytoplasm to allow probe access. Thus, under-digestion leads to suboptimal probe penetration whereas over-digestion leads to destruction of chromosome, cell and tissue structure resulting in the loss of both nucleic acids and morphological detail. When archival wax embedded tissues are used, a dewaxing step is also required. Following proteolysis post-fixation, particularly in aldehyde-based fixatives (e.g. paraformaldehyde), seems useful for the prevention of further loss of material from the slide.

Ad 2: Probe source, labeling and preparation

Many types of probes can be used for in situ hybridization and the appropriate type is determined to a large extent by the application and diagnostic query (Fig. 3). In the section "Applications of FISH in medicine and biology" the choice of probe type will be discussed in relation to application. Here, the different types of probes will briefly be discussed with respect to source and type of labeling.

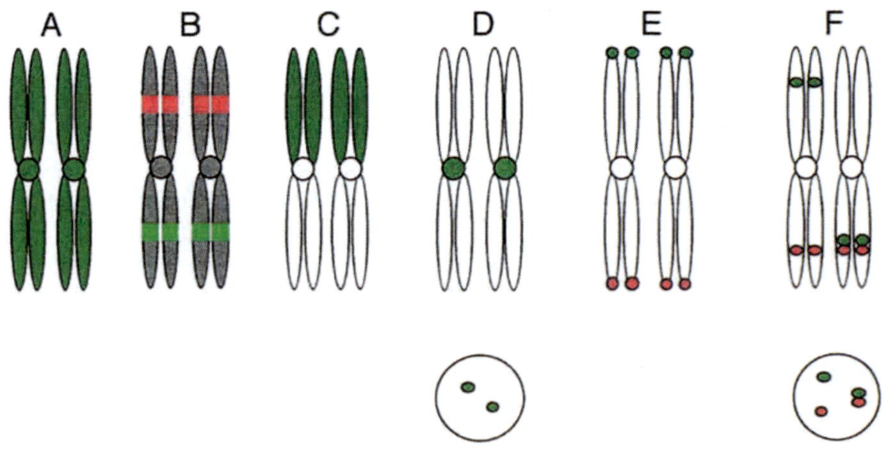

Fig. 3. Schematic illustration of various types of probes used in routine molecular cytogenetic diagnosis. *Upper row*: probes used on metaphase spreads. *Lower row*: probes used on nuclei. **A** Whole chromosome painting probes (one up to 24 differently labeled). **B** Comparative genome hybridization (CGH), **C** Arm-specific painting probes (p- and q-arms are available). **D** Centromeric probes that can be either pan- or chromosome-specific, **E** Telo- and subtelomeric probes that can be either pan- or arm-specific, **F** Locus specific probes (genes or chromosome segments)

A crucial point when deciding on the choice of probe type and label is sensitivity, i.e. one must consider how the target sequence of interest can be detected. This depends on a several factors: (1) size of target sequence (large genes/nucleic acid sequences can bind more probe and hence more label); (2) is the target sequence unique or does it exist in multiple copies (from a few to several thousand copies) at the site of location (the more copies, the more label can be bound); (3) size of probe sequence (theoretically, large probes can carry more label but the maximum probe size in the hybridization mixture is 300 to 500 base pairs as larger probe fragments either have difficulty in penetrating the target sequence or form large aggregates which reduces the amount of available probe and increases background); and 4) indirect labeling techniques offer the possibility for signal amplification.

In Table 2 the different probe types are listed with respect to size of probe sequences, which stretch from very short synthetic oligonucleotides up to composite probes representing a whole chromosome.

The short oligonucleotides (20 bases up to 100 bases) are synthetic DNA, which can be labeled during manufacturing, but because of their size, these sequences can carry only a few fluorochromes. This represents a problem in sensitivity for the detection of unique sequences. However, when dealing with short sequences that are moderately to highly repeated, such as centromeres and telomers, these short probes are very efficient (Koch et al. 1989, Gosden et al. 1991, Hindkjær et al. 1994, Therkelsen et al. 1995).

For detection of single genes or gene regions, cloned probes are often used (Kievits et al. 1990, Rowley et al. 1990, Kuwano et al. 1991). The choice of vector determines the size of the fragment that can be cloned. Cloned DNA fragments in plasmids, cosmids, YACs (yeast artificial chromosomes), PACs (P1 filamentous phage artificial chromosomes) or BACs (bacterial artificial chromosomes) are the most commonly used probes. These are named locus specific probes as the cloned fragment is derived from a specific region of the metaphase.

Often other vectors than plasmids are preferred because of the huge fragments of DNA that can be contained in them. This enables the specific binding of numerous fluorochrome molecules to the region of interest which, in principle, makes detection unproblematic. However, the cloned fragments also contain families of interspersed repeated sequences which at the same time are scattered throughout the genome. Therefore, methods had to be developed to avoid or minimize the possibility of detecting these repeated sequences at other "unspecific" sites in the metaphase (see below).

Table 2. Types of probes and FISH-techniques used in routine molecular cytogenetic analyses

Source of probe	Type of probe	Size of probe-target	FISH technique	Types of chromosomal aberrations identified
Purified chromosomes Purified chromosome regions (e.g. arms- or chromosome bands)	Flow-sorted Microdissected	> 1Mb	Whole chromosome painting (forward and reverse painting) Arm-specific painting FISH-multicolor banding 24-color karyotyping	• Translocations (simple or complex) • Marker/ring chromosome
			ZOO-FISH	
Cloned genes or DNA sequences	YACs, PACs, BACs	< 1 Mb	Locus-specific, e.g. single genes or gene family centromeric sequences telomeric and sub-telomeric sequences	• Microdeletions • Amplifications • Numerical abnormalities
	Cosmids	< 0.1 Mb		
	Phages	< 0.05 Mb		
	Plasmids	< 0.01 Mb		
	RNA probes	< 0.01 Mb	Expression	
Synthetic DNA Synthetic analogs	Oligonucleotides PNA & LNA	< 0.10 kb	Primed in situ labeling (PRINS), e.g. centromeric or telomeric DNA	• Numerical abnormalities • Terminal segment anosomy
Purified DNA from cells	Genomic DNA	> 1 Gb	Comparative genomic hybridization (CGH)	• Amplifications • Deletions

For the detection of whole chromosomes (Pinkel et al. 1986, Lichter et al. 1988, Pinkel et al. 1988, Hulten et al. 1991) or a part thereof (from individual bands (Guan et al. 1993) to arms (Guan et al. 1996)) libraries are generally generated from normal individuals or cell hybrids containing human chromosomes. By fluorescence activated chromosome sorting of metaphases from these sources, chromosome specific libraries for each human chromosome, 1-22, X and Y, have been generated (Deaven et al. 1986, Carter et al. 1992). A sorting experiment can, in general, generate approximately 500 copies of each chromosome. These can then be amplified by DOP-PCR in which a degenerate primer is used in a special PCR-amplification step as described by Telenius et al. (1992). The composite nature of these probe mixtures again ensures that a sufficient number of fluorochromes are bound to the region of interest. Another way of isolating whole chromosomes or parts thereof is by microdissection (Guan et al. 1993, Thangavelu et al. 1994). This can be done by micromanipulators (Meltzer et al. 1992) or laser-equipped devices (Monajembashi et al. 1986). Here only 25–50 copies of each chromosome can be isolated and further amplified by DOP-PCR before fluorescent labeling. Even a single microdissected chromosome can be amplified and used as a chromosome painting probe (Christian et al. 1999, Kjeldsen unpublished data).

The choice of label is governed to some extent by personal preference and, in some situations, by commercial availability. Although isotopic (radioactive) labeling initially was the choice of label because of its sensitivity, it is now largely confined to RNA detection, particularly when non-isotopic measures have been unsuccessful. Radioactive sulfur has been the most frequently used label as it combines specific activity with relatively high morphological resolution.

Non-isotopic ISH sensitivity has been greatly improved and most laboratories are now using non-isotopic labels exclusively. The most commonly used non-isotopic labels are fluorescent dyes such as FITC, rhodamine and Texas Red, which can be coupled to nucleotides, antibodies or macromolecules. There are two basic methods for non-isotopic labeling: (1) fluorescent - direct or indirect; and (2) non-fluorescent. The probes are labeled either directly with nucleotides coupled to a fluorochrome or indirectly with nucleotides coupled to a reporter-molecule, which subsequently can be detected by conventional immunochemical methods. The most commonly used indirect labels are biotin and digoxigenin coupled to nucleotides which - after incorporation - can be detected using fluorescence-conjugated avidin or antibodies, respectively. The choice of direct or indirect labeling is mainly determined by demand for sensitivity as indirect labeling offers the possibility for signal amplification, for example by tyramide (Bobrow et al. 1989).

Labeling of probes is most commonly performed using enzymatic methods (Table 3). DNA probes are conveniently labeled either by nick-translation or randomly primed extension, both of which introduce labeled nucleotides into the probe molecules. The labeling efficiencies, however, are often far from 100%. An alternative labeling method is by chemical cross-linking which gives close to 100% labeling efficiency (Craig et al. 1997). Probe labeling can also be carried out using polymerase chain reaction (PCR) based methods, with either direct incorporation of labeled nucleotides or by using labeled primers. DOP-PCR is often used when flow-sorted or microdissected chromosomes are being labeled.

Oligonucleotides can be labeled during synthesis where 5'-labeled nucleotides can be directly incorporated into the molecule. Enzymatic procedures can also be used for either 3'- or 5'-labeling. One advantage of 3'-labeling is that multiple probe labels can be introduced to form a "tail" thus increasing the sensitivity of detection. A variant of this type of labeling is the primed in situ labeling (PRINS) technique where the labeling of the hybridized oligonucleotide is performed in situ, i.e. an unlabeled oligonucleotide is hybridized to its target DNA sequence in the sample and the labeling is then performed in situ using conjugated nucleotides and a thermostable polymerase (Koch et al. 1989).

For RNA detection (mRNA or viral RNA), RNA probes (riboprobes) are often used because RNA-RNA hybrids are more stable than DNA-RNA hybrids. Riboprobe vectors can be used to generate both sense and antisense probes to allow control of hybridization and the unbound probe can be digested using RNase which do not digest double stranded RNA. Also

Table 3. Labeling strategies of probes

Genomic DNA for CGH	Nick-translation DOP-PCR
Flow sorted or microdissected chromosomes or regions, YACs	DOP-PCR
Cosmids, BACs, PACs	Nick-translation Primed labeling
Plasmids	Nick-translation Primed labeling
Oligonucleotides	PNA
In situ labeling, chemical labeling	Chemical labeling during synthesis
RNA probes	Sense or antisense strand- 5'-label

oligonucleotides, either singly or in a cocktail, have been used to detect high abundance RNA in cells or tissues. RNA probes are generally labeled by in vitro transcription whereby the probe sequence is cloned into a vector containing RNA polymerase promoter sites, and probe molecules are generated using RNA polymerase-mediated incorporation of labeled nucleotides.

Ad 3: Denaturation of probe and sample

When DNA probes are used for the detection of DNA targets within cells or tissues, a denaturation step is required for a successful hybridization to take place. This process can be achieved either by chemical means or by heating above the melting temperature of the DNA double helices.

In practice a combination of the two approaches is used. When double stranded DNA is heated to a temperature above its melting temperature (which is determined by both its length and DNA sequence) the two strands separate. Usually the melting temperature is too high and the time of incubation too long for an efficient denaturation to take place and, at the same time, preserve morphology. To accommodate for this, the inclusion of organic solvents in the denaturation and hybridization has proven efficient. Formamide is the most commonly used of these and it functions by destabilizing the double stranded structure of DNA at a given temperature thereby reducing the effective melting temperature.

The DNA probe and the DNA target molecules can either be denatured separately or by co-denaturation. It has been argued that separate denaturation preserves morphology better, whereas co-denaturation reduces the number of practical steps. For these reasons separate denaturation steps are most often used in molecular cytogenetics and co-denaturation most often used when tissue sections are analyzed.

In principle, riboprobes and oligonucleotide probes are single stranded just like cellular RNA. However, denaturation improves the sensitivity probably by removing secondary structures of RNA probe and target.

Ad 4: Hybridization of probe to sample

When the probe and target molecules are single-stranded, the two types of molecules are brought together under conditions favoring hybridization. This is achieved by ensuring a high probe concentration and lowering the temperature of incubation.

When dealing with probes such as chromosomes, chromosome segments, YACs, BACs and PACs, where many copies of disperse repeats are expected to be present, it is necessary to modify hybridization conditions to avoid cross hybridization to other chromosomal regions containing these repeated sequences. This is done by the so-called chromosomal in situ suppression (CISS) hybridization (Hulten et al. 1991), where the labeled probe is prehybridized shortly with an excess of unlabeled repetitive DNA (Cot-1) before application to the slide. The Cot-1 DNA binds to the probes repeated sequences which in turn cannot hybridize to the chromosomal repetitive DNA leaving the unique sequences contained in the probe single-stranded and free to hybridize to the chromosomal DNA. Recently, a complete set of human chromosome-specific painting probes depleted in repeated sequences have been generated (Bolzer et al. 1999). These probes yield highly specific signals and can be hybridized without a prehybridization step of the labeled probe.

For an efficient hybridization to take place, a number of aspects must be considered:

The specificity of the hybridization (also known as stringency) is a crucial point. For example, if hybridization is carried out at too high a temperature, no probe annealing will take place because both probe and target molecules remain predominantly single-stranded. On the other hand, if the temperature is too low, the probe may anneal at many different unspecific sites on the target DNA which may not be removed in the post-hybridization washing step. The appropriate temperature must be determined by experimentation although some general rules do apply.

Another way to achieve specific annealing is to alter the chemical composition of the hybridization mixture. This can be accomplished in a number of ways: (1) increasing the formamide concentration, thereby destabilizing mismatched hybrids. This has the same effect as increasing the temperature; (2) reducing the salt concentration. This also increases the specificity; (3) increasing the length of the probe. This increases the specificity of the annealing process. When using YACs, cosmids or chromosome-derived probes, the individual labeled molecules must be between 300 and 500 bp. This size is determined by the amount of DNase or length of nick-time when labeling the probe in nicktranslation, (4) increasing the probe concentration. This drives the reaction towards increased formation of probe-target hybrids thereby speeding up the reaction, but may also lead to non-specific background staining (in general, a probe concentration of 1–2 ng/μl is optimal but must be determined empirically), (5) Varying hybridization time. The time required for hybridization depends on the target sequence, i.e. highly repetitive DNA generally requires a short hybridization time (2–4 h) while low or single copy targets require overnight hybridization, and (6) the purity of the DNA probe.

Usually, the hybridization is performed under a sealed coverslip in a moist chamber to minimize the evaporation of water which would otherwise alter the hybridization conditions in an uncontrolled manner.

Ad 5: Post-hybridization washing

After hybridization, unbound probe must be removed. This is carried out by washing the slide in a saline solution (SSC – standard saline citrate) where probe-target hybrids are stable. The specificity of this wash can be regulated by altering one of the following parameters: (1) salt concentrations of washing solutions, (2) formamide concentrations, (3) temperature.

The following are common examples of hybridization and post-hybridization washing: (1) hybridization in 60% formamide, $2\times$SSC at 37°C followed by washing in 60% formamide, $2\times$SSC at 42°C; or (2) hybridization in 50% formamide, $2\times$SSC at 37°C followed by washing in $0.1\times$SSC at 37°C. It should be noted that formamide is a toxic organic solvent and steps involving this compound must be performed under a hood.

Ad 6: Detection

After the final washing step the presence and localization of probe-target hybrids can be demonstrated by detection of the labeled probe. Probes labeled directly with fluorochromes can be visualized immediately by fluorescence microscopy using appropriate filter sets for excitation and emission. Probes labeled indirectly with biotin or digoxigenine can be visualized by methods based on affinity (e.g. streptavidin coupled to a fluorochrome) or immunodetection (fluorochrome-conjugated antibody against digoxigenine), respectively. The streptavidin and antibody can be coupled to either a fluorochrome or to an enzyme-substrate system. Avidin can be used instead of streptavidin but this increases the risk of unspecific background due to the affinity of avidin for cellular components. Fluorescence based approaches are particularly useful when high resolution is required or when more than one sequence is hybridized to the same tissue or cell preparation. In interphase cytogenetics, fluorescence based methods are particular useful as multicolor hybridization techniques can be used (see below).

Before the samples are analyzed, the chromosomes or nuclei are most often counterstained using either DAPI (4',6-diamidino-2-phenylindole)-which emits blue light or propidium iodine which emits red light. DAPI however, is the most commonly used counterstain because probes in general are labeled with FITC (green), rhodamin/Texas Red (red) or both if two probes are hybridized simultaneously.

One important problem when using fluorescent based methods, is bleaching or fading of the signal while being analyzed or left in the light. This problem can be reduced by mounting the sample in so called antifade solutions (Longin et al. 1993).

Non-fluorescent methods are used where high sensitivity is important. Two enzyme-based systems are frequently used, namely the alkaline phosphatase-system and the peroxidase system where the enzymes are coupled to nucleotides, or more commonly, steptavidin or antibodies. For detection, a substrate is then added which by the enzymes action forms a precipitate at the site of annealing. This, in turn, can be detected by light microscopy. The alkaline phosphatase system is considered to be more sensitive than the peroxidase system, which on the other hand tends to give greater resolution.

When isotopic labels are utilized, a dip-slide emulsion technique is used, followed by light microscopy as in the enzyme-based systems.

Refinements of the standard FISH technique

Reverse cytogenetics

The possibilities of isolating chromosomes by flow-sorting (Carter et al. 1992) or by microdissection (Thangavelu et al. 1994) have been used to isolate aberrant, but unidentified chromosomes. After isolation of the aberrant chromosome it is amplified, labeled and used as a probe on metaphases from a normal individual. This technique is called reverse cytogenetics or reverse painting and is used to determine the composition of the aberrant chromosome (Carter et al. 1992).

Multicolor FISH technology

FISH has the capacity to simultaneously visualize different targets in multiple, distinct colors. The simplest form of multicolor FISH is to mix differently labeled probes in a single hybridization solution. However, due to spectral overlap of the fluorochromes, the number of probes identifiable on the basis of a unique fluorescence color is limited to three for the visible part of the spectrum, and to five if far-red and infra-red fluorescing dyes are included. When a more sophisticated approach such as combinatorial labeling is used, the different probes are labeled with combinations of haptens or fluorochromes and the number of identifiable targets can be increased (Nederlof et al. 1990, Ried et al., 1992, Wiegant et al. 1993). In principle, the number of targets identifiable on the basis of a unique, binary combination is 2n-1, where n is the number of different labels. Thus, if five fluorochromes are included, 31 different combinations can be obtained. Another approach is ratio-labeling where varying ratios of different fluorochromes are used for each probe so that it is the fluorescence ratio that identifies the probe (Nederlof et al. 1992). For example five different ratios of red and green fluorescence attached to five different probes allow the discrimination of five different targets with only two fluorochromes. By the recently developed COBRA (Combined Binary Ratio labeling) technique the two methods can be combined and then up to 100 label types can be obtained (Tanke et al. 1999).

The developments of new FISH-techniques have been paralleled by developments in the microscope systems used. While the basic microscopic features have been rather constant, we have witnessed an impressing development in filter technology, where it is now possible to order very precise custom-designed filters. This has, in connection with the development of many new fluorescent dyes, made it possible to perform extensive multicolor analyses. New fluorescent dyes and filters have extended the potential of FISH by simultaneously detecting multiple genomic targets in different colors. Thus, the availability of an increasing number of spectrally distinct fluorochromes can be utilized for the direct or indirect labeling of nucleic acid probes, and then more targets can be visualized in parallel by simply mixing differently labeled probes for a given target.

Fluorescent signals can be viewed both directly in the fluorescence microscope or by using a CCD-camera (charged coupled devices) coupled to the microscope and connected to a computer. Images from each filter set are then captured followed by image analysis in specifically designed software. If the signal to noise ratio is too low, either because of high background or low intensity of signals, the image analysis software can improve the ratio. The images which are in the gray scale can then be overlaid and given artificial colors for presentation. Confocal microscopy utilizes a laser for excitation and photomultipliers for detection of emitted light. Specialized software control the laser, photomultipliers and have similar possibilities for image analysis. One important advantage of using confocal microscopy is that different confocal planes can be captured and processed to generate a three-dimensional image. Such an approach allows for an improved analysis of nuclei or tissues which often contain overlapping signals. However, these analyses are impractical in routine diagnostics at the present technological stage.

24-color karyotyping by spectral analysis (SKY) and multifluor FISH (mFISH)

One of the most recent refinements in FISH-based multicolor techniques for screening whole genomes for both structural and numerical chromosomal aberrations are the following two technically related innovations:

SKY (Schröck et al. 1996) and mFISH (Speicher et al. 1996). Both methods rely on metaphases and allow simultaneous identification of different chromosomes, with each chromosome displayed in a characteristic color. In both methods all 24 differently labeled human chromosomes are hybridized in a single hybridization experiment. With five fluorescent colors at hand, enough combinations of these are available to give 24 artificial colors in the final presentation. The methods of detecting these combinations are, however, different in the two systems.

mFISH is a filter-based approach and employs the sequential image acquisition using five different fluorochrome-specific optical filters, each of which are narrow band pass filters that allow maximum distinction of the fluorescent dyes. Special image analysis software calculates a chromosome segmentation mask based on the DAPI image, followed by measuring the intensities in these predefined areas for all subsequently acquired images. Before these calculations, the software performs an image shift correction. Based on the labeling schemes the program decides whether a single pixel contains signals from one or more fluorochromes and this information forms the basis for chromosome identification and subsequent color assignment.

SKY is based on spectral imaging which is a combination of spectroscopy, CCD imaging and fluorescence microscopy. An interferometer coupled to a CCD camera generates a fluorochrome specific optical path difference that, in turn can be Fourier transformed to provide spectral information.

Consequently, the fluorescence emission spectrum can be recovered simultaneously at all image points with one exposure. Specialized software then classifies the images by identifying pixels with identical spectra and then assigning these pixels the same classification color, which in turn is the basis for the actual karyotyping.

The probe sets for SKY and mFISH are commercially available at a relatively high cost but recently Roberts et al. (1999) have presented new probe construction protocols and strategies for multi-color karyotyping.

Bar-coding of human chromosomes

By using subregional DNA probes in different colors, a pattern that has been termed "chromosome bar code" can be produced. This enables chromosome identification by producing a limited number of bars on each chromosome (Lengauer et al. 1992), providing a potential to screen for intra-chromosomal rearrangements (e.g. deletions and inversions).

In another approach, cross-species color banding probes from flow-sorted, differently labeled gibbon chromosomes are utilized (Wienberg et al. 1990, Wienberg and Stanton 1997). Because of an extensive sequence homology (98%) between gibbon and human DNA and the many chromosomal rearrangements that have taken place during evolution, the hybridization of these probes onto human metaphases results in a specific color banding pattern for each chromosome. This banding pattern has been termed RxFISH and comprises a multi-color set of 110 distinct colored bars or "bands" per haploid chromosome set (Müller et al. 1997a,b).

Yet another bar-code approach is currently being exploited. By selecting a suitable panel of several microdissected overlapping chromosome fragments from each human chromosome and using multicolor labeling, each chromosome will acquire a characteristic bar-code appearance with improved resolution (Chudoba et al. 1999). However, the technique is at present only possible for a limited number of human chromosomes, but is being made avaible for all.

Interphase cytogenetics

The term interphase cytogenetics was proposed by Cremer et al. in 1986 for analysis of interphase nuclei by in situ hybridization.

This method is a useful supplement to other cytogenetic approaches, particularly when time is a crucial factor or for solid tumors in which metaphases are rarely seen or can only be produced with great difficulty. Using combinations of various centromere specific probes, it is possible to screen a large number of cells for numerical chromosomal aberrations by counting the number of signals or spots in each nucleus. When, for example, a probe for an autosome is used, two signals in each nucleus are normally expected and when a cell contains a numerical aberration, either one, three or more signals are expected. However, the interpretation of such signals can be problematic. If, for example, only one spot is seen, does this represent a true monosomy for that chromosome, or does it represent a spatial, but random, co-localization of the two centromeres in that particular cell? Conversely, if three or more spots are seen does this represent a true aneuploidy for that chromosome or does it represent an artifact. Due to varying hybridization efficiencies in the sample and differences in cell cycle state, such normal variations must be encountered. Two rules of thumb are used in practice to minimize these problems. One is that if the two spots are separated by more than the diameter of the largest spots, then they should be counted as two signals. The other rule of thumb is that if >70% of the nuclei falls into one diagnostic category, it can be diagnosed as such (Shaffer et al. 1997, Liehr et al. 1996).

Fiber-FISH

The resolution of FISH for metaphase chromosomes has been assessed to be about 3 Mb which corresponds to a smaller band seen in G-or Q-banded chromosomes making metaphase FISH suited for only coarse mapping of DNA clones. Thus, within a band, clone ordering by FISH to metaphase chromosomes is virtually impossible.

Use of a more relaxed state of the chromatin as in interphase nuclei, provides a dramatic increase in resolution. Interphase FISH resolution has been estimated to be around 100 kb (Lawrence et al. 1990, Trask et al. 1991). The genomic resolution is limited by the spatial resolution of the microscope and determined by the chromatin condensation state. To improve FISH resolution either high-resolution microscopy (Moers et al. 1996) or further decondensation of chromatin has been utilized. As high-resolution microscopy is a rather expensive way of improving resolution power, a more practical approach is to use decondensed chromatin and conventional epi-fluorescence microscopy which can give resolutions up to the theoretical possible (Raap et al. 1996). Several strategies have been pursued to achieve decondensed chromatin in the form of naked DNA fibers and are collectively referred to as Fiber-FISH. For example, SDS-extraction of mammalian cell nuclei immobilized on glass slides (Para and Windle 1993) or extraction conditions adapted from the halo-preparation technique (Wiegant et al. 1992) can produce intact stretches of several hundred kilobase pairs up to over 1 Mb of intact naked DNA fibers at sufficient density.

Early Fiber-FISH results showed that the theoretical resolution could be achieved in practice. A 1 kb of linear DNA measures 0.34 μm in length and optical resolution is 0.2–0.3 μm dependent on objective magnification and numerical aperture. The detection sensitivity limit has been found to be as low as 200 bp (Florijn et al. 1996).

Comparative genomic hybridization (CGH)

Comparative genomic hybridization (CGH) analysis is another powerful refinement of the standard FISH technique, which can be used to screen the whole genome from tissue samples or cell lines providing a comprehensive characterization of changes in the relative copy number of chromosomal material (Kallioniemi et al. 1992). The technique can also be used when no viable cells are available and has even been used to characterize the whole genome from a single cell. However, with CGH it is not possible to identify balanced structural aberrations.

Genomic DNA is purified from the tissue or cell line of interest. This DNA is then labeled by nick-translation using nucleotides coupled to, for example FITC. Control DNA is also labeled by nick-translation using nucleotides coupled to rhodamine. Equal amounts of these two DNA samples are then allowed to competitively hybridize onto a normal metaphase spread of human chromosomes. The red/green color ratio obtained after hybridization indicates either a relative excess or deficiency of target DNA (relative to the control) or equality of hybridization. The color-ratio analysis is generated by specialized software, which analyses images generated by a CCD-camera, coupled to an epifluorescence microscope.

Analytical quality control is a major problem in CGH and perhaps more important here than in conventional FISH. This is due to the fact that the hybridization is analyzed quantitatively. In an effort to overcome this problem a new CGH modification, four-color CGH, was recently developed (Karhu et al., 1999). The modification provides an internal standard in every hybridization which helps in identifying inconsistently hybridized chromosomal regions and when using a special second reference DNA, the dynamic range of hybridization can also be standardized.

The CGH technique also has sensitivity and resolution limitations. For deletion detection the size of the deletion seems to be 5-10 Mb whereas for gain detection, the amplicon size can be 100–200 kb.

DNA Chip technology

A way to improve the resolution of CGH is to artificially create microarrays of individually or appropriately pooled DNA clones or custom synthesized oligonucleotides and use these as the reporting genome instead of the highly condensed metaphase chromosomes (Lichter et al. 1997). In principle, one can choose the CGH resolution that is needed by selecting appropriately spaced DNA sequences in the microarrays. The presence or absence of these sequences in the genome of a patient can then be visualized by using DNA from the patient as a fluorescent probe that is hybridized together with equimolar control genomic DNA to these fixed sequences. The red/green color ratio obtained after hybridization indicates either a relative excess or deficiency of target

DNA (relative to the control) or equality of hybridization. The analysis is performed by specialized software, which analyses images of the microarrays generated by a CCD camera coupled to a fluorescence type microscope.

New developments in probe types

PNA, LNA and padlock probes are new probe types in the FISH field which hold great promise for quantitation and high specificity FISH.

Polypeptide nucleic acids

Polypeptide nucleic acid, PNA, is a synthetic DNA analogue in which the deoxyribose-phosphate backbone is substituted with (2-aminoethyl)-glycine. PNA has the advantage of rapidly forming very stable duplexes with complementary DNA. The neutral peptide backbone provides no electrostatic repulsion to the DNA-phosphate groups therefore favoring the rapid kinetics of PNA-DNA duplex formation and a high thermal stability (Nielsen et al. 1991, Egholm et al. 1993).

Recently, a PNA oligomer was introduced as a FISH probe identifying all 96 telomers in a human metaphase chromosome spread (Lansdorp et al. 1996). The highly efficient hybridization property of the PNA probe contributed to a better FISH stoichiometry, thereby providing a firm basis for measuring relative lengths of telomeres on the basis of fluorescence intensity by digital imaging microscopy. For other repeat sequences such as centromere specific alphoid DNAs, the simple satellite DNAs and Alu-repeats, PNA-FISH probes are currently being developed (Taneja et al. 2001). The better FISH signal intensity and stoichiometry may in the future allow low level repeat and perhaps even unique sequence detection using this new DNA analogue.

Locked nucleic acids

Locked nucleic acid, LNA, is a synthetic DNA analogue that is conformationally restricted containing 2'-O,4-C-methylene LNA nucleoside monomers (Kumar et al. 1998). LNA is capable of recognizing complementary DNA and RNA with high thermal stability which, in the future, may allow for a better FISH signal intensity.

Padlock probes

A strategy to improve discrimination of in situ hybridization of single base mismatches is the so-called padlock probes (Nilsson et al. 1994). These probes are oligonucleotides with probe sequences of about 20 nucleotides on the 3'- and 5'-end with a 30- to 40-mer spacer in between, to which haptens (e.g. biotin and digoxigenine) are coupled for detection purposes. The probe sequences are chosen such that in a hybrid molecule, the 3'- and 5'-ends of the padlock probe are juxtaposed and can then be covalently linked by a ligation step to close the padlock probe. This setup provides three important features with respect to uniqueness of hybridization. First, the formation of a circular molecule allows for extreme stringency conditions in the washing step, secondly, the juxtaposition of the probe ends requires two independent hybridization events, and thirdly the DNA ligase used for ligation is highly dependent on perfect match. The padlock methodology has successfully been used to detect single base mismatches of repeat sequences by fluorescence in situ hybridization on metaphase spreads (Nilsson et al. 1997). However, sensitivity is still an issue but signal amplification by a rolling circle mechanism is a possible way to achieve the sensitivity for single copy sequences (Daubendiek and Kool 1997).

Applications of FISH in medicine and biology

General considerations (Table 4)

The most powerful and important feature of FISH is that the technique allows detection of a particular gene or genetic sequence at the subcellular level in a multi-color mode. In basic science FISH has strongly enhanced gene and genome analysis, enabling a rapid mapping of genes. Diagnostically FISH allows detection of abnormalities that otherwise could not have been identified or only with great difficulty, including identification of pathological chromosome abnormalities in oncology, congenital malformations and single gene disorders. FISH techniques offer the possibility of selecting where in the genome staining shall occur and the technique is therefore used either to complement conventional banding methods or as a substitute for chromosome identification at metaphase or interphase. FISH also has the advantage of being able to detect microbial DNA or RNA sequences without the need for prior propagation of cells or microorganisms.

Table 4. Comparisons of different techniques in cytogenetics

	Advantages	Disadvantages	Solution
Conventional banding (high- or low resolution)	Screening of the whole genome	Difficulties in identifying short chromosome fragments (e.g. markers) and complex rearrangements, especially in low quality chromosome preparations	FISH painting – whole chromosome painting, arm-specific painting, multicolor painting
		Generation of meta- or pro-phases in general requires cell growth. Some times samples cannot be grown (fixed pathology samples) or time is crucial (e.g. late amniocentesis)	Interphase nuclei FISH CGH
		Limited resolution, e.g. microdeletions and other subtle rearrangements	Locus specific FISH probes Telomeric FISH probes
		Cells in metaphase does not always represent the total cell population	Interphase nuclei FISH
Painting FISH	Specific identification of a part of the genome	Some knowledge on identity of the abnormality in question is required – not possible for markers/ring chromosomes or complex rearrangements	Multicolor FISH, e.g. 24-color karyotyping
Interphase FISH	Rapid identification of numerical aberrations in non-dividing cells	Cannot be used to screen for unknown structural aberrations	CGH
CGH	Screening the whole genome for amplifications or deletions in non-viable cells	Resolution is 5–10 Mb Structural balanced rearrangements not identified	Nucleic acid arrays ?
24 color karyotyping	Screening the whole genome in viable cells	Small interstitial rearrangements cannot be identified (amplifications etc)	Nucleic acid arrays ?

However, in many diagnostic queries related to genetic or congenital disorders and to many forms of cancer, a specific approach based on preselected probes is not feasible due to a lack of prior insight regarding the genome region of interest.

The development of 24-color karyotyping and CGH remedies this drawback of classic FISH techniques where an in situ hybridization experiment only allows for a limited fraction of the chromosome complement to be analyzed. 24-color karyotyping and CGH thus represents the next generation of cytogenetic techniques which, like chromosome banding, overviews the whole genome but with far higher sensitivity.

The potential use of FISH techniques is so enormous that it is impossible in a comprehensive way to cover all areas, but in the following sections, some of the more recent and general applications of FISH technology will be briefly reviewed with emphasis on clinical cytogenetics and tumor cytogenetics.

Molecular cytogenetics as a diagnostic tool in clinical cytogenetics and oncology

The vast majority of chromosome analyses in clinical cytogenetics and in oncology are still performed as an analysis of banded metaphases. These banding techniques have not changed much over the past two decades with the result that the minimum size of chromosome material that can be detected, either lost or gained, by banding techniques has not changed very much either. This has left the impression in some quarters, that chromosome analysis in the field of cytogenetics is obsolete and should be replaced by molecular methods based on PCR and gel-analysis. This is, however, far from true, since the standard banding technique is now supplemented with a broad set of FISH-methods based on modern DNA-technology.

Which types of challenges in cytogenetics may optimally be solved by application of FISH technology? The challenges can be divided into two groups:

I. Characterization of specific chromosomal aberrations. The primary goal of employing classic FISH techniques is now mainly analysis or screening for known chromosomal abnormalities (from locus specific events to whole chromosomes) either in a single- or in a multi-color approach.

With the development of the initial FISH techniques, it became possible to identify numerical and structural chromosomal abnormalities using either centromere-specific probes or whole chromosome painting

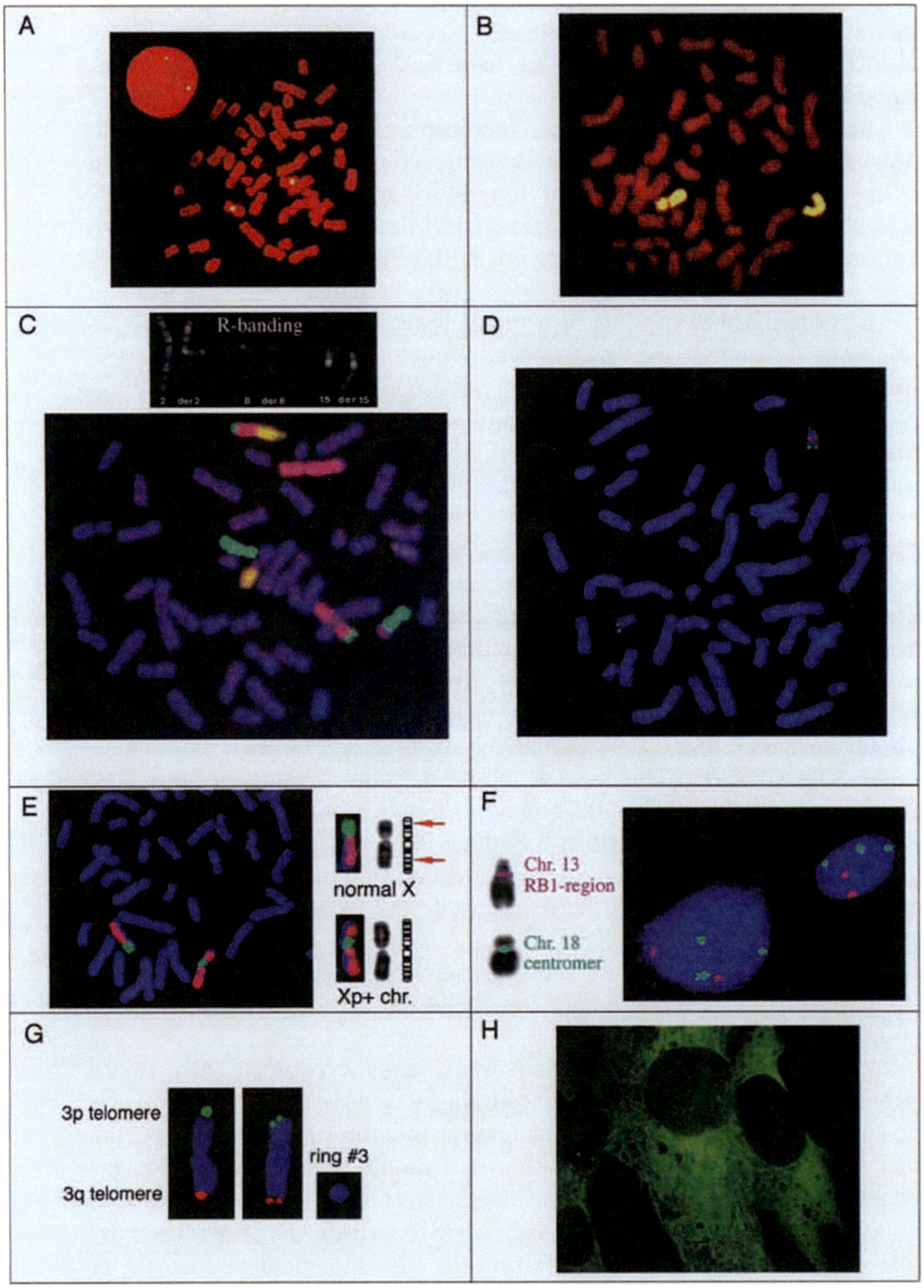
A
B
C
R-banding
2 der2
8 der8
15 der15
D
E
normal X
Xp+ chr.
F
Chr. 13
RB1-region
Chr. 18
centromer
G
3p telomere
3q telomere
ring #3
H

probes in a single-color approach. Figure 4A shows an example of a normal analysis using the centromere-specific probe for chromosome X, and in Fig. 4B the result of a whole chromosome painting using a probe for chromosome 8 is shown.

When complex chromosomal rearrangements are identified by conventional banding techniques, it is possible to get a clue as to which chromosomes are involved and by using a multi-color ratio-labeling approach it later became possible to characterize the chromosomal rearrangement in greater detail (Fig. 4C).

The so-called microdeletion syndromes are an increasingly important group of specific genetic disorders that are characterized by discrete but consistent chromosomal deletions initially identified by high-resolution banding methods (prophase techniques). These methods have an estimated resolution of 2–5 megabase pairs which is barely sufficient to demonstrate such microdeletions, and at the same time these methods are labor-intensive as well as technically difficult. Therefore, even with high quality banding, this type of syndrome can only be identified in approximately 50% of the cases (Dobyns et al. 1991). FISH analyses with suitable cosmid or YAC clones, on the other hand, are ideal for the demonstration of these microdeletion syndromes. The first microdeletion syndrome to be diagnosed with FISH was Miller-Dieker syndrome (Kuwano et al. 1991), which is a rare malformation syndrome, manifested by lissencephaly

Fig. 4. Different examples of FISH techniques used for identification of specific chromosomal aberrations. **A** Detection of highly repeated sequences. Staining of alpha-satellite DNA of chromosome X using cloned alpha-satellite probe labeled indirectly with FITC. Counterstained with propidium iodine. **B** CISS hybridization with a chromosome 8 specific probe indirectly labeled with FITC, propidium iodine used as counterstain. Image captured using a confocal microscope. **C** Three-color ratio labeling hybridization with libraries for chromosomes 2 (red), 8 (green) and 15 (yellow – 50% red and 50% green) identifying a complex translocation that cannot be completely diagnosed by high resolution banding (*upper row*). **D** Locus-specific hybridization at chromosome region 22q11.2 (red) and a control signal (green) for identification of the microdeletion syndrome DiGeorge. **E** Dual color arm-specific hybridization with Xp (red) and Xq (green) showing that Q-banding only identifies a structural abnormality in Xp of unknown origin, but FISH identifies a partial trisomy for Xq and a partial monosomy for Xp. **F** Interphase FISH hybridization identifying an extra chromosome 18 centromere (green) in the nuclei (blue). **G** Telomeric probe hybridization for chromosome 3 (p-arm green and q-arm red) showing two normal chromosome 3, whereas the small marker chromosome (from Fig. 5F) does not contain chromosome 3-derived telomeric DNA. **H** In situ hybridization to permeabilized cell nuclei identifying cytoplasmic mRNA coding for medium chain acyl coenzyme A dehydrogenase (MCAD) (green). Details on experimental conditions can be obtained by contacting the authors.

(smooth brain) and a characteristic face. The syndrome is associated with a microdeletion of chromosome 17p13.3. Since then, FISH has proven to be an invaluable tool for detection of such submicroscopic microdeletions. In fact, FISH has now replaced high-resolution cytogenetic approaches when a specific microdeletion is being investigated. Thus, locus specific probes are now available for diagnosing DiGeorge syndrome (del 22q11.2) (Fig. 4D shows such an example in a dual-color approach), Prader-Willi and Angelman syndromes (del 15q11-13), retinoblastoma (del 13q14), Wolf-Hirschorn syndrome (del 4p13), Williams syndrome (del 7q11.23), Rubinstein-Taybi syndrome (del 16p13.3), X-linked ichthyosis (Xp22.3), and aniridia/Wilms tumor (del 11p13) as examples and more will probably prevail. However, it should be noted that for the listed syndromes microdeletions account for a varying degree of the etiological factors. In for example Prader-Willi and Angelman syndromes, uniparental disomy and imprinting are also important etiological factors that cannot be detected by FISH.

The developments of arm specific probes for each chromosome increased the possibilities for identifying and characterizing intra-chromosomal structural abnormalities that by conventional banding methods can be difficult to ascertain (Fig. 4E).

Interphase cytogenetic diagnosis is currently used for screening prenatal samples (Kuo et al. 1991, Zahed et al. 1992) for numerical aberrations of chromosomes 13, 18, 21, X and Y (Fig. 4F). Time is often a crucial factor and it is therefore of great value that such chromosomal aberrations can be diagnosed within one day using interphase cytogenetics in a multi-color approach. Screening for other aneuploidies is also possible but of limited value due to the low frequency of such abnormalities in prenatal samples. Chromosome microdeletions have also been successfully diagnosed using interphase cytogenetics on peripheral blood smear leukocytes (Novelli et al. 1999). Chromosome microduplications is another group of chromosomal disorders that can be diagnosed by interphase cytogenetics (Shaffer et al. 1997).

Mosaicism is another area where FISH-analysis is superior, as rapid screening of a large number of metaphases or interphase nuclei can be accomplished giving more accurate estimates of the proportion of cells containing the chromosomal abnormality (Kalousek et al 1996). Such information is relevant for better genetic counseling as a high proportion of cells containing an aberrant chromosome directly correlates to a more sever phenotype (Schinzel 1983). Also, in infertility, the rapid and direct screening of large numbers of sperm cells for aneuplodies and structural abnormalities is important (van Hummelen et al. 1996).

In preimplantation genetic diagnosis (PGD) interphase FISH (Delhanty et al. 1993, Verlinsky et al. 1996) also allows for a rapid screening of the most common chromosomal aneuploidies. This offers the opportunity to avoid the implantation of fertilized eggs carrying a numerical chromosome aberration. Investigation of aneuploidies for other chromosomes is also possible and interphase FISH has been employed for identifying chromosome balance in early embryos from patients with reciprocal translocations (Pierce et al. 1998).

In oncology the possibility of using interphase-FISH with translocation-specific probes is of particular interest when the quality of metaphases are poor or if non-viable fixed tumors are used. Furthermore, analysis for specific chromosomal abnormalities in tissues directly without the need for cell propagation may give important information on heterogeneity. The types of chromosome aberrations that can be diagnosed by interphase-FISH are deletions of tumor-suppressor genes e.g. retinoblastoma gene (Chang et al. 1999), amplification of oncogenes e.g. N-myc in neuroblastoma (Hachitanda et al. 1999) and specific types of translocations e.g. Philadelphia chromosome (Yanagi et al. 1999) as well as numerical aberrations (Worsham et al. 1999).

Many malignancies are also characterized by the presence of many cell subpopulations with different genome rearrangements. In these situations molecular cytogenetic analysis is superior to molecular techniques. Standard molecular techniques such as for example PCR would give data that represents the average of aberrations over the whole cell population. When chromosome analysis is undertaken, it is evident that inspection and characterization of the genome of single cells takes place, and thus data on cell diversity is obtained as well. With regard to single cell analysis, interphase-FISH sometimes offers advantages over conventional banding and metaphase FISH. This is due to the fact that these metaphases are obtained by initial cell culturing. When analyzing mixed cell populations there is no guarantee that all subpopulations proliferate to a similar extent during cell culturing. This means that there is a risk of getting very distorted estimates on the distribution of cell subpopulations after culturing. Interphase-FISH offers a simple solution to this problem, since no cellular propagation is involved, therefore allowing for true estimates of cell fractions. Of course, interphase-FISH has severe problems in terms of more detailed screening of the whole genome, but if detection of specific regions is needed, this method offers many advantages, including possibilities for analyzing archive materials or solid tumors that have very low mitotic indexes.

It has recently been found that the subtelomeric regions have the highest number of genes per unit chromosome in the entire genome. Cryptic rearrangements involving these regions is a new area of genetic disorders (involving mental retardation and mild to severe dysmorphism) that involves both balanced and unbalanced rearrangements. FISH using telomeric probes specific for each chromosome-arm can been used to detect such abnormalities in subjects who apparently had normal karyotypes by conventional banding methods (Flint et al 1995, Giraudeau et al. 1997). These terminal segmental aneusomies have been termed telomere syndromes and may account for 10-15% of mentally retarded children having normal banding analyses. Imbalances or rearrangements involving the subtelomeric regions have also been identified in malignancy. Figure 4G shows an example of a ring chromosome 3 that has lost both subtelomeric regions.

II. Characterization of cryptic chromosomal aberrations (whole genome screening). In a historic perspective it was a major breakthrough in 1956 when it was established that man has 46 human chromosomes (Fig. 5A) allowing for a new type of method to screen the whole genome. Refinements of the chromosome preparation techniques allowed a grouping of the chromosomes by size (Fig. 5B) and when banding techniques in the early 1970s were established, it became possible to characterize each individual chromosome (Fig. 5C,D). The development of FISH-techniques (Fig. 4) was encouraged by an ever-increasing number of identified but yet uncharacterized chromosomal aberrations such as simple or complex translocations and the presence of marker and derivative chromosomes. These classic FISH techniques allowed characterization of specific chromosomal rearrangements which was soon followed by the development of multi-color FISH techniques that, in a single hybridization experiment, allowed for both identification and characterization of several types of chromosome aberrations. In a multi-color approach with a few painting probes one must, however, have a clue as to which chromosomes are involved in an aberration, whereas with the more recent refinements of the FISH-techniques such as 24-color karyotyping or CGH-analysis, one can rapidly and reliable screen for many types of chromosomal aberrations and imbalances, without prior knowledge of the involved chromosomes.

In clinical genetics and oncology translocations ranging from simple (involving two chromosomes) to complex rearrangements (involving three or more chromosomes) is another area where 24-color karyotyping has proven valuable (Rao et al. 1998, Veldman et al. 1997). These types of chromosomal rearrangements cannot be identified by CGH-analysis, that on the other hand, is efficient in identifying genomic imbalances such as deletions or amplifications (Kallioniemi et al.1992). Figure 5E shows an example of a trisomy for chromosome 18 detected by CGH-analysis.

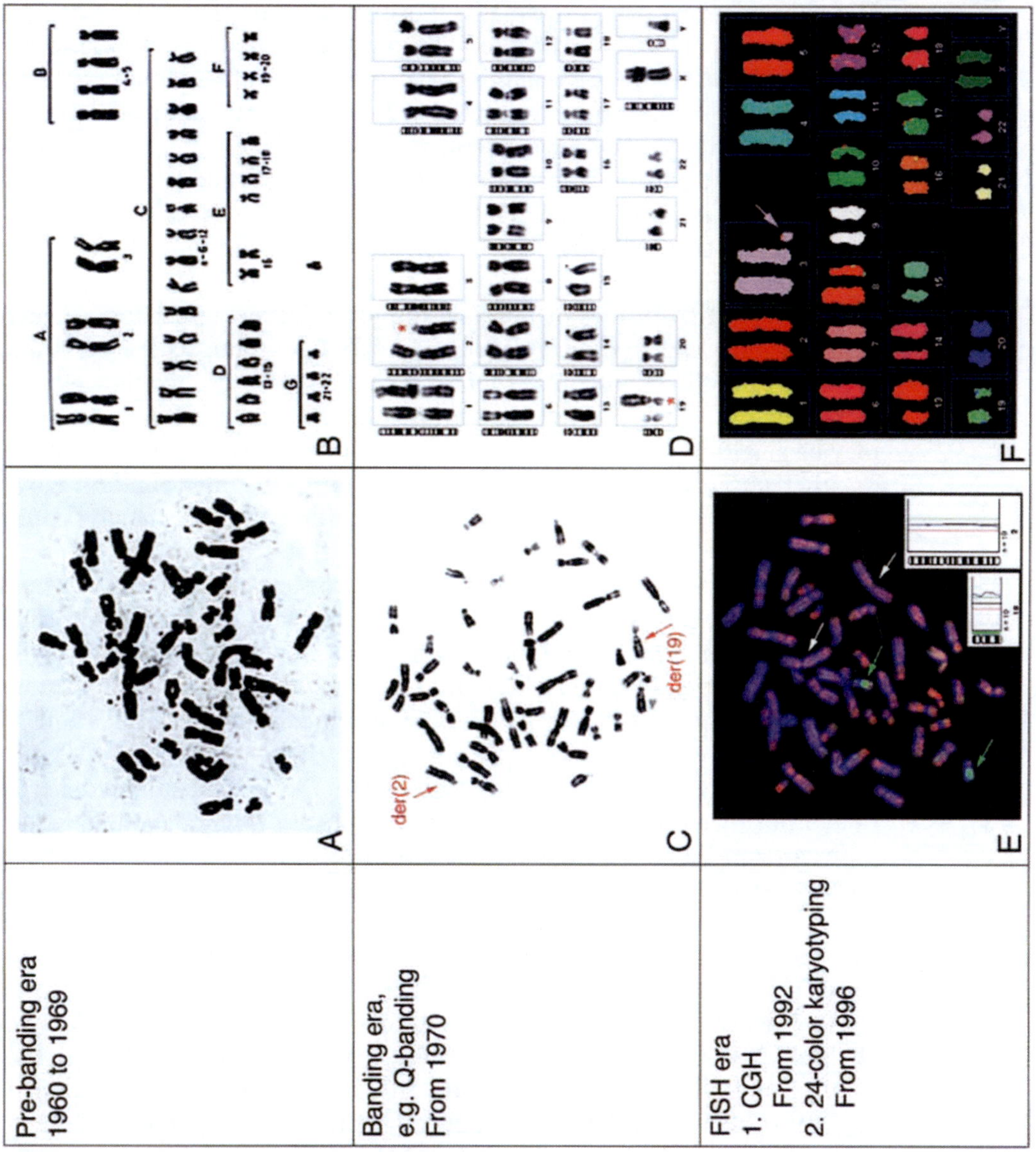

Fig. 5. Technical evolution in whole genome screening techniques for ascertainment of karyotype. **A** Metaphase spread of unbanded chromosomes (Courtesey of Prof. G.B. Petersen) **B** Karyotype based on chromosome size before banding techniques were available. **C** Q-banded metaphase spread showing a t(2;19). **D** Automated karyotyping with chromosome ideograms of the metaphase from C. **E** Comparative genomic hybridization analysis identifying a trisomy 18 (*insert* shows 2 of the 24 computed red-green ratio profiles obtained by the analysis) – chromosome 18 (*green arrows*) and chromosome 2 (*white arrows*). **F** Spectral karyotyping identifying a supernumerary marker chromosome (*gray arrow*) derived from chromosome 3. Details on experimental conditions can be obtained by contacting the authors.

The chromosomal origin of small supernumerary marker (SMC) or ring chromosomes are virtually impossible to establish by conventional cytogenetics, as they contain no banding patterns due to their small size. The incidence of SMCs has been estimated at 3–4 per 1000 patients with mental retardation. The phenotypic consequences of SMCs are difficult to predict especially if the chromosomal origin of the SMCs are unknown. Therefore various FISH techniques have been developed to identify the chromosomal origin of these markers but without any clue as to their identity, the use of classic FISH techniques requires the iterative hybridization of chromosome painting probes or possibly the use of reverse painting. In such cases 24-color karyotyping is both time- and labor-saving as the technique involves analysis of the entire genome in a single hybridization experiment (Fig. 5F). CGH-analysis in conjunction with other FISH techniques may be an alternative way of identifying such chromosomal abnormalities or other numerical aberrations for a whole chromosome (Fig. 5E). However, if a mosaicism is present in 50% or less of the cells, CGH-analysis may not be able to identify the presence of such abnormalities.

In research on archival material of, for example brain tumors and breast cancers, oncogenes and amplicons that are related to certain types of tumors have been identified by CGH-analysis (Du Manoir et al. 1993, Nishizaki et al. 1997). These findings may turn out to be important genetic markers for an estimation of prognosis and stratification of therapy.

In oncology, it is often difficult to ascertain the chromosomal make-up of a tumor by conventional banding methods even where dependable metaphases can be obtained, since the chromosomes often exhibits a poor banding pattern. In spite of this, particular chromosome abnormalities, especially translocations, characteristic for certain types of solid tumors or leukemias, have been established by banding methods and more recently in combination with FISH. Such specific alterations have in some cases correlated with deregulation of translocated genes (Iida et al. 1997) and in other cases new fusion genes are formed (Rubin et al. 1988). The Philadelphia-chromosome in CML represents such a new fusion gene and was the first specific chromosomal rearrangement related to a specific type of leukemia (Nowell and Hungerford 1960). Since then tumor specific non-random chromosome aberrations have been identified. In as many as 80% of the cases, a complete karyotype cannot be established by conventional banding methods even when combined with classic FISH techniques. By including 24-color karyotyping in the analysis, most of these cases can now be characterized (Schröck et al. 1996, Speicher et al. 1996, Veldman et al. 1997, Kerndrup & Kjeldsen 2001) which in turn may further the number of recurrent chromosome abnormalities specific for individual tumors or tumor stages.

To further characterize points of breakage found with whole genome screening techniques, cloned DNA sequences in YACs, PACs, BACS etc. that previously have been mapped to specific regions can be used. These cloned DNA sequences covering all human chromosomes are stored in several places (see referencelist for web-sites) and it is possible to obtain small samples of these clones which then can be amplified, labeled and hybridized to samples for further characterization of the breakpoints.

Chromosomal aberrations identified by conventional banding methods and by FISH techniques can unequivocally be described by an internationally agreed nomenclature, "The International System for Human Cytogenetic Nomenclature" (ISCN). The first conference on this issue was held in Paris in 1971 and the latest revised version ISCN 1995 includes a section for describing abnormalities that have been found by FISH. A standing international committee ensures that the nomenclature is kept up to date, but now a nomenclature for describing findings obtained by the new 24-color karyotyping and CGH-analysis is needed.

Comparative cytogenetics

Chromosomal homologies between different species have been established using an approach termed ZOO-FISH (Solinas-Toldo et al. 1995). The introduction of chromosome painting to the field of comparative cytogenetics has greatly added to the current understanding of chromosome changes that occurred during the evolution of species. The first mammalian karyotype that was painted entirely with the set of chromosome-specific painting probes in a single hybridization experiment was the Indian muntjac (Yang et al. 1995). This deer species has the lowest chromosome number known in mammals (2n=6/7). Since then, all 24 human chromosomes have become available and karyotype analyses of various gibbon species have been performed by assigning the hybridization patterns from all 24 human chromosome-specific painting probes. It was found that the gibbon genome is considerably different from other hominoid primates as all gibbon species are characterized by a highly disrupted chromosomal synteny. Human chromosome specific probes have even been used to paint regions of homology in distant species (Scherthan

of conserved regions in the absence of obvious cytogenetic homologies between human and other species.

RNA expression in situ

Much work in the area of gene expression using FISH technique has been related to tumors, particularly in endocrine systems. Other emerging areas using in situ techniques for RNA expression are in the fields of oncogenes, growth factors, growth factor receptors, adhesion molecules, genes conferring resistance to chemotherapy and cytokines in inflammatory disease (for review McNicol and Farquharson 1997).

The identification of gene expression by the detection of mRNA is usually performed where the levels of protein product are below the limits of immunohistochemical detection. In situ demonstration of RNA expression is of great value in confirming a tumor as the source of hormone production where high circulating hormone levels are present but where the tumor is immunonegative. However, detection of high levels of mRNA does not necessarily imply a high level of functional gene product.

When RNA expression is investigated in situ, it is important to preserve the RNA in tissues. Most authors stress the importance of rapid fixation because of the potential for degradation by the ubiquitous RNases. However, recent reports have shown that RNA can be extracted from postmortem tissues and that hybridization to mRNA can be successfully detected in tissues unfixed for at least 24 h or obtained at autopsy (Gilmore et al. 1993). The half-life of stable RNAs in tissues has been estimated to be 24 h. To check for general preservation of RNA in tissues, the use of probes binding to housekeeping genes or of polydT probes that will hybridize with the polyA tail of mRNAs has been used. However, it should be remembered that the binding pattern of polydT may vary between different tissues and a specific probe signal may be demonstrated even when polydT is negative. Moreover, when using this technique it is important to incorporate controls to assess whether the signal, represents a specific hybridization to the appropriate target sequence.

Expression studies can also be performed in such a way as to allow direct visual evidence of gene expression from a particular chromosome. The technique is technically demanding but refinements of the technique may overcome the present obstacles that restrict a more general use of this FISH variant. An example of a potential use in the future may be in relation to the diagnosis of Beckwith-Weidemann syndrome, as the pattern of expression of H19 and IGF2 genes on the short arm of chromosome 11 in single cells may provide evidence of uniparental disomy associated with this disorder (Weksberg et al. 1993).

In gene therapy monitoring of transcripts from the transgene may be another exciting new area for FISH (Fig. 4H). A general problem in gene therapy at present is instability of the transgene expression, either due to loss of the transgene or due to silencing. FISH-based demonstration of presence/expression of the transgene within target cells would clearly facilitate the development of gene therapy protocols.

Microbiology

Infections can be conceptually characterized as a series of discrete events, beginning with the invasion of the host and continuing with multiplication at the site of entry and subsequent dissemination throughout the organism. The fate of the microorganism and host is determined by the interactions of the microorganism and host cells and by the general and specific immune defenses mounted in response to the infection. Understanding the mechanisms of infections rests on insights derived originally from quantitative assays of infectious microorganisms in organ systems and light and electron microscopic examination of the tissues to visualize the infectious agents and pathological sequelae of the multiplication.

In situ hybridization is one of the more recent and important developments in order to detect, quantitate and locate microorganisms in different cell types at defined anatomical sites (for review McNicol and Farquharson 1997). Because only small amounts of the genome and transcripts of a microorganism need be present for in situ hybridization detection, this technique has opened a whole new line of investigation that is not restricted to a search for the usual indications of infection, e.g. tissue damage, viral particles and proteins. In situ hybridization complements other types of investigation such as immunofluorescence and immunocytochemistry.

Physical mapping of genes on chromosomes

The development of an accurate gene map is one of the main goals of genetic studies in any organism and the localization of genes to specific chromosomes and chromosome segments is one of the most rapidly advancing branches of human genetics. The methodology includes linkage studies, the use of markers or translocated chromosomes in family studies, in vitro fusion of human cells with cells from other species, various molecular techniques and in situ hybridization.

Since ISH was used for the first time in 1981 for localization of a single gene on metaphase chromosomes, the technique has proven to be an invaluable gene-mapping tool (Heiskanen et al. 1996). Further refinements of the use of FISH as a gene-mapping tool has shown that it can be applied to interphase nuclei, mechanically stretched chromosomes (Laan et al. 1995) or DNA fibers (Florijn et al. 1996) to order DNA sequences or genes with increasingly greater resolution. Now individual genes can be mapped as soon as they are cloned.

To determine the chromosome to which a probe has hybridized, a technique for chromosome identification compatible with CISS hybridization has been developed (Klever et al. 1991). It is possible to do Giemsa banding after hybridization but this process requires the relocalization of specific metaphase spreads. Another strategy involves cohybridization with a differently labeled probe (or probe set). Alu DNA BLUR clone has been used because it can give a banding pattern resembling R banding (Baldini and Ward 1991). For simultaneous hybridization strategies, it is important to have different labeling procedures that result in comparable fluorescent signals.

The use of Fiber-FISH has contributed significantly to positional cloning efforts by facilitating physical mapping of probes from the chromosome region of interest by rapid and direct visualization of clone order, and sizes of gaps and overlaps (Heng et al. 1992, Haaf and Ward 1994, Michalet et al. 1997). Even copy numbers of moderately repeated sequences (Shiels et al. 1997) as well as viral integration (Lestou et al. 1996) have been assessed by Fiber-FISH.

Fiber-FISH has proven to be a valuable tool also in contig map construction. The technique also provides a rapid and informative means to study gene rearrangements, because once a contig map of a well-covered disease region has been made, it is a relatively simple task to create a Fiber-FISH color bar code by selecting adequately spaced and/or overlapping clones and proper color assignment of individual DNAs. Gene rearrangements will then be apparent as a disruption of the wild-type bar code thereby easily characterizing either deletion- or translocation breakpoints scattering several hundreds of kilobases at a kilobase resolution (Vaandrager et al. 1996). The ability by Fiber-FISH to scan large genomic regions for gene-rearrangements in a single FISH experiment makes such an approach advantageous over Southern hybridization and PCR-amplification techniques.

Structure of human chromosomes and interphase nuclei

FISH using specific sequences as probes has been extensively used in the study of chromatin structure in chromosomes and interphase nuclei, often supplemented with specialized imaging systems such as confocal microscopy. A new, exciting method for studying spatial positions of sequences is the so-called CO-FISH, which can demonstrate close proximity of sequences. CO-FISH is an acronym for chromosome oriented fluorescence in situ hybridization and has until now mostly been used for detecting obligate chromosome inversions associated with isochromosomes (Bailey et al. 1996). The technique has also been used in the investigation of lateral asymmetry which refers to the unequal fluorescent intensity between adjacent sister chromatids (Goodwin et al. 1996).

The behavior of chromosomes during meiosis has been successfully investigated by FISH techniques, as conventional banding techniques here cannot unequivocally identify individual chromosomes. In this type of experiments FISH allows normal and abnormal chromosomes to be tracked during all stages of meiosis (Goldman et al. 1992, Barlow et al. 1996). FISH has also been found to be an efficient and reliable method to assess replication timing of loci and that allele-specific replication timing is tissue specific for many loci (Selig et al. 1992, Kitsberg et al. 1993).

FISH in combination with immunological methods

The so-called MAC (morphology, antibody, chromosome) technique is an acronym for a technique which allows karyotypic analysis of morphologically and immunotypically classified cells (Knuutila and Keinanen 1985). The technique has also been successfully applied to analyze mitotic cells in hematological malignancies (Tianen et al. 1992) and in the analysis of meiosis I of human spermatocytes (Barlow et al. 1996).

When analyzing subpopulations of cells as in residual disease, (Anastasia et al. 1991) or to predict hematological relapse, (Bernell et al. 1996) the technique FICTION (fluorescence immunophenotyping and interphase cytogenetics) has been used which combines immunophenotyping and interphase FISH to identify the chromosomal status of cells with specific immunophenotypes (Weber-Matthiesen et al. 1993).

Summary

An overall summary of the advantages of FISH techniques is listed in Table 5. The listed advantages clearly make FISH a serious competitor to traditional banding techniques. Whether FISH in the future will replace traditional cytogenetic techniques is at present difficult to say, and will to some extent depend on the future developments in FISH-technology. The following chapters in the book will give numerous examples of these future developments.

Table 5. Several advantages apply to the use of FISH

Sensitivity	In clinical cytogenetics and oncology FISH has the potential to detect cryptic chromosomal deletions and rearrangements, which cannot be detected by conventional banding methods. In microbiology it is possible to detect the presence of microorganisms at an early stage. RNA expression from endogenous or exogenous genes can be detected.
Specificity	By choosing a particular probe or probes and relevant methodology, chromosomal material of unknown or uncertain origin can be identified, and in research the chromosomal localization of specific DNA fragments/sequences can be achieved.
Efficiency	FISH also allows rapid screening for large number of metaphases for one or more target sequences. Specific genetic information is obtained for individual cells and can be correlated to other types of information as FISH can be combined with immunophenotyping of individual cells.
Applicability	Many tissues and types of cells are not directly accessible for conventional cytogenetic methods, but can be examined by FISH analysis on interphase nuclei. Thus, examination of resting cells in interphase may be a much larger and potentially more informative cell population. Also when a new gene or DNA sequence is cloned it is relatively easy to utilize it as a diagnostic tool.
Economy	Because of the simplicity of the basic FISH concept, the technique can now relatively easily be established in almost any laboratory. Furthermore, after establishment of the equipment which is relatively costly, each assay is rather inexpensive. The development of CGH-analysis and 24-color karyotyping has reduced the time and labor-consumption in situations where FISH is used without any knowledge of the chromosomal aberration.

Acknowledgements. We wish to thank. Anja Christoffersen, Kirsten Knudsen, Alice Nielsen and Helle Strømkjær for expert technical assistance. Dr. Gitte Kerndrup, Odense University Hospital is acknowledged for fruitful collaboration with the SKY-system. This work was supported in part by The Karen Elise Foundation, The Danish National Cancer Society, and the Danish Center for Molecular Gerontology.

References

Anastasi J, Vardiman JW, Rudinsky R, Patel M, Nachman J, Rubin CM, Le Beau MM (1991) Direct correlation of cytogenetic findings with cell morphology using in situ hybridization: an analysis of suspicious cells in bone marrow specimens of two patients completing therapy for acute lymphoblastic leukemia. Blood 77:2456-2462

Avery OT, MacLeod CM, McCarty M (1944) Studies on chemical nature of substance inducing transformation of pneomococcal types; induction of transformation by deoxyribonucleic acid fraction isolated from pneumococcus type III. J Exp Biol Med 79:137-158

Bailey SM, Goodwin EH, Meyne J, Cornforth MN (1996) CO-FISH reveals inversions associated with isochromosome formation. Mutagenesis 11:139-144

Baldini A, Ward DC (1991) In situ hybridization banding of human chromosomes with Alu-PCR products: a simultaneous karyotype for gene mapping studies. Genomics 9:770-774

Barlow AL, Hulten MA (1996) Combined immunocytogenetics and molecular cytogenetic analysis of meiosis I human spermatocytes. Chromosome Res 4:562-573

Bernell P, Avidsson I, Jacobsen B, Hast R (1996) Fluorescence in situ hybridization in combination with morphology detects minimal residual disease in remission and heralds relapse in acute leukemia. Br J Haematol 95:666-672

Bobrow MN, Harris TD, Shaughnessy KJ, Litt GJ (1989) Catalyzed reporter deposition, a novel method of signal amplification. Application to immunoassays. J Immunol Methods 125:279-285

Bolzer A, Craig JM, Cremer T, Speicher MR (1999) A complete set of repeat-depleted, PCR-amplifiable, human chromosome-specific painting probes. Cytogenet Cell Genet 84:233-240

Carter NP, Ferguson-Smith MA, Pemyman MT Telenius H, Pelmear AH, Leversha MA, Glancy MT, Wood SL, Cook K, Dyson HM, et al. (1992) Reverse chromosome painting: a method for for rapid analysis of aberrant chromosomes in clinical cytogenetics. J Med Genet 29:299-307

Casperson T, Faber S, Foley GE, Kudynowski J, Modest EJ, Simonsson E, Wagh U, Zech L (1968) Chemical differentiation along metaphase chromosomes. Experimental Cell Research 49:219-222

Caspersson T, Lomakka G, Zech L (1971) The 24 fluorescence patterns of the human metaphase chromosomes – distinguishing characters and variability. Hereditas 67:89-102

Chang H, Bouman D, Boerkoel CF, Stewart AK, Squire JA (1999) Frequent monoallelic loss of D13S319 in multiple myeloma patients shown by interphase fluorescence in situ hybridization. Leukemia 13:105-109

Chaudhuri JP, Vogel W, Voiculescu I, Wolf U (1971) A simplified method of demonstrating Giemsa-band pattern in human chromosomes. Humangenetik 14:83-83

Christian AT, Garcia HE, Tucker JD (1999) PCR in situ followed by microdissection allows whole chromosome painting probes to be made from single microdissected chromosomes. Mamm Genome 10:628-631

Chudoba I, Plesch A, Lorch T, Lemke J, Claussen U, Senger G (1999) High resolution multicolor-banding: a new technique for refined FISH analysis of human chromosomes. Cytogenet Cell Genet 84:156-160

Choo KHA (1994) Methods in Molecular Biology, Vol. 33: In situ hybridization Protocols. Humana Press Inc., Totowa, NJ

Cohen SN, Chang ACY, Boyer HW, Helling RB (1973) Construction of biologically functional bacterial plasmids in vitro. Proc Natl Acad Sci USA 70:3240-3244

Craig JM, Kraus J, Cremer T (1997) Removal of repetitive sequences from FISH probes using PCR-assisted affinity chromatography. Hum Genet 100:472-476

Cremer T, Landegent J, Brückner A, Scholl HP, Schardin M, Hager HD, Devilee P, Pearson P, van der Ploeg M (1986) Detection of chromosome aberrations in the human interphase nucleus by visualization of specific target DNAs with radioactive and non-radioactive in situ hybridization techniques: diagnosis of trisomy 18 with probe L1.84. Hum Genet 74:346-352

Cremer T, Lichter P,Borden J, Ward DC, Manuelidis L (1988) Detection of chromosome aberrations in metaphase and interphase tumor cells by in situ hybridization using chromosome specific library probes. Hum Genet 80:235-246

Daubendiek SL and Kool ET (1997) Generation of catalytic RNAs by rolling transcription of synthetic DNA nanocircles. Nature Biotech 15:273-277

Deaven LL, Van Dilla MA, Bartholdi MF, Carrano AV, Cram LS, Fuscoe JC, Gray JW, Hildebrand CE, Moyzis RK, Perlman J (1986) Construction of human chromosome-specific DNA libraries from flow-sorted chromosomes. Cold Spring Harb Symp Quant Biol 51:159-167

Delhanty JDA, Griffin DK, Handyside AH, Harper J, Atkinson GHG, Pieters MHEC, Winston RML (1993) Detection of aneuploidy and chromosomal mosaicism in human embryos during preimplantation sex determination by fluorescent in situ hybridization, (FISH). Hum Mol Genet 2:1183-1185

Dobyns WB, Curry CJR, Hoyme HE, Turlington L, Ledbetter DH (1991) Clinical and molecular diagnosis of Miller-Dieker syndrome. Am J Hum Genet 48:584-594

du Manoir S, Speicher MR, Joos S, Schröck E, Popp S, Dohner H, Kovacs G, Robert-Nicoud M, Lichter P, Cremer T (1993) Detection of complete and partial chromosome gains and losses by comparative genomic hybridization. Hum Genet 90:590-610

Edwards JH, Harnden DG, Cameron AH, Crosse VM, Wolff OH (1960) A new trisomic syndrome. Lancet 1:787-790

Egholm M, Buchardt O, Christensen L, Behrens C, Freier S, Driver DA, Berg RH, Kim SK, Norden B, Nielsen PE (1993) PNA hybridizes to complementary oligonucleotides obeying the Watson-Crick hydrogen bonding rules. Nature 365:566-568

Flemming W (1882) Zellsubstanz, Kern und Zellteilung. Leipzig.

Flint J, Wilkie AOM, Buckle VJ, Winter RM, Holland AJ, McDermid HE (1995) The detection of subtelomeric chromosomal rearrangements in idiopathic mental retardation. Nat Genet 9:132-140

Florijn RJ, van de Rijke FM, Vrolijk J, Blonden LAJ, Hofker MH, den Dunnen JT, Tanke HJ, van Ommen GJB, Raap AK (1996) Exon mapping by fiber-FISH or LR-PCR. Genomics 38:277-282

Ford CE, Jacobs PA, Lajtha LG (1958) Human somatic chromosomes. Nature, London; 181:1565-1568

Ford CE, Jones KW, Polani PE, Almeida JCDE, Briggs JH (1959) A sex-chromosomal anomaly in a case of gonadal dysgenesis (Turner's syndrome). Lancet 1:711-713

Gerhard DS, Kawasaki ES, Bancroft FC, Szabo P (1981) Localization of a unique gene by direct hybridization in situ. Proc Natl Acad Sci USA 78:3755-3759

Gilmore JH, Lawler CP, Eaton AM, Mailman RB (1993) Postmortem stability of dopamine D1 receptor mRNA and D1 receptors. Brain Res Mol Brain Res 18:290-296

Giraudeau F, Aubert D, Young I, Horsley S, Knight S, Kearney L, Vergnaud G, Flint J (1997) Molecular cytogenetic detection of a deletion of 1p36.3. J Med Genet 34:314-317

Goldman ASH, Hulten MA (1992) Chromosomes in situ suppression hybridization in human male meiosis. J Med Genet 29:98-102

Goodwin EH, Meyne J, Bailey SM Quigley D (1996) On the origin of lateral assymetry. Chromosoma 104:345-347

Gosden J, Hanratty D, Starling J, Fantes J, Mitchell A, Porteus D (1991) Oligonucleotide-primed in situ DNA synthesis (PRINS): a method for chromosome mapping, banding, and investigation of sequence organization. Cytogenet Cell Genet 57:100-104

Guan X-Y, Trent JM, Meltzer PS (1993) Generation of band-specific painting probes from a single microdissected chromosome. Hum Mol Genet 2:1117-1121

Guan X-Y, Zhang H, Bittner M, Jiang Y, Meltzer PS, Trent JM (1996) Chromosome arm painting probes. Nat Genet 12:10-12

Haaf T and Ward DC (1994) High resolution ordering of YAC contigs using extended chromatin and chromosomes. Hum Mol Genet 3:629-633

Hachitanda Y, Saito M, Mori T, Hamazaki M (1997) Application of fluorescence in situ hybridization to detect N-myc (MYCN) gene amplification on paraffin-embedded tissue sections of neuroblastomas Med Pediatr Oncol 2:135-138

Harper M & Saunders G (1981) Localization of single copy DNA sequences on G-banded human chromosomes by in situ hybridization. Chromosoma 83:431-439

Heiskanen M, Peltonen L, Palotie A (1996) Visual mapping by high resolution FISH. Trends Genet 12:379-382

Heng HHQ, Squire J, Tsui LC (1992) High resolution mapping of mammalian genes by in situ hybridization to free chromatin. Proc Natl Acad Sci USA 89:9509-9513

Hindkjær J, Koch J, Mogensen J, Kølvraa S, Bolund (1994) Primed InSitu (PRINS) labeling of DNA. In: Choo KHA (ed.) In situ hybridization Protocols. Methods in Molecular Biology, vol 33. Humana Press Inc., Totowa, NJ, pp 95-107

Hsu TC (1952) Mammalian chromosomes in vitro. I. The karyotype of man. J Hered 43:167-172

Hulten MA, Gould CP, Goldman ASH, Waters JJ (1991) Chromosome in situ suppression hybridization in clinical cytogenetics. J Med Genet 28:577-582

Iida S, Rao PH, Butler M, Corradini P, Boccadoro M, Klein B, Chaganti RS, Dalla-Favera R (1997) Deregulation of MUM1/IRF4 by chromosomal translocation in multiple myeloma. Nat Genet 17:226-230

Kallioniemi A, Kallioniemi O-P, Sudar D, Rutovitz D, Gray JW, Waldman F, Pinkel D (1992) Comparative genomic hybridization for molecular cytogentic analysis of solid tumours. Science 258:818-821

Kalousek DK, Langlois S, Robinson WP, Telenius A, Bernard L, Barrett IJ, Howard-Peebles PN, Wilson RD (1996) Trisomy 7 CVS mosaicism: pregnancy outcome, placental and DNA analysis in 14 cases. Am J Med Genet 65:348-352

Karhu R, Rummukainen J, Lörch T, Isola J (1999) Four-Color CGH: A new method for quality control of comparative genomic hybridization. Genes, Chromosomes Cancer 24:112-118

Kelly TJ, Smith HO (1970) A restriction enzyme from Hemophilus influenzae. II. J Mol Biol 51:393-409

Kerndrup G & Kjeldsen E (2001) Acute leukemia cytogenetics: an evaluation of combining G-band karyotyping with multicolor spectral karyotyping. Cancer Genet Cytogenet 1:7-11

Kievits T, Dauwerse JG, Wiegant J, Devilee P, Breuning MH, Cornelisse CJ, van Ommen GJB et al (1990) Rapid subchromosomal localization of cosmids by nonradioactive in situ hybridization. Cytogenet Cell Genet 53:134-136

Kitsberg D, Selig S, Brandeis M, Simon I, Keshet I, Driscoll DJ, Nicholls RD, Cedar H (1993) Allele-specific replication timing of imprinted gene regions. Nature 364:459-463

Klever M, Grond-Ginsbach C, Scherthan H, Schroeder-Kurth TM (1991) Chromosomal in situ suppression hybridization after Giemsa banding. Hum Genet 86:484-486

Knuutila S, Keinanen M (1985) Chromosome banding technique for morphologically classified cells. Cytogenet Cell Genet 39:70-72

Koch JE, Kølvraa S, Petersen KB, Gregersen N, Bolund L (1989) Oligo-nucleotide-priming methods for the chromosome-specific labelling of alpha satellite DNA in situ. Chromosoma 98:259-265

Koch JE, Hindkjær J, Kølvraa S, Bolund L (1995) Construction of a panel of chromosome-specific oligonucleotide probes (PRINS-primers) useful for identification of individual human chromosomes in situ. Cytogenet Cell Genet 71:142-147

Kumar R, Singh SK, Koshkin AA, Rajwanshi VK, Meldgaard M, Wengel J (1998) The first analogues of LNA (locked nucleic acids): phosphorothioate-LNA and 2'thio-LNA. Bioorg Med Chem Lett 8:2219-2222

Kuo W-L, Tenjin H, Segraves R, Pinkel D, Golbus MS, Gray J (1991) Detection of aneuploidy involving chromosomes 13, 18, 21 by fluorescence in situ hybridization (FISH) to interphase and metaphase amniocytes. Am J Hum Genet 49:112-119

Kuwano A, Ledbetter SA, Dobyns WB, Emanuel BS, Ledbetter DH (1991) Detection of deletions and cryptic translocations in Miller-Dieker syndrome by in situ hybridization. Am J Hum Genet 49:707-714

Laan M, Kallioniemi O-P, Hellsten E, Alitalo K, Peltonen L, Palotie A (1995) Mechanically stretched chromosomes for high resolution FISH mapping. Genome Res 5:13-20

Lander ES, Linton LM, Birren B, Nusbaum C, Zody MC, Baldwin J, Devon K, Dewar K, Doyle M, FitzHugh W, Funke R, Gage D, Harris K, Heaford A, Howland J, Kann L, Lehoczky J, LeVine R, McEwan P, McKernan K, Meldrim J, Mesirov JP, Miranda C, Morris W, Naylor J, Raymond C, Rosetti M, Santos R, Sheridan A, Sougnez C, Stange-Thomann N, Stojanovic N, Subramanian A, Wyman D, Rogers J, Sulston J, Ainscough R, Beck S, Bentley D, Burton J, Clee C, Carter N, Coulson A, Deadman R, Deloukas P, Dunham A, Dunham I, Durbin R, French L, Grafham D, Gregory S, Hubbard T, Humphray S, Hunt A, Jones M, Lloyd C, McMurray A, Matthews L, Mercer S, Milne S, Mullikin JC, Mungall A, Plumb R, Ross M, Shownkeen R, Sims S, Waterston RH, Wilson RK, Hillier LW, McPherson JD, Marra MA, Mardis ER, Fulton LA, Chinwalla AT, Pepin KH, Gish WR, Chissoe SL, Wendl MC, Delehaunty KD, Miner TL, Delehaunty A, Kramer JB, Cook LL, Fulton RS, Johnson DL, Minx PJ, Clifton SW, Hawkins T, Branscomb E, Predki P, Richardson P, Wenning S, Slezak T, Doggett N, Cheng JF, Olsen A, Lucas S, Elkin C, Uberbacher E, Frazier M, Gibbs RA, Muzny DM, Scherer SE, Bouck JB, Sodergren EJ, Worley KC, Rives CM,

Gorrell JH, Metzker ML, Naylor SL, Kucherlapati RS, Nelson DL, Weinstock GM, Sakaki Y, Fujiyama A, Hattori M, Yada T, Toyoda A, Itoh T, Kawagoe C, Watanabe H, Totoki Y, Taylor T, Weissenbach J, Heilig R, Saurin W, Artiguenave F, Brottier P, Bruls T, Pelletier E, Robert C, Wincker P, Smith DR, Doucette-Stamm L, Rubenfield M, Weinstock K, Lee HM, Dubois J, Rosenthal A, Platzer M, Nyakatura G, Taudien S, Rump A, Yang H, Yu J, Wang J, Huang G, Gu J, Hood L, Rowen L, Madan A, Qin S, Davis RW, Federspiel NA, Abola AP, Proctor MJ, Myers RM, Schmutz J, Dickson M, Grimwood J, Cox DR, Olson MV, Kaul R, Raymond C, Shimizu N, Kawasaki K, Minoshima S, Evans GA, Athanasiou M, Schultz R, Roe BA, Chen F, Pan H, Ramser J, Lehrach H, Reinhardt R, McCombie WR, de la Bastide M, Dedhia N, Blocker H, Hornischer K, Nordsiek G, Agarwala R, Aravind L, Bailey JA, Bateman A, Batzoglou S, Birney E, Bork P, Brown DG, Burge CB, Cerutti L, Chen HC, Church D, Clamp M, Copley RR, Doerks T, Eddy SR, Eichler EE, Furey TS, Galagan J, Gilbert JG, Harmon C, Hayashizaki Y, Haussler D, Hermjakob H, Hokamp K, Jang W, Johnson LS, Jones TA, Kasif S, Kaspryzk A, Kennedy S, Kent WJ, Kitts P, Koonin EV, Korf I, Kulp D, Lancet D, Lowe TM, McLysaght A, Mikkelsen T, Moran JV, Mulder N, Pollara VJ, Ponting CP, Schuler G, Schultz J, Slater G, Smit AF, Stupka E, Szustakowski J, Thierry-Mieg D, Thierry-Mieg J, Wagner L, Wallis J, Wheeler R, Williams A, Wolf YI, Wolfe KH, Yang SP, Yeh RF, Collins F, Guyer MS, Peterson J, Felsenfeld A, Wetterstrand KA, Patrinos A, Morgan MJ; International Human Genome Sequencing Consortium. (2001) Initial sequencing and analysis of the human genome. Nature 409:860-921

Langer PR, Waldrop AA, Ward DC (1981) Enzymatic synthesis of biotin-labeled polynucleotides: Novel nucleic acid affinity probes. Proc Natl Acad Sci USA 75:6633-6637

Lansdorp PM, Verwoerd NP, van de Rijke FM, Dragowska V, Little M-T, Dirks RW, Raap AK, Tanke HJ (1996) Heterogeneity in telomere length of human chromosomes. Hum Mol Genet 5:685-691

Lawrence JB, Singer RH, McNeil JA (1990) Interphase and metaphase resolution of different distances within the human dystrophin gene. Science 249:928-932

Lejeune J, Gautier M, Turpin R (1959) Etude des chromosomes somatiques de neuf enfants mongoliens. Compt Rend 248:1721-1722

Lengauer C, Speicher MR, Popp S, Jauch A, Tniwaki M, Nagaraja R, Riethman H, Donis-Keller H, D'Urso M, Schlessinger D, Cremer T (1993) Chromosomal bar codes produced by multicolor fluorescence in situ hybridization with multiple YAC clones and whole chromosome painting probes. Hum Mol Genet 2:505-512

Lestou VS, Strehl S, Lion T, Gadner H, Ambros PF (1996) High resolution FISH of the entire integrated Epstein-Barr virus genome on extended human DNA. Cytogenet Cell Genet 74:211-217

Levy ER, Herrington CS (1995) Non-isotopic methods in molecular biology: a practical approach. Oxford Oxford University Press

Lichter P, Cremer T, Borden J, Manuelidis L, Ward DC (1988) Delineation of individual human chromosomes in metaphase and interphase cells by in situ suppression hybridization using chromosome specific library probes. Hum Genet 80:224-234

Lichter P, Cremer T, Tang C-JC, Watkins PC, Manuelidis L, Ward DC (1988) Rapid detection of human chromosome 21 aberrations by in situ hybridization. Proc Natl Acad Sci USA 85:9664-9668

Liehr T, Rautenstrauss B, Grahl H, Bathke KD, Ekici A, Rauch A, Rott H-D (1996) Mosaicism for the Charcot-Marie-Tooth disease type 1A duplication suggests somatic reversion. Hum Genet 98:22-28

Longin A, Souchier C, Ffrench M, Bryon PA (1993) Comparison of anti-fading agents used in fluorescence microscopy: image analysis and laser confocal microscopy study J Histochem Cytochem 41: 1833-1840

Malcolm S, Barton P, Murphy CS, Ferguson-Smith MA (1981) Chromosomal localization of a single copy gene by in-situ hybridization: human beta globin genes on the short arm of chromosome 11. Ann Hum Genet 45:135-141

McNicol AM, Farquharson MA (1997) In situ hybridization and its diagnostic applications. J Pathol 1823:250-261

Meltzer PS, Guan XY, Burgess A, Trent JM (1992) Rapid generation of region specific probes by chromosome microdissection and their application. Nat Genet 1:24-28

Michalet X, Ekong R, Fougerousse F, Rousseaux S, Schurra C, Hornigold N, von Slegtenhorst M, Wolfe J, Povey S, Beckman JS, Bensimon A (1997) Dynamic molecular combing: stretching the whole human genome for high resolution studies. Science 277:1518-1523

Mitelman F (Ed.) (1995) ISCN. An international system for human cytogenetic nomenclature. Basel Karger.

Moers MHP, Kalle WHJ, Ruiter AGT, Wiegant JCAG, Raap AK, Greve J, de Grooth BG, van Hulst NF (1996) Fluorescence in situ hybridization on human metaphase chromosomes detected by near-field scanning optical microscopy. J Microsc 182:40-45

Molecular Cytogenetics and PositionalCloning Center in Berlin: http://www.mpimg-berlin-dahlem.mpg.de/-cytogen/

Monajembashi S, Cremer C, Cremer T, Wolfrum J, Greulich KO (1986) Microdissection of human chromosomes by a laser microbeam. Exp Cell Res 167:262-265

Morris A, Boyd E, Dhanjal S, Lowther GW, Aitken DA, Young J, Menzies AL, Imrie SJ, Connor JM (1999) Two years' prospective experience using fluorescence in situ hybridization on uncultured amniotic fluid cells for rapid prenatal diagnosis of common chromosomal aneuploidies. Prenat Diagn 19:546-551

Müller S, O'Brien PCM, Ferguson-Smith M, Wienberg J (1997a) A novel source of highly specific chromosome painting probes for human karyotype analysis derived from primate homologues. Hum Genet 101:149-153

Müller S, Rocchi M, Ferguson-Smith M, Wienberg J (1997b) Toward a multicolor chromosome bar code for the entire human karyotype by fluorescence in situ hybridization. Hum Genet 100:271-278

Nederlof PM, van de Flier S, Wiegant J, Raap AK, Tanke HJ, Ploem JS, van der Ploeg M (1990) Multiple fluorescence in situ hybridization. Cytometry 11:126-131

Nederlof PM, van de Flier S, Vrolijk J, Tanke HJ, Ploem JS, Raap AK (1992) Fluorescence ratio measurements of double-labeled probes for multiple in situ hybridization by digital imaging microscopy. Cytometry 13:839-845

Nielsen PE, Egholm M, Berg RH, Buchardt O (1991) Sequence-selective recognition of DNA by strand displacement with a thymine-substituted polyamide. Science 254:1497-1500

Nilsson M, Malmgren H, Samiotaki M, Kwiatkowski M, Chowdhary BP, Landegren U (1994) Padlock probes: circularizing oligonucleotides for localized DNA detection. Science 265:2085-2088

Nilsson M, Krjeci K, Koch J, Kwiatkowski M, Gustavsson P, Landegren U (1997) Padlock probes reveal single-nucleotide differences, parent of origin and in situ distribution of centromeric seqeunces in human chromosomes 13 and 21. Nat Genet 16:252-255

Nishizaki T, Chew K, Chu L, Isola J, Kallioniemi A, Weidner N, Waldman FM (1997) Genetic alterations in lobular breast cancer by comparative genomic hybridization. Int J Cancer 74:513-517

Novelli A, Sabani M, Caiola A, Digilio MC, Giannotti A, Mingarelli R, Novelli G, Dallapiccola B (1999) Diagnosis of DiGeorge and Williams syndromes using FISH analysis of peripheral blood smears. Mol Cell Probes 4:303-307

Nowell PC, Hungerford DA (1960) A minute chromosome in human chronic granulocytic leukemia. Science 132:1497-1499

OMIM - On line Mendelian Inheritance in Man: http://www.ncbi.nih.gov/Omim/

Para I, Windle B (1993) High resolution visual mapping of stretched DNA by fluorescent in situ hybridization. Nat Genet 5:17-21

Patau K, Smith DW, Therman E, Inhorn SL, Wagner HP (1960) Multiple congenital anomaly caused by an extra autosome. Lancet 1:790-793

Pierce KE, Fitzgerald LM, Seibel MM, Zilberstein M (1998) Preimplantation genetic diagnosis of chromosome balance in embryos from a patient with a balanced reciprocal translocation. Molec Hum Reprod 4:167-172

Pinkel D, Straume T, Gray JW (1986) Cytogenetic analysis using quantitative, high-sensitivity, fluorescence hybridization. Proc Natl Acad Sci U S A 83:2934-2938

Pinkel D, Landegent J, Collins C, Fuscoe J, Segraves R, Lucas J, Gray JW (1988) Fluorescence in situ hybridization with human chromosome specific libraries: detection of trisomy 21 and translocation of chromosome 4. Proc Natl Acad Sci USA 85:9138-9142

Rao PH, Cigudosa JC, Ning Y, Calasanz MJ, Iida S, Tagawa S, Michaeli J, Klein B, Dalla-Favera R, Jhanwar SC, Ried T, Chaganti RS (1998) Multicolor spectral karyotyping identifies new recurring breakpoints and translocations in multiple myeloma. Blood 92:1743-1748

Raap AK, Florijn RJ, Blonden LAJ, Wiegant JCAG, Vaandrager JW, Vrolijk J, den Dunnen J, Tanke HJ, van Ommen GJB (1996) Fiber-FISH as a DNA mapping tool. Methods Enzymol 9:67-73

Ried T, Baldini A, Rand TC, Ward DC (1992) Simultaneous visualization of seven different DNA probes by in situ hybridization using combinatorial fluorescence and digital imaging microscopy. Proc Natl Acad Sci USA 89:1388-1392

Roberts I, Wienberg J, Nacheva E, Grace C, Griffin D, Coleman N (1999) Novel method for the production of multiple colour chromosome paints for use in karyotyping by fluorescence in situ hybridisation. Genes Chromsom Cancer 25:241-250

Roux W (1883) Über die Bedeutung der Kerntheilungsfiguren, Engelmann, Leipzig.

Rowley JD, Diaz MO, Espinosa R, Patel YD, van Melle E, Ziemin S, Taillon-Miller P et al (1990) Mapping chromosome band 11q23 in human acute leukemia with biotinylated probes: identification of 11q23 translocation breakpoints with a yeast artificial chromosome. Proc Natl Acad Sci USA 87:9358-9362

Rubin CM, Carrino JJ, Dickler MN, Leibowitz D, Smith SD, Westbrook CA (1988) Heterogeneity of genomic fusion of BCR and ABL in Philadelphia chromosome-positive acute lymphoblastic leukemia. Proc Natl Acad Sci U S A 85:2795-2799

Saiki RK, Scharf S, Faloona F, Mullis KB, Horn GT, Erlich HA, Arnheim N (1985) Enzymatic amplification of beta-globin genomic sequences and restriction site analysis for diagnosis of sickle cell anemia. Science 230:1350-1354

Schinzel A (1983) Catalog of unbalanced chromosome aberrations in man. Walter de Gruyter and Co. Berlin, pp 1.7

Schröck E, du Manoir, S, Veldman T, Schoell B, Wienberg J, Fergueson-Smith MA, Ning Y, Ledbetter D, Bar-Am I, Soenksen D, Garini Y, Ried T (1996) Multicolor spectral karyotyping of human chromosomes. Science 273:494-497

Schröck E, Veldman T, Padilla-Nash H, Ning Y, Spurbeck J, Jalal S, Shaffer LG, Papenhausen P, Kozma C, Phelan MC, Kjeldsen E, Schonberg S.A, O'Brien P, Biesecker L, du Manoir S., Ried T (1997) Spectral karyotyping refines cytogenetic diagnostics of constitutional chromosomal abnormalities. Hum Genet 101:255-262

Selig S, Okumura K, Ward DC, Cedar H (1992) Delineation of DNA replication time zones by fluorescence in situ hybridization. EMBO J 11:1217-1225

Shaffer LG, Kennedy GM, Spikes AS, Lupski JR (1997) Diagnosis of CMT1A duplications and HNPP deletions by interphase FISH: Implications for testing in the cytogenetics laboratory. Am J Med Genet 69:325-331

Shiels C, Coutelle C, Huxley C (1997) Analysis of ribosomal and alphoid repetitive DNA by fiber-FISH. Cytogenet Cell Genet 76:20-22

Sinsheimer RL (1977) Recombinant DNA. In: "Annual Review of Biochrmistry" vol 46. Annual Reviews, Palo Alto, CA, pp 415-438

Smith HO, Wilcox KW (1970) A restriction enzyme from Hemophilus influenzae. I. Purification and general properties. J Mol Biol 51:379-391

Solinas-Toldo S, Lengauer C, Fries R (1995) Comparative map of human and cattle. Genomics 27:489-496

Solinas-Toldo S, Lampel S, Stilgenbauer S, Nickolenko J, Benner A, Dohner H, Cremer T, Lichter P (1997) Matrix-based comparative genomic hybridization: biochips to screen for genomic imbalances. Genes Chromosomes Cancer 20:399-407

Speicher MR, Ballard SG, Ward DC (1996) Karyotyping human chromosomes by combinatorial multi-fluor FISH. Nat Genet 12:368-375

Taneja KL, Chavez EA, Coul J, Lansdorp PM (2001) Multicolor fluorescence in situ hybridization with peptide nucleic acid probes for enumeration of specific chromosomes in human cells. Genes Chromosomes Cancer 30:57-63

Tanke HJ, Wiegant J, van Gijlswijk RP, Bezrookove V, Pattenier H, Heetebrij RJ, Talman EG, Raap AK, Vrolijk J (1999) New strategy for multi-colour fluorescence in situ hybridisation: COBRA:COmbined Binary RAtio labelling. Eur J Hum Genet 7:2-11.

Telenius H, Pelear AH, Tunnacliffe A, Carter NP, Behmel A, Gergueson-Smith MA, Nordenskjøld M, Pfranger R Ponder BAJ (1992) Cytogenetic analysis by chromosome painting using DOP-PCR amplified flow sorted chromosomes. Genes Chromosomes Cancer 4:257-263

Thangavelu M, Pergament E, Espinosa R III, Bohlander SK (1994) Characterization of marker chromosomes by microdissection and fluorescence in situ hybridization. Prenat Diagn 14:583-588

Therkelsen AJ, Nielsen A, Koch JE, Hinkjær J, Kølvraa S (1995) Staining of human telomeres with primed in situ labeling (PRINS). Cytogenet Cell Genet 68:115-118

Tiainen M, Popp S, Parlier V, Emmerich P, Bellomo MJ, Ruutu T, Cremer T, Knuutila S (1992) Chromosomal in situ suppression hybridization of immunologically classified mitotic cells in hematologic malignancies. Genes Chromosomes Cancer 4:135-140

Tjio JH, Levan A The chromosome number of man. Herditas 1956; 42:1-6.

Trask BJ, Massa H, Kenwrick S, Gitschier J (1991) Mapping of human chromosome Xq28 by two-color fluorescence in situ hybridization of DNA sequences in interphase cell nuclei. Am J Hum Genet 48:1-15

Vaandrager JW, Schuuring E, Zwikstra E, de boer CJ, Kleiverda KK, van Kierken JHJM, Kluin-Nelemans HC, van Ommen GJB, Raap AK, Kluin PM (1996) Direct visualization of dispersed 11q13 chromosomal translocations in mantle cell lymphoma by multicolor DNA fiber fluorescence in situ hybridization. Blood 88:1177-1182

Van Hummelen P, Lowe XR, Wyrobek AJ (1996) Simultaneous detection of structural and numerical abnormalities in sperm of healthy men by multicolor fluorescence in situ hybridization. Hum Genet 98:608-615

Veldman T, Vignon C, Schröck E, Rowley JD, Ried T (1997) Hidden chromosome abnormalities in haematological malignancies detected by multicolour spectral karyotyping. Nat Genet 15:406-410

Venter JC, Adams MD, Myers EW, Li PW, Mural RJ, Sutton GG, Smith HO, Yandell M, Evans CA, Holt RA, Gocayne JD, Amanatides P, Ballew RM, Huson DH, Wortman JR, Zhang Q, Kodira CD, Zheng XH, Chen L, Skupski M, Subramanian G, Thomas PD, Zhang J, Gabor Miklos GL, Nelson C, Broder S, Clark AG, Nadeau J, McKusick VA, Zinder N, Levine AJ, Roberts RJ, Simon M, Slayman C, Hunkapiller M, Bolanos R, Delcher A, Dew I, Fasulo D, Flanigan M, Florea L, Halpern A, Hannenhalli S, Kravitz S, Levy S, Mobarry C, Reinert K, Remington K, Abu-Threideh J, Beasley E, Biddick K, Bonazzi V, Brandon R, Cargill M, Chandramouliswaran I, Charlab R, Chaturvedi K, Deng Z, Di Francesco V, Dunn P, Eilbeck K, Evangelista C, Gabrielian AE, Gan W, Ge W, Gong F, Gu Z, Guan P, Heiman TJ, Higgins ME, Ji RR, Ke Z, Ketchum KA, Lai Z, Lei Y, Li Z, Li J, Liang Y, Lin X, Lu F, Merkulov GV, Milshina N, Moore HM, Naik AK, Narayan VA, Neelam B, Nusskern D, Rusch DB, Salzberg S, Shao W, Shue B, Sun J, Wang Z, Wang A, Wang X, Wang J, Wei M, Wides R, Xiao C, Yan C, Yao A, Ye J, Zhan M, Zhang W, Zhang H, Zhao Q, Zheng L, Zhong F, Zhong W, Zhu S, Zhao S, Gilbert D, Baumhueter S, Spier G, Carter C, Cravchik A, Woodage T, Ali F, An H, Awe A, Baldwin D, Baden H, Barnstead M, Barrow I, Beeson K, Busam D, Carver A, Center A, Cheng ML, Curry L, Danaher S, Davenport L, Desilets R, Dietz S, Dodson K, Doup L, Ferriera S, Garg N, Gluecksmann A, Hart B, Haynes J, Haynes C, Heiner C, Hladun S, Hostin D, Houck J, Howland T, Ibegwam C, Johnson J, Kalush F, Kline L, Koduru S, Love A, Mann F, May D, McCawley S, McIntosh T, McMullen I, Moy M, Moy L, Murphy B, Nelson K, Pfannkoch C, Pratts E, Puri V, Qureshi H, Reardon M, Rodriguez R, Rogers YH, Romblad D, Ruhfel B, Scott R, Sitter C, Smallwood M, Stewart E, Strong R, Suh E, Thomas R, Tint NN, Tse S, Vech C, Wang G, Wetter J, Williams S, Williams M, Windsor S, Winn-Deen E, Wolfe K, Zaveri J, Zaveri K, Abril JF, Guigo R, Campbell MJ, Sjolander KV, Karlak B, Kejariwal A, Mi H, Lazareva B, Hatton T, Narechania A, Diemer K, Muruganujan A, Guo N, Sato S, Bafna V, Istrail S, Lippert R, Schwartz R, Walenz B, Yooseph S, Allen D, Basu A, Baxendale J, Blick L, Caminha M, Carnes-Stine J, Caulk P, Chiang YH, Coyne M, Dahlke C, Mays A, Dombroski M, Donnelly M, Ely D, Esparham S, Fosler C, Gire H, Glanowski S, Glasser K, Glodek A, Gorokhov M, Graham K, Gropman B, Harris M, Heil J, Henderson S, Hoover J, Jennings D, Jordan C, Jordan J, Kasha J, Kagan L, Kraft C, Levitsky A, Lewis M, Liu X, Lopez J, Ma D, Majoros W, McDaniel J, Murphy S, Newman M, Nguyen T, Nguyen N, Nodell M, Pan S, Peck J, Peterson M, Rowe W, Sanders R, Scott J, Simpson M, Smith T, Sprague A, Stockwell T, Turner R, Venter E, Wang M, Wen M, Wu D, Wu M, Xia A, Zandieh A, Zhu X (2001) The sequence of the human genome. Science 291:1304-1351

Verlinsky Y, Cieslak J, Freidine M, Ivakhnenko V, Wolf G, Kovalinska L, White M, Lifchez A, Kaplan B, Moise J, Valle J, Ginsberg N, Strom C, Kuliev A (1996) Polar body diagnosis of common aneuploidies by FISH. J Asist Reprod Genet 13:157-162

Watson JD, Crick FHC (1953) Molecular structure of nucleic acids. A structure for deoxyribose nucleic acid. Nature 171:37-738

Watson JD, Crick FHC (1953) Genetical implications of the structure of deoxyribose nucleic acid. Nature 171:964-967
Weber-Matthiesen K, Pressl S, Schlegelberger B, Grote W (1993) Combined immunophenotyping and interphase cytogenetics on cryostat sections by the new FICTION method. Leukemia 7:646-649
Weksberg R, Teshima I, Williams BR, Greenberg CR, Pueschel SM, Chernos JE, Fowlow SB, Hoyme E, Anderson IJ, Whiteman DA, et al (1993) Molecular characterization of cytogenetic alterations associated with Beckwith-Weidemann syndrome (BWS) phenotype refines the localization and suggests the gene for is imprinted. Hum Mol Genet 2:549-556
Wieghant J, Kalle W, Mullenders L, Brookes S, Hoovers JMN, Dauwerse JG, van Ommen GJB, Raap AK (1992) High resolution in situ hybridization using DNA halo preparations. Hum Mol Genet 1:587-591
Wiegant J, Wiesmeijer CC, Hoovers JMN, Schuuring E, d'Azzo A, Vrolijk J, Tanke HJ, Raap AK (1993) Multiple and sensitive fluorescence in situ hybridization with rhodamine-, fluorescein-, and coumarin-labeled DNA's. Cytogenet Cell Genet 63:73-76
Wilkinson DG (1992) In situ hybridization: a practical approach. Oxford: Oxford University Press.
Yanagi M, Shinjo K, Takeshita A, Tobita T, Yano K, Kobayashi M, Terasaki H, Naoe T, Ohnishi K, Ohno R (1999) Simple and reliably sensitive diagnosis and monitoring of Philadelphia chromosome-positive cells in chronic myeloid leukemia by interphase fluorescence in situ hybridization of peripheral blood cells. Leukemia 13:542-552
Yang F, Carter NP, Shi L, Ferguson-Smith MA (1995) A comparative study of the karyotype of the muntjacs by chromosome painting. Chromosoma 103:642-652
Yunis JJ (1976) High resolution of human chromosomes. Science 191:1268-1271
Yunis JJ (1981) Mid-prophase human chromosomes. The attainment of 2000 bands. Hum Genet 56:293-298
Zahed L, Murer-Orlando M, Vekemans M (1992) In situ hybridization studies for the detection of common aneuploidies in CVS. Prenat Diagn 12:483-493

Websites

//genome.ucsc.edu/ Assembly of draft human genome
//genome.wustl.edu/gsc//human/Mapping/ Links to clones and accession no. maps of the human genome
//www.molgen.mpg.de/~cytogen/ Cloned DNA sequences covering all human chromosomes
//www.ncbi.nlm.nih.gov/genome/guide/ Views of chromosomes and maps and loci with links to NCBI
//www.ncbi.nlm.nih.gov/genome/Entez/ Human genes and disease (OMIM) and medical literature (PubMed)

Microscopy and Imaging Systems

ANDREAS PLESCH

Introduction

Any FISH technique remains useless without visualization of the probes. In many routine FISH applications visual interpretation of the FISH images by looking down the microscope may be sufficient. An imaging system with its image enhancement capabilities will usually facilitate the analysis and add the convenience of documentation and archiving of the results. In a number of modern FISH techniques, however, quantitative digital image analysis is indispensable to obtain the relevant information. Each step in the imaging process (or each element in the chain of imaging components) has the inherent risk of influencing (generally deteriorating) the image. The following chapter summarizes the fundamental aspects and limitations of the imaging process.

Basic concepts

Optical resolution

Any lens or optical instrument suffers from a physical phenomenon called diffraction. Diffraction is due to the wave nature of light and will make an infinitely small point-like object appear blurred (Fig. 1). The apparent size of the blurred image depends on the wavelength of the light and of the ratio of focal length and diameter of the lens and is given by the following equation:

$$d = 2.44\ \lambda\ f/D \qquad (1)$$

✉ Andreas Plesch, MetaSystems GmbH, Robert-Bosch-Str. 6, Altlussheim, 68804, Germany (*phone* +49-6205-39610; *fax* +49-6205-32270;
e-mail aplesch@metasystems.de)

where lambda is the wavelength, f the focal length of the lens, and D the clear diameter (aperture) of the lens. As D/(2f) is also called the numerical aperture (N.A.) of the lens, Eq. (1) can be written as

$$d = 1.22\ \lambda\ /\ \text{N.A.} \tag{2}$$

Example: when observed in green light (lambda=520 nm) with an immersion objective of N.A. 1.3, the apparent diameter of the image of a point object (of a very small hybridization signal) becomes d [µm]= 1.22×0.52 µm /1.3 = 0.49 µm. This is the smallest apparent diameter of any, even the smallest, object.

Even the tiniest object (FISH signal) will be visible if its intensity is sufficiently high. Its apparent size will depend on the numerical aperture of the lens. The lower the N.A. the larger (more blurred) the object will appear.

Resolution or more precisely the optical resolution limit means the closest distance between two very small objects (point sources) that still allows them to be distinguished as two separate objects in the image. In order to be resolved the distance between two point objects (FISH signals) should exceed half their apparent diameter d, which corresponds to a minimum distance of 0.25 µm (object plane). This minimum distance between two point objects will lead to a noticeable dip in the resulting intensity profile of two adjacent signals and will thus allow distinguishing of the partially overlapping peaks (Fig. 2). The minimum distance is also called resolution element or *resel*. At the resolution limit the relative depth or contrast of the dip defined as c=(Imax-Imin)/ Imax is 0.2 or 20%.

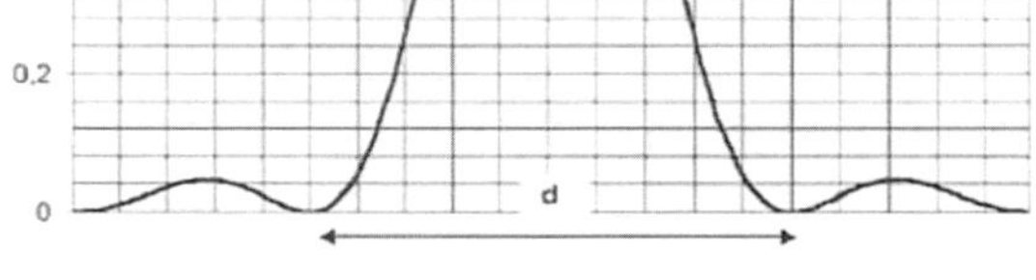

Fig. 1. Intensity distribution of the image of a point source with its apparent size *d*

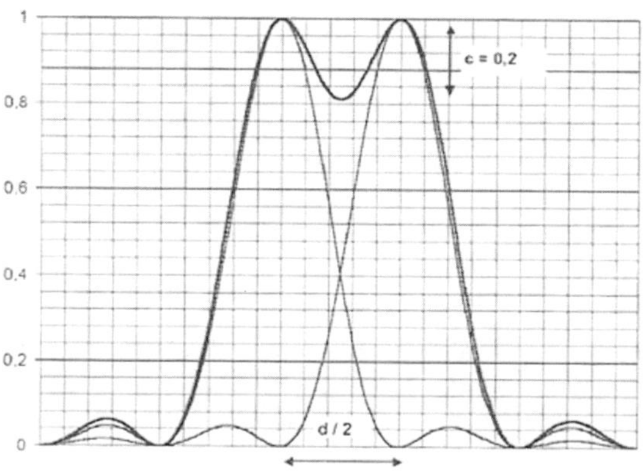

Fig. 2. Two point sources at a separation corresponding to the resolution limit. The resulting intensity profile clearly shows two peaks with a contrast of 0.2

It is the numerical aperture of the lens that determines the resolution limit, not the magnification. Magnification will only adapt the image size to the image sensor [eye, film, CCD (charged coupled device) sensor].

Contrast

Up to now we assumed the ideal situation of two bright objects on a completely black background. In reality, there will be some background fluorescence due to several reasons that will be discussed later. Figure 3 shows the situation of two signals lying on top of the background intensity I_0. The effective contrast of the dip has been reduced to 0.1. It will be more difficult to distinguish the two signals.

The presence of fluorescence background will decrease the effective contrast of two adjacent signals, making it harder to distinguish them.

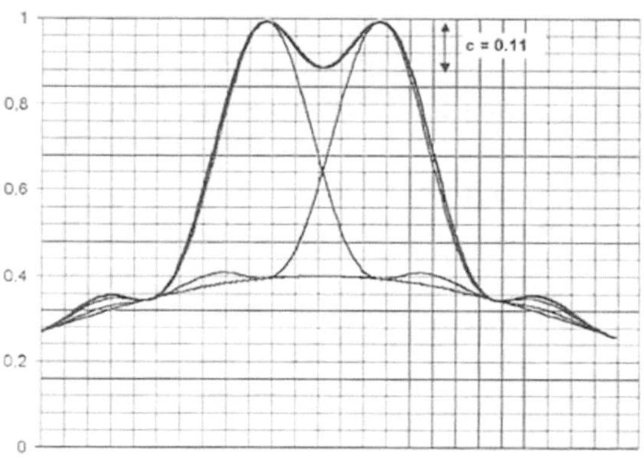

Fig. 3. Contrast drops to 0.1 in the presence of a background level of 40%

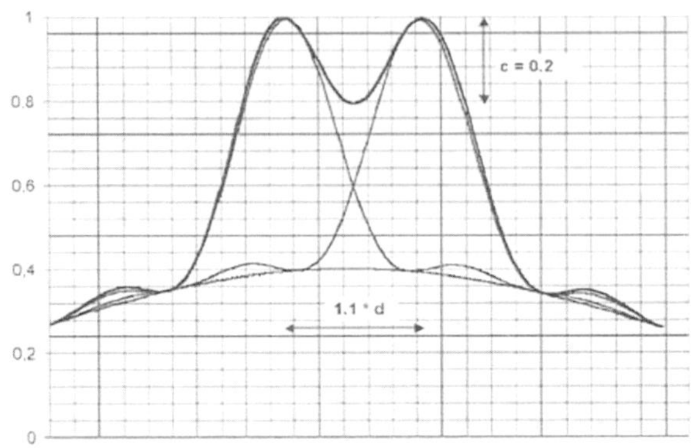

Fig. 4. In the presence of background the contrast of 0.2 is reached at a signal separation that is 10% larger than in the absence of background

If we now increase the distance between the signals by 10%, inter-signal contrast is up to 0.2 again (Fig. 4), which was the contrast at the resolution limit of two signals without background. This illustrates the close interdependence of resolution and contrast.

Optical resolution and contrast are closely related. Reduced image contrast due to background effectively reduces the resolution.

Spatial resolution of image sensor

The image sensor converts the optical image to an electronic signal that can then be digitized. CCD (charged coupled device) arrays are commonly used as an image sensor. A CCD consists of an array of discrete light sensing rectangular elements, called pixels. For the sake of simplicity we will consider a one-dimensional CCD array. All the light reaching an individual pixel will be converted to an electronic charge representing the average intensity of that pixel. Local intensity variations within an individual pixel will not be detected. It is obvious that smaller, more closely spaced pixels allow the resolving of smaller image details. To make use of the available image resolution, the pixel spacing of the sensor has to be smaller than half the resolution limit. If this sampling theorem is fulfilled, then at least one pixel will "see" the dip between the individual peaks reflecting the two objects or signals. This is illustrated in Fig. 5. CCD1 does not resolve the two objects while the electric signal of CCD2 reflects the central dip and thus discriminates between the two objects.

In order to make use of the optical resolution, the image sensor must fulfill the sampling theorem.

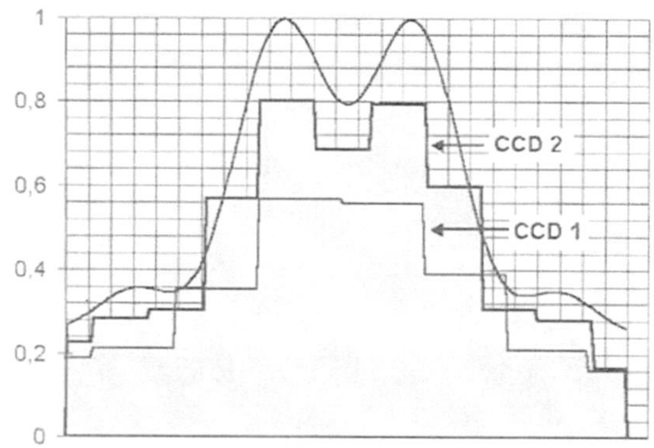

Fig. 5. CCD camera resolution: CCD1 has pixels too large to fulfill the sampling theorem and does not resolve the two spots. CCD2 fulfills the sampling theorem and discriminates between the two objects

Dynamic range of the image sensor

The capability to distinguish between different intensity levels is of central importance in image sampling. The maximum number of distinguishable intensity levels is called dynamic range. For our purpose it will be sufficient to consider the sensor and the digitizer as one unit. The digitizer converts the analog signal of the CCD sensor to a digital signal. The digitizer can be a separate component (in combination with an analog or video CCD camera) or integrated in the camera (digital CCD camera). Digital signals can only take discrete levels whose number depends on the number of bits representing a signal level. Common digitizers use 8 bits which allows discrimination between 2^8 or 256 discrete signal levels or gray levels. An 8 bit digital signal has thus a maximum dynamic range of 256, and in the absence of image noise, the smallest intensity variation that can be detected is 1:256 or 0.4%. This holds only, if the complete dynamic range of the digitizer is used. If the exposure time is too short, the signal will use only a fraction of the full dynamic range and the minimum detectable contrast will be reduced accordingly. This underlines the importance of the automatic exposure control of any FISH imaging system. Theoretically, the larger the dynamic range of the sensing system, the smaller the image contrast that can be resolved.

The smallest detectable contrast in an image depends on the dynamic range of the image sensor and digitizer.

The imaging system

To get an idea how these fundamental concepts are related to real imaging systems, we have to take a closer look at the main functional units that imaging systems are composed of. Each component will potentially degrade the quality of the image:

- illuminating system of the microscope – including fluorescence lamp, fluorescence filter set, and optical path
- magnifying system of the microscope – the objective lens
- camera – including the digitizer
- image processing system

The illuminating system

The fluorescence lamp

Usually, high-pressure mercury lamps are used in FISH applications, as their spectral characteristics provide efficient excitation of many dyes – from ultraviolet excitation (DAPI) to deep red (Cyanine 5 and Cyanine 5.5). Mercury lamps of 100 W as well as of 50 W are available for microscopy. The difference in light intensity reaching the object (slide) is less than a factor of 2. Some microscopists prefer the 50-W lamp as photo bleaching is less of a problem during lengthy manual adjustments. The use of infrared dyes like Cyanine 7 requires a Xenon lamp with its high intensity in the far-red and infrared spectral range. Perfect stabilization of the lamp power supply is normally not an issue as the integration times required will effectively smooth out residual ripples of the supply current. While the fluorescence lamp does not influence the optical resolution, incorrect lamp alignment will severely reduce the illumination intensity, lead to an inhomogeneous illumination within the image and reduce image contrast.

Fluorescence filters

Fluorescent dyes are characterized by their excitation and emission spectra. The emission spectrum of a fluorochrome is shifted towards longer wavelengths as compared to the excitation spectrum. The spectral distance between excitation and emission maxima is called Stokes shift. Fluorescence filter sets are used to separate the illuminating light (excitation) and fluorescence light (emission). Figure 6 shows the principle. The excitation filter selects the spectral band to illuminate the dye. The dichroic mirror reflects the excitation light down to the specimen. Dichroic mirrors are designed to reflect light at shorter wavelengths (e.g. blue light) while they will transmit longer wavelengths (e.g. green and red). The specimen emits fluorescence but also reflects some of the incident light. As the intensity of the reflected excitation light may be orders of magnitude higher than the fluorescence signal, it must not be transmitted into the observation path of the microscope. Most of the reflected light will not pass the dichroic mirror but will be reflected back towards the lamp. Any residual excitation light will be blocked by the emission filter that follows the dichroic mirror.

Fluorescence filters are always a compromise of selectivity and throughput. Very narrow excitation and emission filters will be highly selective but the fluorescence intensity may be too low for manual scanning of the slides and for focussing. If the filters are too wide autofluorescence and optical cross talk may obscure the FISH signal, particularly in the case of multi-fluorochrome techniques. The image contrast will depend critically on the correct choice of filter sets.

Single band filter sets provide the best compromise of selectivity and optical throughput. They may, however, introduce a registration error or pixel shift when the subsequent capture of several dyes requires filter block changes within the same image. Although registration errors may be corrected automatically or interactively in the digital image,

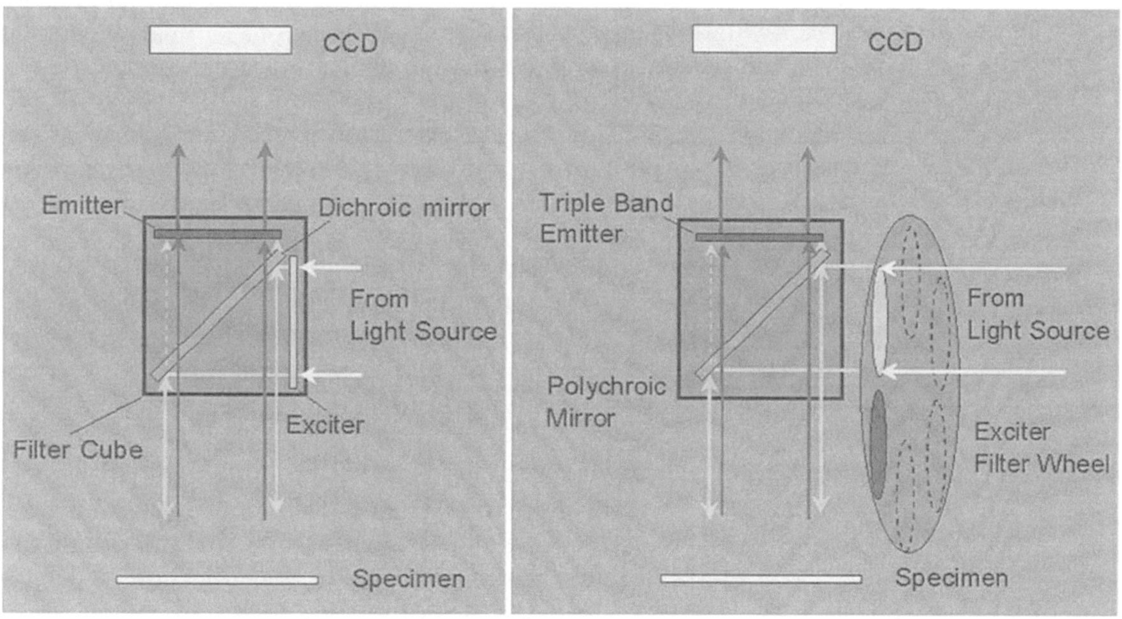

Fig. 6. Principle of a single band filter set (*left*) and triple band filter set (*right*) with individual excitation filters

they can cause problems in certain applications if they are not reproducible. Single band filter sets are not an appropriate solution if the number of fluorochromes is greater than the number of filter block positions of a microscope. In this case single band filter sets are combined with one or several dual band filter sets.

Multiple band filter combinations (Fig. 6) use dichroic mirrors and emission filters with two or more separate spectral bands of transmission/reflection. They can be combined with multiple band excitation filters (for simultaneous observation of several dyes) or with individual single band exciter filters (for subsequent observation of the dyes). Changing the excitation filter without moving the dichroic mirror and emitter filter does not introduce registration problems, no matter whether the excitation filter change is performed manually (filters mounted in filter slider) or by a motorized excitation filter wheel.

The optical path

The design of the illumination and detection optical paths finally is crucial, as any significant light losses between lamp and object, on the one hand, and between object and sensor, on the other, will require longer exposure times that may increase the thermal sensor noise (dark current) to an unacceptable level. Losses in the observation path are always more critical than in the excitation path. Photo-bleaching or fading of delicate specimens can ultimately limit the total number of photons that are converted to fluorescence. The signal may fade before sufficient light has reached the sensor. In such a situation an increased exposure time will not compensate for light losses in the observation path.

Stray light arising from optical elements or mechanical surfaces can deteriorate the image contrast significantly. A recent optical design (Axioplan2 Imaging from Carl Zeiss Jena) improves the signal contrast by using a light trap to catch the fraction of the illumination light that passes the dichroic mirror and is reflected and scattered back into the observation light path. A simple manipulation that is always worthwhile trying, to improve the contrast, is lowering the condenser or swiveling the condenser front lens out of the optic path to reduce reflections off the lens surface back through the specimen into the observation light path.

Magnifying system

Usually, the objective lens of the microscope is the only magnifying system to consider. In FISH imaging, immersion objectives with high numeric aperture (N.A) are used. The high numerical aperture guarantees high optical resolution and light throughput. Typical magnifications used for FISH imaging are 100× and 63×. They provide a high enough magnification with a field of view wide enough to cover complete metaphases. For

Table 1. Size of the resolution element (resel) for commonly used objective lenses for green light of 520 nm (emission maximum of FITC). The maximum pixel size of a CCD camera that is required to use the theoretical resolution is also given.

Objective	100×/1.3 Oil	63×/1.4 Oil	40×/0.75
Resel (object plane)	240 nm	230 nm	420 nm
Resel (image plane)	24 μm	14 μm	17 μm
CCD pixel size required	< 12 μm	< 7 μm	< 8.5 μm

interphase FISH, a 40× lens may also be useful. Low absorption in the ultraviolet and blue range of the spectrum is important, as DAPI which is commonly used for counter staining DNA is excited around 360 nm. Field flatness correction and chromatic correction are less important than in conventional wide field photographic documentation. CCD sensors cover a smaller field of view, typically 10 to 20% of the total field of view. Therefore the camera will not see any decrease in sharpness caused by field curvature in the far periphery. Longitudinal chromatic aberration can be overcome by refocusing the individual fluorochromes during the sequential image acquisition process. In mFISH applications even an objective with very high chromatic correction may require refocusing to achieve optimum sharpness through the spectral range from far blue (DAPI) to far-red (Cy5).

In Table 1 the theoretical resolution limit of commonly used objective lenses is given for green light (520 nm).

Image sensor

The most common image sensor used for FISH imaging is the CCD camera. CCD cameras are available as black and white and as color cameras. For several reasons, color CCDs are not suitable for digital FISH imaging applications:

- The amount of fluorescence of counterstain and hybridization signal often differs by orders of magnitude. It is thus necessary to use individual exposure (integration) times for the individual color channels. Color cameras use the same exposure time for the red (R), green (G) and blue (B) channels.
- The RGB-filters integrated in color CCD cameras do not match the emission spectra of the fluorochromes that are usually used in FISH applications.

- One-chip color cameras use one third of the available pixels per color channel (R,G,B) and provide insufficient resolution. 3-chip color cameras use three separate CCD sensors to achieve a resolution equivalent to black and white CCDs. 3-chip CCD cameras are very expensive.

For low light level applications like FISH, CCD cameras need the capability of long-term exposure or on-chip integration. In this operation mode the individual sensor elements will collect the incident light for a certain time, the exposure or integration time, before the pixels are read out.

Increased sensor noise is an unwanted side effect of long-term exposure. In addition to the electric charge generated by the incident light, thermally generated charges will show up as bright pixels making very long exposure times impractical. Cooling the CCD sensor reduces thermal noise and allows the use of very long integration times. Cooling does not increase the sensitivity of the CCD. A cooled CCD will require the same integration time as an uncooled camera, but it will add less noise to the image.

CCD cameras come in different chip sizes and resolutions. The chip size is usually defined by the diagonal in inches, e.g. 2/3"or 1/2". Typical chip resolutions are 768×576 pixels and 1280×1024 pixels; typical pixel sizes are 11 μm and 6.7 μm (Table 2).

Table 2. Specifications of common CCD cameras

Camera type	M300JAI	M1 JAI	C 5985H Hamamatsu	ORCA Hamamatsu
Chip size (inch)	2/3	2/3	1/2	2/3
Cooled Chip	–	–	x	x
Pixels	756 x 581	1280 x 1024	756 x 572	1280 x 1024
Pixel size (μm)	11	6.7	8.6	6.7
Read-out speed (images/s)	25 (analog)	12 (analog)	25 (analog)	9 (digital)
Dynamic range of CCD	min. 800:1	min 630:1	min 1000:1	1300:1
Digitization depth (bit)	8	8	8	10/ 12
Dynamic range of image	256:1	256:1	256:1	1024:1/4096:1
Relative cost	1	3	4	8

What is the minimum CCD resolution that is needed to use the full optical resolution of a good microscope objective? Do we need a mega pixel CCD or will a standard CCD do? Let us consider two objective lenses commonly used in cytogenetics: a 100×/1.3 Oil, and a 63×/1.4 Oil objective. The optical resolution limit is 240 nm and 230 nm, respectively (in the object plane). This corresponds to 100×240 nm =24 μm and 63×230 nm=14 μm, respectively, on the CCD sensor (image plane), due to the magnification between object and image plane. To fulfill the sampling theorem, the pixel size must be smaller than 12 μm for the 100× objective, and smaller than 7 μm for the 63× objective.

A standard 2/3" video CCD camera having 756×581 pixels and a pixel size of 11 μm will fulfill the sampling theorem when using the 100× objective. At 63× magnification, the same camera will reduce the image resolution below the theoretical optical limit of the objective. The mega pixel cameras with their pixel size of 6.7 μm will make use of the full resolution of the 63× and 40× objectives. We conclude that a standard video CCD camera is appropriate to exploit the full theoretical optical resolution of a 100× immersion objective lens.

The signal to noise ratio of the CCD defines the dynamic range. CCD noise is mainly photon noise due to the statistical nature of the photon-electron conversion process, read-out noise and (particularly for low signal intensities requiring long exposure times) dark current. The digitization depth is directly related to the contrast resolution or number of gray levels that are available in the digital image. Apparently, digitization to 8 bit or 256 gray levels reduces the camera dynamic range. Does this mean that 10 bit or 12 bit digitization is necessary? In most applications 8-bit digitization (256 gray levels) will be sufficient. However, if FISH signals with very different intensities are to be captured in the same image (labeled with the same fluorochrome), some information will be lost. Depending on the exposure time selected, one of the signals will be correctly exposed while the other will either be saturated or will disappear in the background noise. The problem can be solved if two acquisitions are performed with appropriate integration times. Multiplied with the respective integration times, the 256 image gray levels of two separate exposures allow even quantitative analysis of signals that differ by a factor of 1000 and more.

Image processing system

The image processing system comprises the hard- and software necessary to control the camera and digitizer, and the microscope (if motorized) with attachments like external excitation filter wheel. It further includes the software to process the raw images and to enhance (for qualitative interpretation) or to analyze (for measuring and quantitative interpretation) the image information. The image display finally provides the interface to the user. For optimum results the following requirements should be met:

Image acquisition Focusing is a crucial step in image capture. The system should provide on-screen focussing of the digital live image. The exposure time used for focussing is a compromise between displayed signal intensity and refresh rate of the displayed image. Automatic or interactive contrast stretching of the displayed image is helpful. Fast data transmission is of practical importance to obtain pseudo real-time display. An efficient software-supported focussing procedure will avoid unnecessary lengthy exposure of the slide and excessive signal fading.

Long-term exposure is almost always necessary, except for a very strong counterstain. The system must ensure correct exposure by controlling the integration time of the camera for all fluorochromes (color channels) of each individual image automatically. If done properly, automatic integration time control will allow for making use of the full dynamic range of FISH signals with an 8-bit digitizer. An algorithm that allows the setting of the number of saturated points in the image, avoids underexposure of the image due to a few bright pixels that may be caused by defective sensor elements of the CCD. The option of reducing the size of the measuring window (similar to a "spot measurement" in photography) is extremely helpful to exclude a bright artifact from the exposure control which would otherwise result in too short an integration time and consequently underexposed signals.

Software-controlled filter change is convenient and further reduces total acquisition time. Automatic capture of a focus stack of images, which can be used to calculate an extended focus image, requires automatic control of the microscope focus during the capture sequence. This function is handy for conventional interpretation and documentation of interphase nuclei or tissue sections; it is indispensable for automatic interphase FISH scanning applications.

Image pre-processing

Registration errors (also called pixel shift) due to the emission filters will always be a potential problem, particularly in multi fluorochrome applications. If the pixel shift is reproducible, the software should automatically correct registration errors after having once calibrated the respective shifts for each filter cube. In applications with similar information in each individual color channel, e.g. comparative genomic hybridization, automatic image registration can be performed by cross-correlation of the color channels.

Raw images usually show some fluorescence background. Manual or interactive thresholding followed by automatic contrast normalization will result in enhanced images that display the relevant information at maximum contrast.

Particularly for quantitative analyses like CGH (metaphase as well as array-based), mFISH, multi color banding (mBAND), or telomere quantification a non-uniform background will add a variable bias to the result. An effective, locally adaptive background correction algorithm is absolutely necessary to obtain reliable data. For qualitative FISH analysis and for image documentation, local or "region of interest" operations for thresholding, contrast stretching, etc. are helpful.

Quantitative image analysis

Data analysis of FISH images can be divided into two classes: 1. Measurement of image parameters and 2. Classification of objects. Calculation of CGH ratio profiles and quantitation of telomere size by measuring telomere signal intensity are applications of the first class. Classification means extracting appropriate data, called features, from the objects of interest and using these features to recognize or identify objects. In mFISH and mBAND analyses for example, spectral features are analyzed to identify chromosomes or fragments of chromosomes. In scanning applications morphometric features, often combined with spectral information, are used to find metaphases or to pin-point rare fluorescent labeled cells in a huge number of unlabeled normal cells. The analysis strategies employed are highly application-dependent; their detailed discussion is beyond the scope of this chapter.

Practical examples

Figure 7 shows an interphase nucleus with several small hybridization signals (BAC clones) directly labeled with Texas Red. Only the Texas Red information is displayed in black and white, the DAPI counter stain is not relevant for the following considerations and has been omitted.

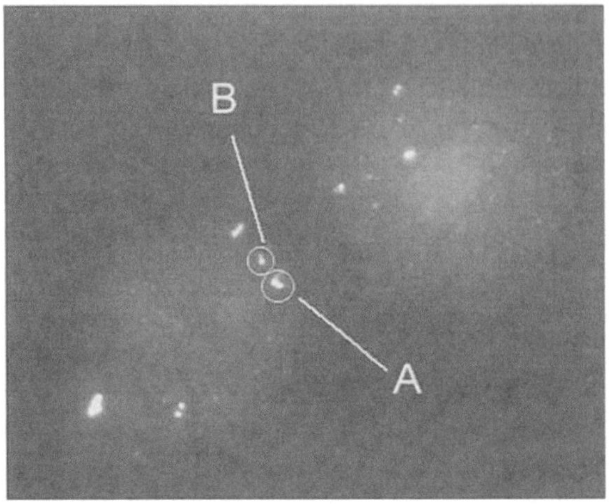

Fig. 7. Small hybridization spots on interphase nuclei. Only Texas Red fluorescence is shown. Taken with Plan-Neo-fluoar 100×/1.3 Oil and a standard CCD (M300) with an integration time of 3.8 s. *Regions A* and *B* show spots close to the resolution limit

The image shows bright hybridization spots along with some unspecific signals and an elevated fluorescence background within the nuclei. The signal to background ratio is approximately 4.

The situation for whole chromosome painting signals (FITC-labeled chromosomes of a metaphase hybridized with the "24XCyte" mFISH kit from MetaSystems) is shown in Fig. 8a. The intensity ratio of the painted and unpainted chromosomes is 4 again. Is the dynamic range possibly limited by the dark current of the uncooled CCD sensor (M300) used for capturing? This is not the case, as the background intensity between the chromosomes is significantly lower than the fluorescence intensity of the unlabeled chromosomes. And even this inter-chromosomal background is not only due to dark current, but is caused to a great extent by auto fluorescence within the cytoplasm, by the very small fraction of reflected excitation light that passes the emission filter and by ambient light. Painting probes labeled with Texas Red usually yield a higher dynamic range at comparable exposure times. The difference in the dynamic range is probably due to different behavior of the fluorochromes leading to increased non-specific FITC fluorescence and to increased auto fluorescence caused by the shorter wavelength of the FITC excitation. Figure 8b illustrates the influence of dark current on the image quality. The fluorescence lamp was dimmed and the same metaphase was captured again. Focusing had to be performed before reducing the light level, as the signals were much too dark to be visible in the microscope or on the screen. The required exposure time was 125 s. Of course, the dynamic range fell to 2.5 and the intra-chromosomal background as well as the overall noise level has been in-

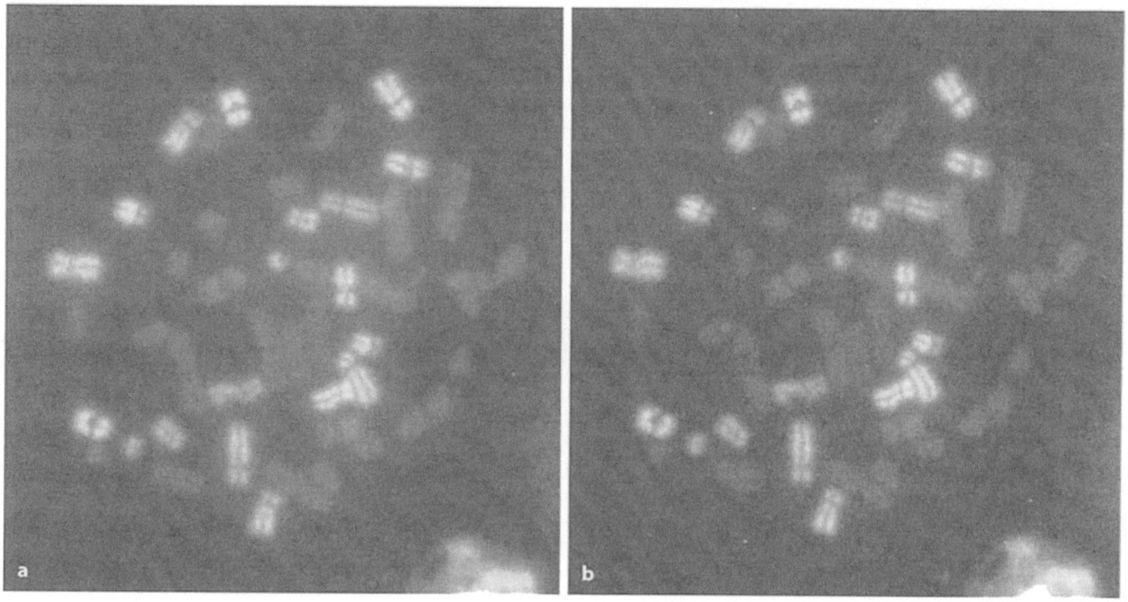

Fig. 8. **a** FITC-labeled chromosome paints (24Xcyte mFISH kit from MetaSystems). Plan-Neofluoar 100×/1.3 Oil, M300 standard CCD, integration time 2.4 s. **b** Same metaphase captured at very low illumination light level. Integration time was 125 s

creased. However, we obtain a decent, analyzable image, with a background that is still dominated by residual fluorescence of labeled nucleotides that are unspecifically bound to DNA rather than by CCD dark current. This is an important result.

- The dynamic range of typical FISH signals is limited by the sample itself and does not exceed the dynamic range that an 8-bit digital system can handle.
- In most applications an uncooled standard video CCD is suitable for long-term exposure of signals down to, and even below the limit of visibility, if the exposure control is properly done.

To learn about the limit of resolution we will take a closer look at the two regions marked in Fig. 7. Region A includes two spots of about 0.8 μm diameter, separated by 0.5 μm. Region B shows two spots of similar size, separated by 0.4 μm. Spots smaller than 0.8 μm are not found in the image. The theoretical minimum size, however, was calculated to be 0.58 μm and 0.54 μm for the 100×/1.3 and the 63×/1.4 objectives, respectively. This theoretical limit is not reached due to several factors: (1) a

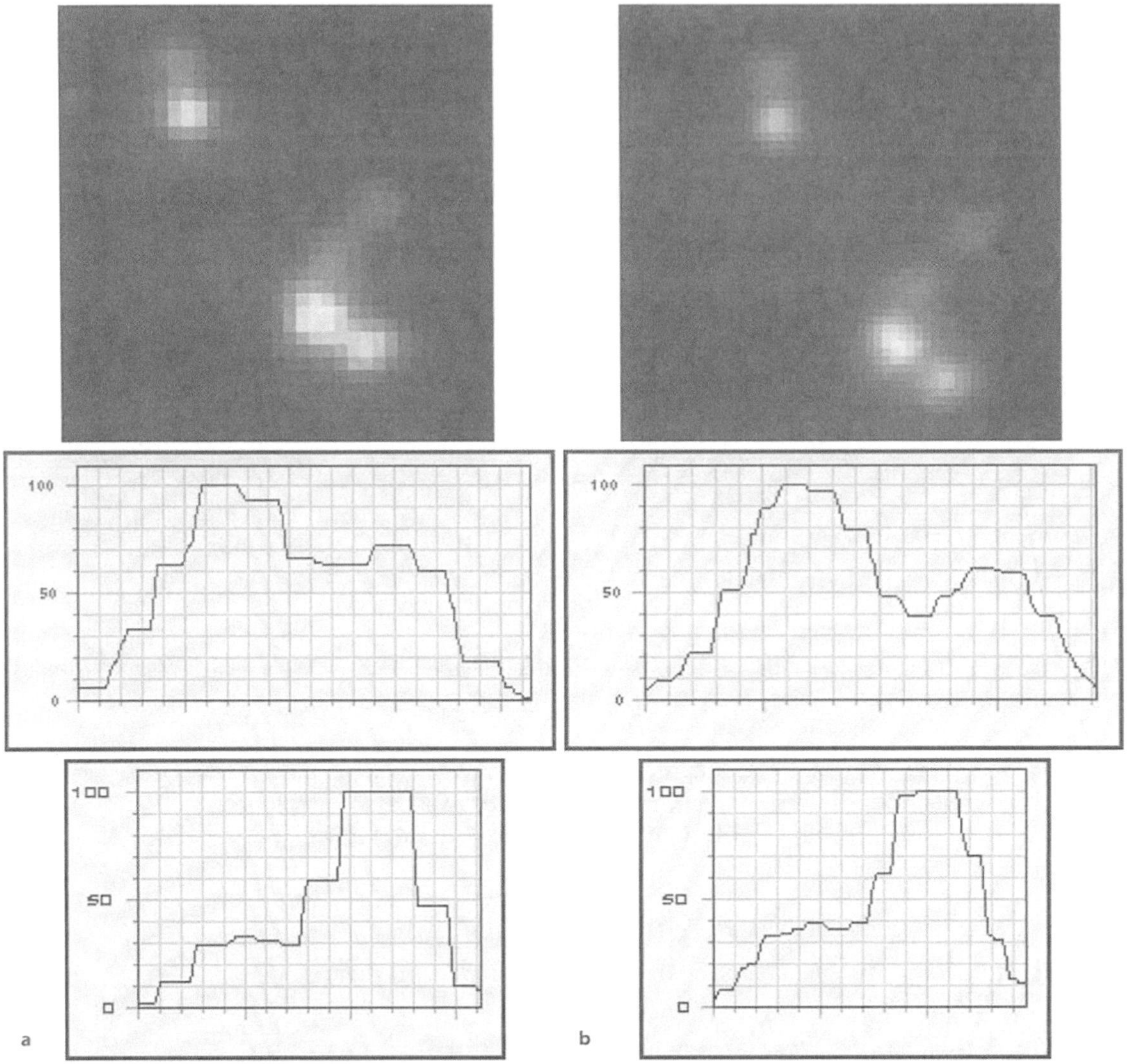

Fig. 9. Comparison of images (top) and intensity profiles of signals of region *A* (*middle*) and region *B* (*bottom*). **a** Plan-Neofluar 100×, M300 camera; **b** Plan-Neofluar 100×, M1 camera; **c** Plan-Apochromat 63x, M300 camera; **d** Plan-Apochromat 63×, M1 camera. For explanations refer to text.

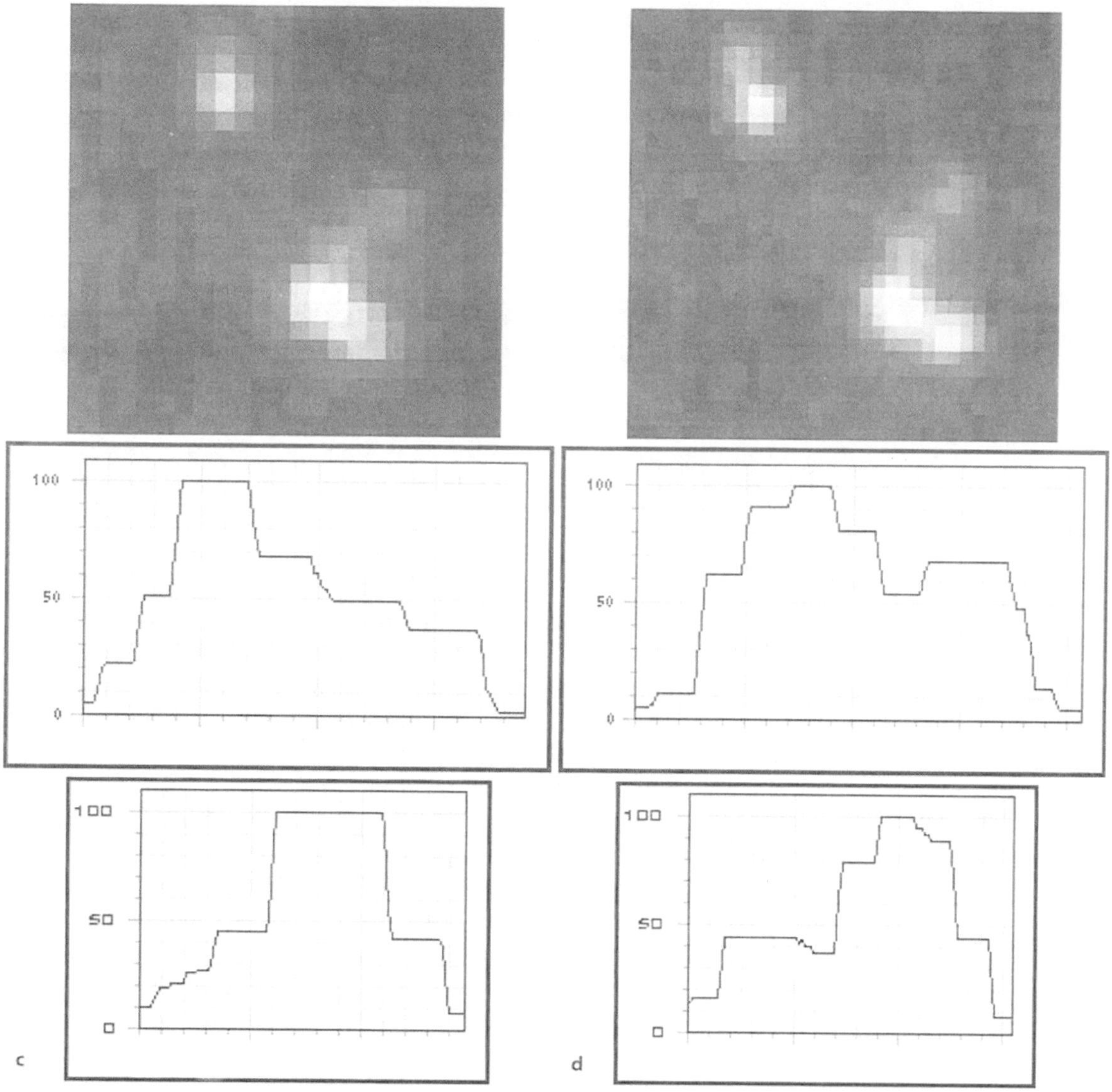

Fig. 9c, d

real objective, even a highly corrected Plan-Apochromat, will exhibit some residual optical aberrations. (2) Light scattering will occur within the specimen. (3) Particularly for small, faint signals, focusing may be difficult. Each factor will contribute to signal broadening. The real optical resolution limit is about 50% below the diffraction limited theoretical resolution.

Figure 9 shows the results of two different CCD cameras (Fig. 9a,c: standard CCD M300 with 11 μm pixels; Fig. 9b,d: mega pixel CCD M1 with 6.7 μm pixels) and two different objectives (Fig. 9a,b: Plan-Neofluar 100x/1.3 Oil; Fig. 9c,d: Plan-Apochromat 63×/1.4 Oil). The raw images were digitally zoomed to approximately the same size (upper row of Fig. 9). Zooming was performed without interpolation in order to recognize the individual pixels. Intensity profiles of the signals are displayed for regions A (middle row of Fig. 9) and B (bottom row of Fig. 9), respectively. For better comparison of their shape the profiles have been normalized. Consequently, they range from 0 (which corresponds to the fluorescence background intensity outside the spots) to 100 (which corresponds to the maximum intensity of the spots). At 100× magnification (Fig. 9a,b) both cameras resolve the two signals of region A (middle row of Fig. 9a,b), although the mega pixel CCD provides slightly better contrast. This is reflected by the more pronounced dip of the profile. Due to the poor signal intensity of one of the two signals of region B (bottom row of Fig. 9a,b) the darker region separating the signals is at the very limit of visibility.

With the 63× Plan-Apochromat and the standard CCD, the centers of the signals of region A are approximately 2.5 pixels apart. While the signals should be resolved if they were of approximately equal intensity, they appear as one larger signal (middle row of Fig. 9c) due to the low intensity of one of the spots with only 40% of the intensity of the second spot. The fluorescence background further reduces the contrast. This situation corresponds roughly to the simulation of Fig. 10. The mega pixel CCD resolves the spots of region A (middle row of Fig. 9d), while the signals of region B are again at the limit of resolution due to their low and uneven intensity (bottom row of Fig. 9c,d).

Although intensity profiles are helpful to appreciate subtle differences of signal intensities and very valuable to explain theory and compare theoretical to real results, they do not exactly correlate to the visual interpretation of the image. The information content of the image as a two-dimensional scene is higher than a one-dimensional profile. The human visual system performs a lot of image analysis and pattern recognition that the human observer is not aware of. For example, the presence of two spots is quite evident from the morphology of the signals in region A. From their

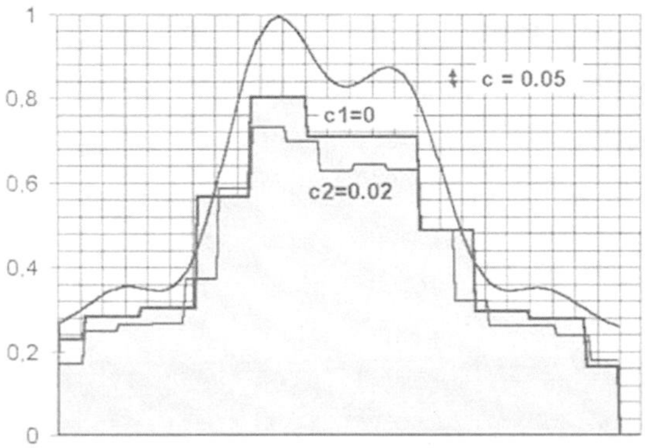

Fig. 10. Intensity profile of two signals having different intensities. Their separation is close to the resolution limit of the standard CCD. However, due to the reduced signal contrast c of only 0.05 the camera signal does not resolve the two peaks (c1=0). The mega pixel camera provides a slightly better result (relative contrast c2 of camera signal is 0.02)

contour the spots of region A will be interpreted as two individual spots although their intensity profile does not exhibit a second peak.

Conclusion

A FISH imaging system is more than the sum of its components. However, one of the components will always define the ultimate limit. It is of particular importance to include the FISH preparation as a major (if not *the* major) limiting component in the imaging process. Resolution and image quality is determined by the signal to background ratio, which is in most practical situations limited by the sample itself. No imaging system is able to add information to the raw image data. It can just help to visualize the information and remove unwanted distracting information. Moderately priced standard hardware components can provide completely satisfactory results if the software of the imaging system ensures that they are used to their full potential. Ironically, more sophisticated equipment can even compromise the results for merely practical reasons (as was the case in early mega pixel CCD cameras that did not provide a sufficiently fast display rate for precise focussing). The most important advice to be given is to check any imaging system under realistic conditions with the same type of preparations that the system will ultimately be used for. The worst thing to do is to make a decision based on the specification sheets of the system's individual hardware components.

Part B

FISH on Native, Human Tissues

FISH on Chromosome Preparations of Peripheral Blood

THOMAS LIEHR and UWE CLAUSSEN

Introduction

Peripheral blood is the most frequently used tissue for chromosome studies, as it is easy to obtain, can simply be brought into short-term culture and metaphase spreads can be prepared in high quality within a short time. Such metaphase spreads, as well as the huge amounts of previously superfluous interphase nuclei in cytogenetic preparations, are a material very well suited to FISH analyses. All kinds of FISH probes can be used to analyze metaphase spreads, while the interphase nuclei can actually be analyzed successfully only by satellite and (sub-)telomere probes, plasmids, cosmids, BACs, YACs or P1-clones and not – at least not in routine-approaches – by whole or partial chromosome painting probes. However, the quality of metaphases is important in routine diagnostics, as well as in research approaches (e.g. gene mapping). Thus, the preparation procedure of the chromosomes themselves is discussed in more detail in the following.

The currently used method of cell culture and chromosome preparation is a result of the combination of many single step procedures introduced into the protocol by different researchers. An aliquot of peripheral blood is added to culture medium, mixed with 10–20% fetal calf serum, penicillin/streptomycin to avoid contaminating prokaryotic cell growth and phytohaemagglutinin. The latter is a mitogen, stimulating in vitro growth of T lymphocytes (Nowell 1960; Moorhead et al. 1960). After 72 h of incubation at 37°C/5% CO_2, mimicking the conditions in human veins, cells are harvested. The addition of colcemid (= deacetylmethylcolchicine) acts as a mitotic spindle inhibitor, so that a block is introduced

✉ Thomas Liehr, Institut für Humangenetik, Kollegiengasse 10, Jena, 07743, Germany (*phone* +49-3641-935533; *fax* +49-3641-935502; *e-mail* i8lith@mti-n.mti.uni-jena.de)
Uwe Claussen, Institut für Humangenetik, Kollegiengasse 10, Jena, 07743, Germany

between meta- and anaphase (Ford and Hamerton 1956, Tjio and Levan 1956). The so called air-drying method for chromosome preparation of Moorhead et al. (1956) includes in the following order, a hypotonic treatment with 0.075 M KCl (according to Hsu 1952, Hughes 1952), a so called fixation and several washing steps with Carnoy's fixative (= methanol/glacial acetic acid 3:1) and finally dropping of the suspension on a slide surface.

Even though such preparations of human chromosome have been made for more than 40 years, their structure and the process of chromosome spreading are not completely understood up to now. Recent studies have revealed that this spreading process is not based on a "bursting" procedure, as suggested for years, but that fixed lymphocytes at the metaphase stage spread very slowly on the slide (Hliscs et al. 1997). Moreover, this surprisingly slow process leads to a spreading, which is humidity dependent – a fact described (Spurbeck et al., 1996), but previously not well understood. Moreover, chromosomes are elongated due to a stretching or swelling process, arising during the metaphase spreading (Hliscs et al. 1997, Claussen et al. 1999).

A standard preparation of human lymphocytes and a two-color FISH approach including the possibility to enhance the DAPI-banding pattern by BrdU (according to Heng and Tsui 1993) is described here.

Outline

1. Heparinized blood is added to cell culture medium, cell growth is stimulated by phytohaemagglutinin, and the short-term culture is incubated for 72 h at 37°C/5%CO_2. For the last 16 h BrdU can be added to the culture to enhance the DAPI banding pattern (see point 3). Cells are harvested after the addition of colcemid, subjected to a hypotonic treatment, fixed and washed in methanol/glacial acetic acid (3:1). The resulting suspension is dropped on slides, which are air dried and incubated over night at room temperature (see comment 1).

2. For a FISH approach, these slides are treated with RNase and pepsin prior to denaturation.

3. Denaturation of the slides is performed for about 2–3 min at 75°C. Biotinylated and/or digoxigenated probes are denatured for 5 min and hybridized on the slides for 3 nights at 37°C in a humid chamber. After a postwashing series, detection of the biotinylated probe is performed with an FITC-avidin system leading to a green signal. The di-

goxigenated probe with anti-digoxigenin-rhodamine leads to red signals. After counterstaining of the nuclei with DAPI and addition of antifade solution, slides can be evaluated on a fluorescence microscope.

Procedure

Preparation of peripheral blood lymphocytes suspension

1. Add 1 ml of heparinized blood to 9 ml of cell culture medium (e.g. RPMI 1640 medium/ 20% fetal calf serum/ 300 μg/ml L-glutamine/ 1 U/ml penicillin/ 1 μg/ml streptomycin/ 0.1 ml phytohaemagglutinine {Gibco BRL, 061-00576C}), mix the suspension carefully and incubate for 72 h at 37°C/5%CO_2 (see comments 2 and 3).
2. 16 h before harvesting, add 180 μg of BrdU to the 10 ml of the cell culture (comment 4).
3. 30 min before harvesting of the cells 1 μg of colcemid is added to a 10 ml culture flask, the flask is mixed gently and incubated at 37°C/ 5%CO_2.
4. The fluid is put into a 15-ml tube – sterile conditions no longer need to be observed.
5. Centrifuge the solution at room temperature (= RT) for 8 min at 1000 rpm and discard the supernatant by sucking it off carefully with a glass pipette (1 ml of supernatant is left in the tube to avoid loss of material).
6. For hypotonic treatment the pellet is resuspended in 10 ml 0.075 M KCl (37°C) and incubated at 37°C for 20 min.
7. Slowly add 0.6 ml of Carnoy's fixative (= methanol/glacial acetic acid 3:1) (4°C) and mix the solution carefully.
8. Repeat step 5.
9. Resuspend the pellet in 10 ml of fixative (4°C) and incubate at 4°C for 20 min.
10. Repeat step 5.
11. Resuspend the pellet in 5 ml of fixative (4°C) and repeat step 5.
12. Repeat step 11 twice.

13. According to the density of the suspension, the pellet is finally resuspended in 0.3–1 ml of fixative (suck off as much of the suspension as necessary after step 12).

14. Place 1 to 2 drops of the suspension, each on a clean and humid slide with a glass pipette and let the slide dry at RT.

15. After an incubation over night at RT, the slides can be subjected to the pretreatment (see below), stored dust free at RT for several weeks or frozen at -20°C for several months.

Slide pretreatment

1. Slides with the chromosome suspension are incubated in 2×SSC for 5 min at RT (in a 100 ml Coplin jar on a shaker).

2. Remove slides from the Coplin jar, add 100 μl of RNase solution/slide and cover with a 24×50 mm coverslip (RNase solution: per slide 100 μl 2×SSC plus 1 μl of RNase stock (= 5 μg/μl)).

3. Incubate the slides in a humid chamber for 15 min at 37°C (see comment 5).

4. Put slides back into the Coplin jar with 100 ml 2×SSC (RT) and remove the coverslips with forceps. Leave slides in 2×SSC solution for 3 min with gentle agitation.

5. Discard 2×SSC and replace it with 100 ml 1×PBS (RT) for 5 min (shaker).

6. Replace 1×PBS with 100 ml prewarmed pepsin-buffer (37°C) and incubate the slides for 10 min at 37°C, without agitation (see comment 5). (Pepsin-buffer: Add 1 ml of 1M HCl to 99 ml of distilled water and incubate at 37°C for about 20 min; then add 50 μl of the pepsin stock solution 10% (w/v) (Serva 31855) and leave the Coplin jar at 37°C; make fresh as required).

7. Replace fluid with 100 ml 1×PBS/$MgCl_2$, incubate at RT for 5 min with gentle agitation. $MgCl_2$ will block the enzymatic activity of pepsin. (1×PBS/$MgCl_2$ = 5% (v/v) 1M $MgCl_2$ in 1×PBS)

8. Postfix cells on the slide surfaces by replacing 1×PBS/$MgCl_2$ with 100 ml of formalin-buffer (3 ml acid free formaldehyde [37%; Roth 4979.1] in 100 ml 1xPBS) for 10 min (RT, with gentle agitation).

9. Formalin-buffer is replaced by 100 ml 1×PBS for 2 min (RT, with gentle agitation).
10. Finally, slides are dehydrated by an ethanol series (70%, 90%, 100%, 3 min each) and air dried (see comment 6).

Fluorescence in situ hybridization (FISH)

1. Add 100 µl denaturation-buffer to the slides and cover with (24×50 mm) coverslips. (Denaturation-buffer: 70% (v/v) deionized formamide, 10% (v/v) filtered double distilled water, 10% (v/v) 20×SSC, 10% (v/v) phosphate buffer; make fresh as required; phosphate buffer: prepare 0.5 M Na_2HPO_4 and 0.5 M NaH_2PO_4, mix these two solutions (1:1) to get pH 7.0, then aliquot and store at -20°C).
2. Incubate slides on a warming plate for 2–3 min at 75°C (see comment 7).
3. Remove the coverslips immediately with forceps and place slides in a Coplin jar filled with 70% ethanol (4°C) to conserve target DNA as single strands.
4. Dehydrate slides in ethanol (70%, 90%, 100%, 4°C, 3 min each) and air dry.
5. For each slide to be hybridized, treat the labeled DNA-probes according to manufacturers instructions or add 1.0 µl of biotin or digoxigenine-labeled chromosome specific satellite probes (e.g. Oncor Inc., Vysis, Inc.) plus 0.6 µl of 1 µg/µl COT1-DNA (Gibco) to 20 µl of the hybridization-buffer in a 1.5-ml microtube, vortex and spin down. {Hybridization-buffer: Dissolve 2 g dextran sulfate in 10 ml 50% deionized formamide/2×SSC/50mM phosphate buffer for 3 h at 70°C. Aliquot and store at -20°C} (see comment 8).
6. Denature probe-solution at 75°C for 5 min and cool immediately on ice to conserve probe DNA in single strands for satellite probes or handle the probe according to manufactures instructions (e.g. do a prehybridization step at 37°C for 20 min for any kind of library, YACs, BACs, cosmids or P1-clones).
7. Add 20 µl of probe-solution onto each denatured slide, put 24×50 mm coverslips on the drops and seal with rubber cement (Fixogum; Marabu).
8. Incubate slides for three nights at 37°C in a humid chamber (see comment 9).

9. Take the slides out of 37°C, remove rubber cement with forceps and coverslips by letting them swim off in 4×SSC/0.2%Tween (RT, 100-ml Coplin jar) (see comment 10).

10. Postwash the slides 3× 5 min in formamide-solution (45°C) followed by 3×5 min in 2×SSC (37°C) in a 100-ml Coplin jar, with gentle agitation.

11. Put the slides in 4×SSC/0.2% Tween (100 ml, RT), for a few seconds.

12. Add 50 µl of solution I to each slide, cover with 24×50 mm coverslips and incubate at 37°C for 30 min in a humid chamber. (Solution I: FITC-avidin (CAMON Vector Laboratories A2011)/4×SSC + 0.2% Tween + 5% BSA (1:300); make fresh as required).

13. Remove the coverslips and wash 3×3 min in 4×SSC/0.2% Tween (RT, with gentle agitation).

14. Add 50 µl of solution II to each slide, cover with 24×50 mm coverslips and incubate at 37°C for 75 min in a humid chamber. (Solution II: Biotinylated antiavidin (CAMON Vector Laboratories BA0300)/ Anti-digoxigenin-rhodamine (Boehringer Mannheim, 1207750)/4× SSC + 0.2% Tween + 5% BSA (1:20:100); make fresh as required).

15. Repeat step 13.

16. Repeat steps 12 and 13.

17. Counterstain the slides with DAPI-solution (100 ml in a Coplin jar, RT) for 8 min. (DAPI-solution: Dissolve 5 µl of DAPI (4,6-diamidino-2-phenylindol.2HCl stock-solution; Serva 18860) in 100 ml 4×SSC/ 0.2% Tween; make fresh as required).

18. Wash slides several times in water for a few seconds and air dry.

19. Add 15 µl of antifade Vectashield (CAMON Vector Laboratories H1000), cover with coverslips and look at the results under a fluorescence microscope. As examples of successful FISH experiments see Figure 1.

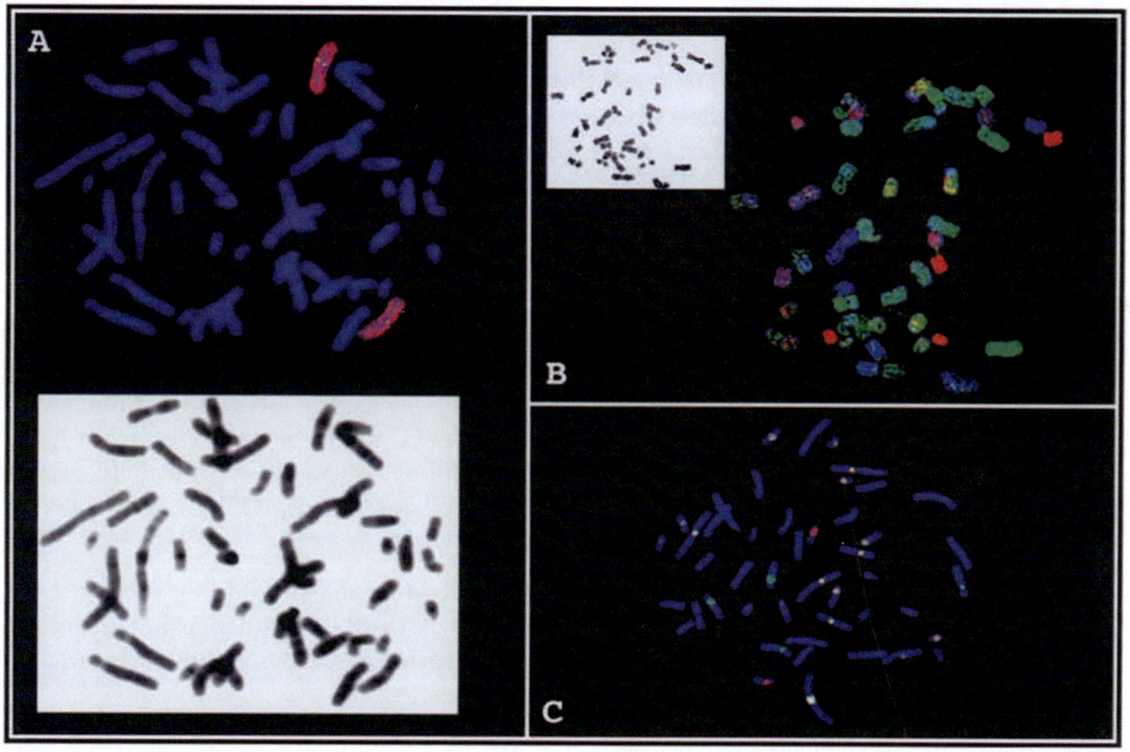

Fig. 1. Results of FISH experiments performed on chromosome preparations of healthy probands. Images were captured on a Zeiss Axioplan microscope using the ISIS digital FISH imaging system (MetaSystems, Altlussheim, Germany) using an XC77 CCD camera with on-chip integration (Sony). **A** The results of a FISH experiment using a whole chromosome painting probe for chromosome 12 (red) and of an mdm2-specific probe (green) in 12q13-14 are shown. In the lower part of A the inverted DAPI-picture, which corresponds to the GTG-banding pattern of the human chromosomes, is shown and it can be easily recognized that it is really chromosome 12 which has been painted. **B** The result of a multi-color FISH experiment is presented. Apart from all 24 different whole chromosome painting (wcp) probes, partial chromosome painting (pcp) probes for the long arms of nearly all human chromosomes have been applied. The wcp probes have been labeled as described in Senger et al. (1998) and the pcp probes have been labeled with a sixth fluorochrome (diethylaminocumarine). In the left upper part of B the inverted DAPI picture of the metaphase is shown. **C** A multi-color FISH result using 7 different centromeric DNA-probes labeled with three different fluorochromes is shown (see also Nietzel et al. 2001 and Chapter 32)

Comments

1. In this chapter only the preparation of peripheral blood metaphase spreads is described. For more specific preparation protocols (e.g. for bone marrow or solid tumors) see more specialized handbooks (e.g. Wegner 1999).
2. It is a well-known fact that EDTA- or sodiumacetate-treated blood samples can not be successfully cultured in short-term culture.
3. Sterile cell culture conditions have to be kept, so as not to contaminate the short-term culture.
4. BrdU is a thymidine analogue which is readily incorporated into chromosomes. Cultures containing BrdU should be protected from light, otherwise chromosome breaks will occur. Moreover, the yield of metaphase may be reduced compared to cultures without BrdU, as this chemical is cytotoxic.
5. RNase and pepsin pretreatment conditions should be tested in each laboratory on a single slide first. Both RNase and pepsin concentrations can be too stringent, resulting in clean slides without any remaining metaphase spreads or nuclei.
6. The pretreated slides can be hybridized immediately or stored at RT for up to 3 weeks. If longer storage is necessary, slides are stable at -20°C for several months.
7. Denaturation times of 2–3 min only will still allow the metaphase chromosomes to be evaluated. If only interphase nuclei are to be analyzed, longer denaturation times of up to 6 min can lead to stronger and better estimable results.
8. In the current protocol the most common indirect probe-ligands, biotin and digoxigenine and one example of a detection system for them are described. However, there are many other possibilities for directly labeled ligands (like SpectrumGreen, SpectrumOrange, SpectrumAqua, etc.) or detection possibilities for the indirect ones (like Cy5, Cy5.5, TexasRed, etc.). The corresponding protocols for detection are supplied by the companies and distributors and the chosen detection system mainly depends on the available filter sets of the own fluorescence microscope.
9. Incubation can be stopped – if necessary – after 24 or 96 h, as well. While in the first case weaker signals are possible, in the second

case some cross-hybridization problems may arise. The incubation time has to be tested and adjusted according to the probe type.

10. During the washing steps, it is important to prevent the slide surfaces drying out, otherwise background problems may arise.

Acknowledgements. The continuos support of the Carl Zeiss GmbH (Jena, Germany) is gratefully acknowledged.

References

Claussen, U, Lemke, J, Michel, S, Chudoba, I, Mühlig, P, Westermann, M, Claussen, J, Rubtsov, N, Grummt U-W (1999) The concept of chromosome condensation and de-condensation during mitosis has to be replaced by the concept of chromosome region specific swelling. Cytogenet Cell Genet 1999; Vol 85: p 166

Ford, CE, Hamerton, JL (1956) Ford, CE, Hamerton, JL (1956) The chromosomes in man. Nature 178:1020-1030

Heng, HHQ, Tsui L-C (1993) Modes of DAPI banding and simultaneous in situ hybridization. Chromosoma, 102:325-332

Hliscs, R, Mühlig, P, Claussen, U (1997) The spreading of metaphases is a slow process which leads to a stretching of chromosomes. Cytogenet Cell Genet, 76:167-171

Hsu TC (1952) Mammalian chromosomes in vitro. 1. karyotype of man. J of Heredity, 43:167-172

Hughes A (1952) Some effects of abnormal toxicity on dividing cells in chick tissue cultures. Quart J of Mikroscop Sci, 93:207

Moorhead, PS, Nowell, PC, Mellman, WJ (1960) Chromosome preparations of leukocytes cultured from human peripheral blood. Exp Cell Res 20: 613-616

Nietzel A, Rocchi M, Heller A, Starke H, Wlodarska I, Beensen V, Claussen U, Liehr 7 (2001) A new multicolor-FISH approach for the characterization of marker chromosomes: Centromere-specific multicolor-FISH (cenM-FISH). Hum Genet, 108: 199-204

Nowell, PC (1960) Phytohemagglutinin: An initiator of mitosis in cultures of normal human leukocytes. Cancer Res 20:462-466

Senger, G, Chudoba, I, Plesch A (1998) Multicolor-FISH – the identification of chromosome aberrations by 24 colors. BIOforum, 9/98: 99-503

Spurbeck, JL, Zinsmeister, AR, Meyer, KJ, Jatal, SM (1996) Dynamics of chromosome spreading. Am J Med Genet, 61:387-393

Tjio, JH, Levan, A (1956) The chromosome number of man. Hereditas, 42:6-9

Wegner, RD (1999) Diagnostic Cytogenetics. Springer, Berlin Heidelberg

FISH on Uncultured Blood or Bone Marrow

THOMAS LIEHR

Introduction

Interphase cytogenetics (= assay to obtain some "cytogenetic information" from interphase nuclei) has been made possible since the advent of the fluorescence in situ hybridization (FISH) technique (Langer et al., 1981, Pinkel et al., 1986). It is an approach widely used in diagnostics since reliable and well estimable probes have been made available for this technique; i.e. commercially available satellite probes for the centromeric regions of the human chromosomes to study numerical chromosomal changes, and cosmids, YACs, BACs or P1-clones suitable for the detection of translocations, inversions, deletions, duplications or amplifications (for summary see e.g. Sawyer et al., 1992; Wolfe and Herrington, 1997).

Peripheral blood or bone marrow aspirate are normally subjected to a short-term culture for 24 to 72 h to obtain metaphase spreads for standard cytogenetic banding analyses (according to Verma and Babu, 1989). However, in some exceptional cases, direct preparation of uncultured blood or bone marrow should be given priority (Liehr et al. 1995).

In some cases clinicians send EDTA- or sodium acetate-treated instead of heparinized blood samples for cytogenetic analysis. It is a well-known fact that such blood or bone marrow samples cannot be successfully cultured in short-term culture. However, for some FISH analyses, interphase nuclei are sufficient, as in the detection of microdeletion or microduplication syndromes (i.e. CATCH 22, hereditary neuropathy with liability to pressure palsies = HNPP, Miller-Diecker syndrome, Cri du Chat syndrome, Williams-Beuren syndrome, Prader-Willi syndrome or Charcot-Marie-Tooth disease Type 1A and others), numerical chromosome aberrations detectable by centromeric probes (see Figure 1), or in control of

✉ Thomas Liehr, Institut für Humangenetik, Kollegiengasse 10, Jena, 07743, Germany (*phone* +49-3641-935533; *fax* +49-3641-935502; *e-mail* i8lith@mti-n.mti.uni-jena.de)

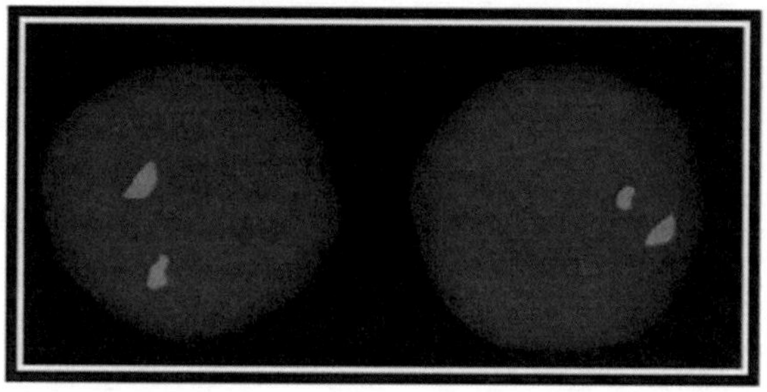

Fig. 1. Interphase nuclei prepared from EDTA-treated blood, prepared according to the protocol described in this chapter. Alphoid probes specific for chromosome 8 (red) and 17 (green) have been applied. Probes were derived from Oncor digoxigenated (D8Z2) and biotinylated (D17Z1) and detected as described using rhodamine and FITC. Two normal nuclei – with respect to the number of centromeric signals of chromosomes 8 and 17 – are shown. Images were captured on a Zeiss Axioplan microscope using the ISIS digital FISH imaging system (MetaSystems, Altlussheim, Germany) using an XC77 CCD camera with on-chip integration (Sony)

the course of a leukemic process with a known aberration (e.g. bcr-abl rearrangement).

Even heparinized blood samples could be prepared directly, for example in cases to be tested for Charcot-Marie-Tooth disease Type 1A, as the disease causing duplication is only 1.5 Mbp in size. This kind of duplication can be detected exclusively due to locus-specific probes in interphase nuclei and never in metaphase spreads, neither by GTG banding nor by FISH analysis. Thus, the time-consuming and cost-intensive procedure of short-term culture and metaphase preparation could be relinquished in such cases (Liehr et al., 1995, 1997; for a summary of Charcot-Marie-Tooth disease, see Patel and Lupski, 1994).

Moreover, for bone marrow samples of leukemia, it has to be remembered, that the detected percentage of subpopulations even after a 24-h short-term culture does not necessarily represent the in vivo situation. Thus, the interphase cell analysis of directly prepared bone marrow aspirate gives more valid information to the clinician in repeated analysis of bone marrow aspirates during the course of a leukemic process or in the control of reverse sex bone marrow transplantation (Liehr and Gebhart, unpublished data).

Outline

1. 1 ml of uncultured EDTA-, sodium acetate-treated or heparinized blood or bone marrow aspirate is washed in cell culture medium, subjected to a hypotonic treatment and fixed and washed in methanol/glacial acetic acid (3:1). The resulting suspension is dropped on slides, which are incubated over night at room temperature.
2. For a FISH approach, these slides are treated with RNase and pepsin prior to denaturation.
3. Denaturation of the slides is performed for about 5 min at 75°C. Biotinylated and/or digoxigenated probes are denatured for 5 min and hybridized on the slides for 3 nights at 37°C in a humid chamber. After a postwashing series, detection of the biotinylated probe is performed with an FITC-avidin system which leads to a green signal and of the digoxigenated probe with anti-digoxigenin-rhodamine leading to red signals. After counterstaining of the nuclei with DAPI and addition of antifade solution, slides can be evaluated on a fluorescence microscope.

Procedure

Preparation of uncultured blood or bone marrow

1. Add 1 ml of blood or bone marrow (EDTA-, sodium acetate-treated or heparinized) to 9 ml of cell culture medium (e.g. RPMI 1640 medium) and mix the solution carefully.
2. Centrifuge the solution in a 15-ml tube at room temperature (= RT) for 8 min at 1000 rpm and discard the supernatant by sucking it off carefully with a glass pipette (1 ml of supernatant is left in the tube to avoid loss of material).
3. For hypotonic treatment the pellet is resuspended in 10 ml 0.4% KCl (37°C) and incubated at 37°C for 20 min.
4. Slowly add 0.6 ml of fixative (= methanol/glacial acetic acid 3:1) (4°C) and mix the solution carefully.
5. Repeat step 2.
6. Resuspend the pellet in 10 ml of fixative (4°C) and incubate at 4°C for 20 min.

7. Repeat step 2.
8. Resuspend the pellet in 5 ml of fixative (4°C) and repeat step 2.
9. Repeat step 8 twice.
10. According to the density of the suspension, the pellet is finally resuspended in 0.3–1 ml of fixative (suck off as much of the suspension as necessary after step 9).
11. Bring 1 to 2 drops of the suspension, each, on a clean and humid slide with a glass pipette and let the slide dry at RT.
12. After an incubation over night at RT, the slides can be subjected to the pretreatment (see below), stored dust free at RT for several weeks or frozen at -20°C for several months.

Slide pretreatment

1. Slides with the interphase nuclei are incubated in 2×SSC for 5 min at RT (in a 100-ml Coplin jar on a shaker).
2. Remove slides from the Coplin jar, add 100 µl of RNase solution/slide and cover with a 24×50 mm coverslip (RNase solution: per slide 100 µl 2×SSC plus 1 µl of RNase stock (= 5 µg/µl)).
3. Incubate the slides in a humid chamber for 15 min at 37°C (see comment 1).
4. Put slides back into the Coplin jar with 100 ml 2×SSC (RT) and remove the coverslips with forceps. Leave slides in 2×SSC solution for 3 min with gentle agitation.
5. Discard 2×SSC and replace it with 100 ml 1×PBS (RT) for 5 min (shaker).
6. Replace 1×PBS with 100 ml prewarmed pepsin-buffer (37°C) and incubate the slides for 10 min at 37°C, without agitation (see comment 1). (Pepsin-buffer: Add 1 ml of 1 M HCl to 99 ml of distilled water and incubate at 37°C for about 20 min; then add 50 µl of the pepsin stock solution 10% (w/v) (Serva 31855) and leave the Coplin jar at 37°C; make fresh as required).
7. Replace fluid with 100 ml 1×PBS/$MgCl_2$, incubate at RT for 5 min with gentle agitation. $MgCl_2$ will block the enzymatic activity of pepsin (1×PBS/$MgCl_2$ = 5% (v/v) 1M $MgCl_2$ in 1×PBS).

8. Postfix nuclei on the slide surfaces by replacing $1\times PBS/MgCl_2$ with 100 ml of formalin-buffer (3 ml acid free formaldehyde [37%; Roth 4979.1] in 100 ml 1xPBS) for 10 min (RT, with gentle agitation).
9. Formalin-buffer is replaced by 100 ml $1\times$PBS for 2 min (RT, with gentle agitation).
10. Finally, slides are dehydrated by an ethanol series (70%, 90%, 100%, 3 min each) and air dried (see comment 2).

Fluorescence in situ hybridization (FISH)

1. Add 100 µl denaturation-buffer to the slides and cover with (24×50 mm) coverslips. (Denaturation-buffer: 70% (v/v) deionized formamide, 10% (v/v) filtered double distilled water, 10% (v/v) $20\times$SSC, 10% (v/v) phosphate buffer; make fresh as required; phosphate buffer: prepare 0.5 M Na_2HPO_4 and 0.5 M NaH_2PO_4, mix these two solutions (1:1) to get pH 7.0, then aliquot and store at -20°C).
2. Incubate slides on a warming plate for 5 min at 75°C (see comment 3).
3. Remove the coverslips immediately with forceps and place slides in a Coplin jar filled with 70% ethanol (4°C) to conserve target DNA as single strands.
4. Dehydrate slides in ethanol (70%, 90%, 100%, 4°C, 3 min each) and air dry.
5. For each slide to be hybridized, treat the labeled YAC-, BAC- or cosmid-DNA according to manufacturers instructions or add 1.5 µl of biotin or digoxigenine labeled chromosome specific satellite probes (e.g. Oncor Inc., Vysis, Inc.) plus 0.6 µl of 1 µg/µl COT1-DNA (Gibco) to 20 µl of the hybridization-buffer in a 1.5 ml microtube, vortex and spin down. (Hybridization-buffer: Dissolve 2 g dextran sulfate in 10 ml 50% deionized formamide/$2\times$SSC/50 mM phosphate buffer for 3 h at 70°C. Aliquot and store at -20°C).
6. Denature probe-solution at 75°C for 5 min and cool immediately on ice to conserve probe DNA in single strands for satellite probes or handle the probe according to the manufacturers instructions (e.g. do a prehybridization step at 37°C for 20 min for YACs, BACs, cosmids, P1-clones).

7. Add 20 µl of probe-solution onto each denatured slide, put 24×50 mm coverslips on the drops and seal with rubber cement (Fixogum; Marabu).
8. Incubate slides for three nights at 37°C in a humid chamber (see comment 4).
9. Take the slides out of 37°C, remove rubber cement with forceps and then the coverslips by letting them swim off in 4×SSC/0.2%Tween (RT, 100-ml Coplin jar) (see comment 5).
10. Postwash the slides 3×5 min in formamide-solution (45°C) followed by 3×5 min in 2×SSC (37°C) in a 100-ml Coplin jar, with gentle agitation.
11. Put the slides in 4×SSC/0.2% Tween (100 ml, RT), for a few seconds.
12. Add 50 µl of solution I to each slide, cover with 24×50 mm coverslips and incubate at 37°C for 30 min in a humid chamber. [Solution I: FITC-avidin (CAMON Vector Laboratories A2011)/4×SSC + 0.2% Tween + 5% BSA (1:300); make fresh as required].
13. Remove the coverslips and wash 3×3 min in 4×SSC/0.2% Tween (RT, with gentle agitation).
14. Add 50 µl of solution II to each slide, cover with 24×50 mm coverslips and incubate at 37°C for 75 min in a humid chamber. [Solution II: Biotinylated antiavidin (CAMON Vector Laboratories BA0300)/anti-digoxigenin-rhodamine (Boehringer Mannheim, 1207750)/4×SSC + 0.2% Tween + 5% BSA (1:20:100); make fresh as required].
15. Repeat step 13.
16. Repeat steps 12 and 13.
17. Counterstain the slides with DAPI-solution (100 ml in a Coplin jar, RT) for 8 min. [DAPI-solution: Dissolve 5 µl of DAPI (4,6-diamidino-2-phenylindol.2HCl stock-solution; Serva 18860) in 100 ml 4×SSC/0.2% Tween; make fresh as required].
18. Wash slides several times in water for a few seconds and air dry.
19. Add 15 µl of antifade Vectashield (CAMON Vector Laboratories H1000), cover with coverslips and look at the results under a fluorescence microscope.

Comments

1. RNase and pepsin pretreatment conditions should be tested in each laboratory on a single slide first. Both RNase and pepsin concentrations can be too stringent, resulting in clean slides without any remaining nuclei.
2. The pretreated slides can be hybridized immediately or stored at RT for up to 3 weeks. If longer storage is necessary slides are stable at -20°C for several months.
3. In other FISH-protocols with denaturation times of 2–3 min only, the maintenance of available metaphase chromosomes is the main aspect, which is of no significance in the actual protocol.
4. Incubation can be stopped – if necessary – after 48 or 96 h, as well. While in the first case weaker signals are possible, in the second case some cross-hybridization problems may arise.
5. During the washing steps it is important to prevent the slide surfaces drying out, otherwise background problems may arise.

Acknowledgements. Supported by the Wilhelm Sander-Stiftung (99.105.1) and the EU (QLRT-1999-31590). The continuos support of the Carl Zeiss GmbH (Jena, Germany) is gratefully acknowledged.

References

Langer, PR, Waldrop, AA, Ward, DC (1981) Enzymatic synthesis of biotin-labeled polynucleotides: Novel nucleic acid affinity probes. Proc Natl Acad Sci USA, 78:6633-6637

Liehr, T, Thoma, K, Kammler, K, Gehring, C, Ekici, A, Bathke, KD, Grehl, H, Rautenstrauss, B (1995) Direct preparation of uncultured EDTA-treated or heparinized blood for interphase FISH analysis. Appl Cytogenet, 21:185-188

Liehr, T, Grehl, H, Rautenstrauss, B (1997) Molecular diagnosis of PMP22-associated neuropathies using fluorescence in situ hybridization (FISH) on archival peripheral nerve tissue preparations. Acta Neuropathol, 94:266-271

Patel, PI, Lupski, JR (1994) Charcot-Marie-Tooth disease: a new paradigm for the mechanism of inherited disease. Trends in Genetics, 10:128-133

Pinkel, D, Straume, T, Gray, JW (1986) Cytogenetic analysis using quantitative, high sensitivity, fluorescence hybridization. Proc Natl Acad Sci USA, 83:2934-2938

Sawyer, JR, Johnson, MP, Miller, OJ (1992) Traditional and molecular cytogenetics. J Reprod Med 37, 485-498.

Verma, RS, Babu, A (1989) Human chromosomes: Manual of basic techniques. Pergamon Press, New York.
Wolfe, KQ, Herrington, CS (1997) Interphase cytogenetics and pathology: a tool for diagnostics and research. J Pathol, 181:359-361

FISH on Blood or Bone Marrow Smears

Thomas Liehr

Introduction

Cytogenetic diagnostics of leukemia subtypes are nowadays done routinely by conventional banding analysis as well as by fluorescence in situ hybridization (FISH). A lack of metaphase spreads in cytogenetic preparations in many cases can be compensated, using interphase cytogenetics with specific probes like commercially available satellite probes for the centromeric regions of the human chromosomes, cosmids, YACs, BACs or P1-clones (for summary see e.g. Sawyer et al., 1992; Wolfe and Herrington, 1997). Special approaches, i.e. the control of bone marrow transplants, can be performed on interphase nuclei without the necessity of time consuming cytogenetic preparation at all. Moreover, for blood or bone marrow samples of leukemia, it has to be remembered, that the percentage of subpopulations even after a 24-h short-term culture does not necessarily represent the in vivo situation. Thus, the interphase cell analysis of uncultured bone marrow aspirate gives the clinician more valid information in repeated analysis of bone marrow aspirates during the course of a leukemic process or in the control of reverse sex bone marrow transplantation (Liehr and Gebhart, unpublished data).

Interphase nuclei from peripheral blood or bone marrow can be prepared directly according to Liehr et al. (1995) (see Chapter 4) or – more quickly – by making a blood or bone marrow smear on a slide (Gebhart et al., 1995). Furthermore, the latter technique has the advantage that the cellular structure can be maintained and tumor cells can be distinguished from nontumor cells (e.g. Haferlach et al., 1998). In the protocol reported in the following, this can be achieved when replacing a slide pretreatment step with a 10-min incubation of the slides in 2×SSC at room temperature.

✉ Thomas Liehr, Institut für Humangenetik, Kollegiengasse 10, Jena, 07743, Germany (*phone* 49-3641-935533; *fax* +49-3641-935502; *e-mail* i8lith@mti-n.mti.uni-jena.de)

Outline

1. 100 to 200 µl of uncultured EDTA-, sodium acetate-treated or heparinized blood or bone marrow aspirate is dropped on a clean dry slide, spread over the slide surface and air dried.
2. For a conventional FISH approach without the goal of maintaining the cellular structure of the analyzed cells/nuclei, these slides are treated with RNase and pepsin prior to denaturation.
3. Denaturation of the slides is performed for about 5 min at 75°C. Biotinylated and/or digoxigenated probes are denatured for 5 min and hybridized on the slides for 3 nights at 37°C in a humid chamber. After a postwashing series, detection of the biotinylated probe is performed with an FITC-avidin system which leads to a green signal and of the digoxigenated probe with anti-digoxigenin-rhodamine leading to red signals. After counterstaining of the nuclei with DAPI and addition of antifade solution, slides can be evaluated under a fluorescence microscope.

Procedure

Blood or bone marrow smear

1. Drop 100 to 200 µl of blood or bone marrow (EDTA-, sodium acetate-treated or heparinized) on the one end of a dry clean slide (see comment 1).
2. Spread the fluid over the whole slide surface with the small edge of a 24×60 mm coverslip. The edge of the coverslip is dipped into the blood or bone marrow and moved slowly – once only – over the slide and without touching the slide surface, which could disrupt the cells.
3. Let the blood or bone marrow dry out at room temperature (= RT) for approx. 12 h before introduction into the slide pretreatment (see below).

Slide pretreatment

During the whole slide pretreatment procedure, the slides lose the red erythrocytes and they become more and more transparent. At the end (step 10), the nuclei of the blasts and lymphocytes are the only remaining cellular components on the slide surface. If the cellular structure is to be maintained and tumor cells are to be distinguished from nontumor cells, this slide pretreatment step must be replaced by a 10-min incubation of the slides in $2\times$SSC at room temperature, prior to the FISH.

1. Slides with the blood or bone marrow smear are incubated in $2\times$SSC for 5 min at RT (in a 100-ml Coplin jar on a shaker).
2. Remove slides from the Coplin jar, add 100 µl of RNase solution/slide and cover with a 24×50 mm coverslip [RNase solution: per slide 100 µl $2\times$SSC plus 1 µl of RNase stock (= 5 µg/µl)].
3. Incubate the slides in a humid chamber for 15 min at 37°C (see comment 2).
4. Put slides back into the Coplin jar with 100 ml $2\times$SSC (RT) and remove the coverslips with forceps. Leave slides in $2\times$SSC solution for 3 min with gentle agitation.
5. Discard $2\times$SSC and replace it with 100 ml $1\times$PBS (RT) for 5 min (shaker).
6. Replace $1\times$PBS with 100 ml prewarmed pepsin-buffer (37°C) and incubate the slides for 10 min at 37°C, without agitation (see comment 2). (Pepsin-buffer: Add 1 ml of 1 M HCl to 99 ml of distilled water and incubate at 37°C for about 20 min; then add 50 µl of the pepsin stock solution 10% (w/v) (Serva 31855) and leave the Coplin jar at 37°C; make fresh as required).
7. Replace fluid with 100 ml $1\times$PBS/$MgCl_2$, incubate at RT for 5 min with gentle agitation. $MgCl_2$ will block the enzymatic activity of pepsin ($1\times$PBS/$MgCl_2$ = 5% (v/v) 1M $MgCl_2$ in $1\times$PBS).
8. Postfix nuclei on the slide surfaces by replacing $1\times$PBS/$MgCl_2$ with 100 ml of formalin-buffer (3 ml acid free formaldehyde [37%; Roth 4979.1] in 100 ml 1xPBS) for 10 min (RT, with gentle agitation).
9. Formalin-buffer is replaced by 100 ml $1\times$PBS for 2 min (RT, with gentle agitation).
10. Finally, slides are dehydrated by an ethanol series (70%, 90%, 100%, 3 min each) and air dried (see comment 3).

Fluorescence in situ hybridization (FISH)

1. Add 100 µl denaturation-buffer to the slides and cover with (24×50 mm) coverslips. [Denaturation-buffer: 70% (v/v) deionized formamide, 10% (v/v) filtered double distilled water, 10% (v/v) 20×SSC, 10% (v/v) phosphate buffer; make fresh as required; phosphate buffer: prepare 0.5 M Na_2HPO_4 and 0.5 M NaH_2PO_4, mix these two solutions (1:1) to get pH 7.0, then aliquot and store at -20°C].
2. Incubate slides on a warming plate for 5 min at 75°C (see comment 4).
3. Remove the coverslips immediately with forceps and place slides in a Coplin jar filled with 70% ethanol (4°C) to conserve target DNA as single strands.
4. Dehydrate slides in ethanol (70%, 90%, 100%, 4°C, 3 min each) and air dry.
5. For each slide to be hybridized, treat the labeled YAC-, BAC- or cosmid-DNA according to the manufacturers instructions or add 1.5 µl of biotin or digoxigenin labeled chromosome specific satellite probes (e.g. Oncor Inc., Vysis, Inc.) plus 0.6 µl of 1 µg/µl COT1-DNA (Gibco) to 20 µl of the hybridization-buffer in a 1.5-ml microtube, vortex and spin down. (Hybridization-buffer: Dissolve 2 g dextran sulfate in 10 ml 50% deionized formamide/2×SSC/50mM phosphate buffer for 3 h at 70°C. Aliquot and store at -20°C).
6. Denature probe-solution at 75°C for 5 min and cool immediately on ice to conserve probe DNA in single strands for satellite probes or handle the probe according to the manufacturers instructions (e.g. do a pre-hybridization step at 37°C for 20 min for YACs, BACs, cosmids, P1-clones).
7. Add 20 µl of probe-solution onto each denatured slide, put 24×50 mm coverslips on the drops and seal with rubber cement (Fixogum; Marabu).
8. Incubate slides for three nights at 37°C in a humid chamber (see comment 5).
9. Take the slides out of 37°C, remove rubber cement with forceps and the coverslips by letting them swim off in 4×SSC/0.2%Tween (RT, 100-ml Coplin jar) (see comment 6).

10. Postwash the slides 3×5 min in formamide-solution (45°C) followed by 3×5 min in 2×SSC (37°C) in a 100-ml Coplin jar, with gentle agitation.
11. Put the slides in 4×SSC/0.2% Tween (100 ml, RT), for a few seconds.
12. Add 50 μl of solution I to each slide, cover with 24×50 mm coverslips and incubate at 37°C for 30 min in a humid chamber. [Solution I: FITC-avidin (CAMON Vector Laboratories A2011)/4×SSC + 0.2% Tween + 5% BSA (1:300); make fresh as required].
13. Remove the coverslips and wash 3×3 min in 4×SSC/0.2% Tween (RT, with gentle agitation).
14. Add 50 μl of solution II to each slide, cover with 24×50 mm coverslips and incubate at 37°C for 75 min in a humid chamber. [Solution II: Biotinylated antiavidin (CAMON Vector Laboratories BA0300)/anti-digoxigenin-rhodamine (Boehringer Mannheim, 1207750)/4×SSC + 0.2% Tween + 5% BSA (1:20:100); make fresh as required].
15. Repeat step 13.
16. Repeat steps 12 and 13.
17. Counterstain the slides with DAPI-solution (100 ml in a Coplin jar, RT) for 8 min. [DAPI-solution: Dissolve 5 μl of DAPI (4,6-diamidino-2-phenylindol 2 HCl stock-solution; Serva 18860) in 100 ml 4×SSC + 0.2% Tween; make fresh as required].
18. Wash slides several times in water for a few seconds and air dry.
19. Add 15 μl of antifade Vectashield (CAMON Vector Laboratories H1000), cover with coverslips and look at the results under a fluorescence microscope.

Comments

1. Sodium citrate-treated blood or bone marrow is best suited for the preparation of a "smear-slide" as it spreads the best. However, blood or bone marrow treated with the other anticoagulants can be used, as well.
2. RNase and pepsin pretreatment conditions should be tested in each laboratory on a single slide first. Both RNase and pepsin concentrations can be too stringent, resulting in clean slides without any remaining nuclei.

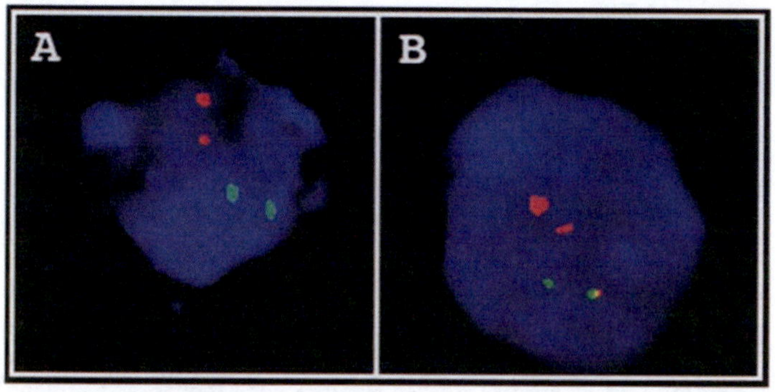

Fig. 1. Result of a FISH experiment performed on the bone marrow smear of a patient suffering from chronic myelogenous leukemia (CML). In most of the CML patients the Philadelphia chromosome is present, which results from a translocation t(9;22)(q34;q11.2) that fuses the abl proto-oncogene on chromosome 9 to the bcr (breakpoint cluster region) on chromosome 22. To determine the frequency of Philadelphia-positive cells in the bone marrow interphase, FISH was performed using the LSI bcr/abl ES translocation probe of Vysis. The bcr-region is depicted in green, the abl-region in red. In **A**, a Philadelphia-negative cell with two red and two green signals is shown. In **B**, a Philadelphia-positive cell with one green signal (for the normal chromosome 22), one big red signal (for the normal chromosome 9), one green and red fusion signal, resulting in a yellow mix-color (for the Philadelphia-chromosome) and one small red signal (for the translocated chromosome 9) are visible. For details see Vysis product catalog 1999/2000, p. 37 and Sinclair et al. (1997). Images were captured with the ISIS digital FISH imaging system (MetaSystems, Altlussheim, Germany) using an XC77 CCD camera with on-chip integration (Sony)

3. The pretreated slides can be hybridized immediately or stored at RT for up to 3 weeks. If longer storage is necessary, slides are stable at -20°C for several months.
4. In other FISH-protocols with denaturation times of 2-3 min only, the maintenance of available metaphase chromosomes is the main aspect, which is of no significance in the actual protocol.
5. Incubation can be stopped – if necessary – after 48 or 96 h, as well. While in the first case weaker signals are possible, in the second case some cross-hybridization problems may arise.
6. During the washing steps it is important to prevent the slide surfaces drying out, otherwise background problems may arise.

Acknowledgements. Supported by the Wilhelm Sander-Stiftung (99.105.1) and the EU (ICA2-CT-2000- 10012 and QLRT-1999-31590). The continuos support of the Carl Zeiss GmbH (Jena, Germany) is gratefully acknowledged.

References

Gebhart, E, Liehr, T, Harrer, P, Reichardt, S, Schmitt, G, Thoma, K, Gramatzki, M, Trautmann, U (1995) Determination by interphase FISH of the clonality of aberrant karyotypes in human hematopoietic neoplasias. Leukemia and Lymphoma, 17:295-302

Haferlach, T, Loffler, H, Nickenig, C, Ramm-Petersen, L, Meeder, M, Schoch, R, Schlegelberger, B, Schnittger, S, Schoch, C, Hiddemann, W (1998) Cell lineage specific involvement in acute promyelocytic leukaemia (APL) using a combination of May-Grunwald-Giemsa staining and fluorescence in situ hybridization techniques for the detection of the translocation t(15;17)(q22;q12). Br J Haematol 103:93-99

Liehr, T, Thoma, K, Kammler, K, Gehring, C, Ekici, A, Bathke, KD, Grehl, H, Rautenstrauss, B (1995) Direct preparation of uncultured EDTA-treated or heparinized blood for interphase FISH analysis. Appl Cytogenet, 21:185-188

Sawyer, JR, Johnson, MP, Miller, OJ (1992) Traditional and molecular cytogenetics. J Reprod Med 37:485-498

Sinclair, PB, Green, AR, Grace, C, Nacheva, EP (1997) Improved sensitivity of BCR-ABL detection: a triple-probe three-color fluorescence in situ hybridization system. Blood 90:1395-1402

Wolfe, KQ, Herrington, CS (1997) Interphase cytogenetics and pathology: a tool for diagnostics and research. J Pathol, 181:359-361

Epithelial Cells from Buccal Smears and Urine

ANITA RAUCH

Introduction

Aneuploidy is usually investigated by karyotyping lymphocytes from peripheral blood. However, in several aneuploidy syndromes, tissue mosaicism is frequently found. In Pallister-Killian syndrome, for example, the supernumerary isochromosome 12p is usually not seen in blood lymphocytes, although it occurs in up to 100% of fibroblasts. Therefore it is necessary to karyotype other tissues than blood. Traditionally, fibroblasts from a skin biopsy were used to analyze suspected mosaic karyotypes. However, this rather invasive procedure is often refused by patients. Therefore, interphase FISH on cells from buccal smears (Ohashi et al. 1993) or urinary sediment provides a useful noninvasive alternative method to analyze numerical chromosomal aberrations, if tissue mosaicism is suspected.

In order to achieve a high hybridization efficiency, which also allows accurate analysis even of low percentage mosaicism by FISH with single copy probes, we recommend pretreatment of cells with hypotonic solution, RNase and pepsin. For hypotonic treatment we are using a solution containing dithioerythritol, which was used by Otto to investigate Chinese hamster cell lines by flow karyotyping (Otto 1988). In addition, application of directly labeled probes reduces background fluorescence, which is especially disturbing if caused by bacteria in a signal imitating manner.

✉ Anita Rauch, Institut für Humangenetik, Schwabachanlage 10, Erlangen, 91054, Germany (*phone* +49-9131-8522318; *fax* +49-9131-209297;
e-mail arauch@humgenet.uni-erlangen.de)

Buccal smears

The oral cavity is covered by a mostly nonkeratinized squamous epithelium. Keratinization occurs only at the gingiva and lingual villi. While the oral mucosa is regenerated through mitotic activity of the basal layer, superficial cell layers are permanently peeled off and found in the saliva. Therefore epithelial cells from the oral mucosa are easily accessible by a simple buccal smear without any pain.

Urinary cells

Urine sediment contains many exfoliative cells from the epithelial layer of the kidney pelvis, ureter, bladder, urethra, vagina, and prostate, i.e. from the urothel. Urine cells are usually investigated for aneuploidy in bladder cancer or for chromosomal mosaicism in mental-retardation syndromes. Urine cells can be cultured similarly to amniotic fluid to obtain metaphase spreads as described in the Springer cytogenetic lab manual (Körner et al. 1999). However, it is difficult to avoid contamination from bacteria or yeast. Much easier and less time-consuming is the preparation of urine cells for interphase FISH. This was successfully invented for the diagnosis and follow-up of bladder cancer (Meloni et al. 1993). In clinical cytogenetics, it provides an ideal method to study tissue mosaicism of aneuploidy, especially in trisomy and tetrasomy 20q (Pfeiffer et al. 1998), where the aneuploidy is often restricted to urothel which is also found in amnion fluid (Koskull et al. 1984).

The major prerequisite for analysis of urine cells is to prevent the cells from being damaged by toxic urine components. Therefore the urine sample is ideally prepared immediately after urination, or at least within 3 h.

Outline

The procedure of interphase FISH on epithelial cells from buccal smears or urinary sediment is outlined in Fig. 1.

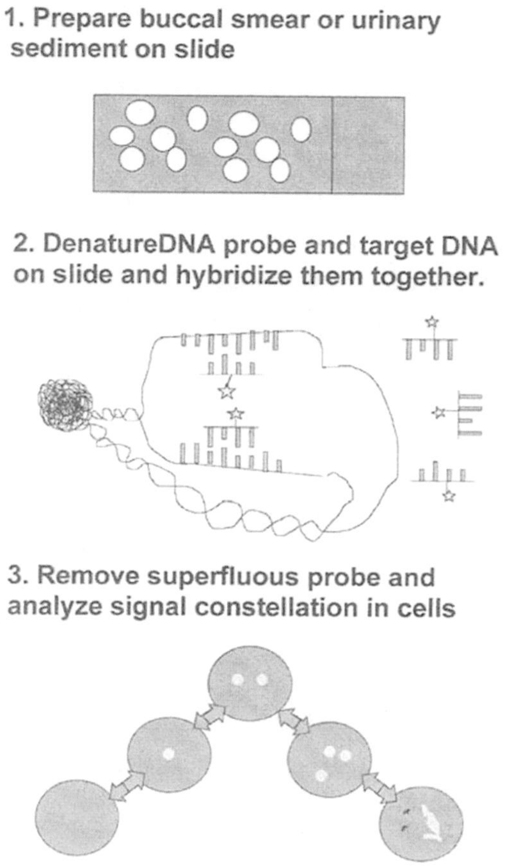

Fig. 1. Outline of interphase FISH from epithelial cells

Materials

Slide preparation

- Quadriperm trays (Invitro Systems & Services)
- Otto's solution:
  - 100 ml Tris-HCl buffer 10 mmol/l, pH 7.4: 1.21 g Tris-amino-methane and 1 l aqua dest., adjust with HCl to pH 7.4
  - 220 mg KCl (30 mmol/l)
  - 100 mg $MgCl_2 \times 6\ H_2O$ (5 mmol/l)
  - 46 mg dithioerythritol (3 mmol/l)

- Fixative:
  - Acetic acid/methanol 1:4 (e.g. 50 ml acetic acid plus 150 ml methanol, prepare freshly)
- RNase working solution (100 ng/μl): 1 μl stock solution plus 99 μl 2×SSC
  - Stock solution: dissolve 10 mg RNase A (Roche) in 1 ml 2×SSC, cook for 10 min and store in portions of 100 μl in -20°C
- Pepsin stock solution 10%: Dissolve 1g pepsin (Fluka) in 10 ml aqua bidest and store in portions of 50 μl at -20°C
- 2×SSC: 50 ml 20×SSC plus 450 ml aqua dest
- 20×SSC: 87.65 g NaCl plus 44.1 g Na-citrate plus 500 ml aqua bidest, adjust pH to 7.0 with HCl 32%, store at room temperature
- PBS: 50 ml 10×PBS plus 450 ml aqua dest
- 10×PBS: 80 g NaCl plus 2 g KCl plus 11.5 g Na_2HPO_4 plus 2 g KH_2PO_4 add 1 l aqua bidest, adjust pH to 7.0 with NaOH, store at room temperature
- 1 M $MgCl_2$: dissolve 20.3 g $MgCl_2$ in 100 ml aqua bidest, store at room temperature
- PBS-$MgCl_2$: 10 ml 1 M $MgCl_2$ plus 190 ml PBS
- 1% Formaldehyde: 2.7 ml formaldehyde (37%, Merck) plus 97.3 ml PBS-$MgCl_2$

Denaturation

- 70% formamide/2×SSC×: per slide: 70 μl deionized formamide plus 10 μl phosphate buffer plus 10 μl 20× ×SSC plus 10 μl aqua dest (may be stored at -20°C)
- Phosphate buffer: dissolve 0.7 g sodium dihydrogen phosphate monohydrate in 10 ml aqua bidest, dissolve 0.89 g sodium dihydrogen phosphate dihydrate in 10 ml aqua bidest and mix both, store at -20°C in portions of 1 ml.
- Deionized formamide: ad 2.5 g ion exchanger (mixed bed resin) to 50 ml formamide, stir for 2 h, let the ion exchanger settle down and aliquot formamide in 1 ml portions, store at -20°C.

Hybridization mix

- Commercially available, directly labeled probes should be used with provided hybridization mix following supplier's instructions.
- Self-labeled probes (e.g. with Roche Corp. Nick Translation Mix according to supplier's instructions) use with 20% dextran sulfate: dissolve 2 g dextran sulphate in 5 ml deionized formamide plus 1 ml phosphate buffer plus 1 ml 20×SSC plus 3 ml aqua bidest, store at -20°C.

Washing solutions

- 1×SSC: add 25 ml 20×SSC to 475 ml aqua bidest, adjust pH to 7.0-7.5 with 5 drops of 1 M HCl. Store at room temperature
- 2×SSC/0.1% NP-40: add 50 ml 20×SSC to 450 ml aqua bidest, ad 0.5 ml NP-40, adjust pH to 7.0-7.5 with 1 drop of HCl 32%

Counterstaining

- Dapi-solution: per slide, mix 1 ml 2×SSC/0.1% NP-40 with 1 µl Dapi stock solution (plus 1 µl propidium iodide stock solution if there is not a red fluorochrome being used to detect DNA probe)
- Dapi stock solution: dissolve 10 mg DAPI (Serra) in 10 ml aqua dest and store aliquots of 50 µl at -20°C (the aliquot at use can be stored at 4°C)
- Propidium iodide stock solution: dissolve 10 mg propidium iodide (Sigma) in 10 ml aqua bidest and store aliquots of 50 µl at -20°C (the aliquot at use can be stored at 4°C)
- Antifade: Mounting medium for fluorescence (Vectashield)

Procedure

Preparation of buccal smears

1. Scrape the inner surface of the buccal mucosa and saliva with a wooden tongue depressor and smear it onto a slide.
2. Let the slide air dry.
3. Transfer the slide into a Quadriperm tray.
4. Carefully overlay the slide with Otto's solution and incubate for 30-60 min at 37°C.
5. Remove Otto's solution carefully and overlay the slide with fixative, incubate at room temperature for 5 min.

6. Remove fixative and incubate another 5 min at room temperature.
7. Remove fixative once again and incubate another 5 min at room temperature.
8. Remove fixative and let the slide air dry.
9. If the slide is not further used within a few days, store it at -20°C.

Preparation of urinary cells

1. Transfer the not necessarily sterile, but fresh urine sample into a 50 ml tube.
2. Centrifuge at 172 *g* for 8 min (if r=16 cm 1000 U/min).
3. Remove the supernatant leaving a few millimeters.
4. Take the sediment with the remaining supernatant in a pipette and drop it onto a clean slide.
5. Let the slide air dry.
6. Put the slide into a Quadriperm tray.
7. Carefully overlay the slide with Otto' solution and incubate for 30-60 min at 37°C.
8. Remove Otto's solution carefully and overlay the slide with fixative, incubate at room temperature for 5 min.
9. Remove fixative and incubate another 5 min at room temperature.
10. Remove fixative once again and incubate another 5 min at room temperature.
11. Remove fixative and let the slide air dry.
12. If the slide is not further used within a few days, store it at -20°C.

RNase/Pepsin pretreatment

RNase

1. Cover slide with prepared epithelial cells with 100 µl RNase working solution and a coverslip and incubate in a metal tray in a 37°C water bath for 15 min.

2. Remove coverslip and wash 3×3 min in 2×SSC at room temperature in a Coplin jar on a shaker.

3. Add 50 µl of pepsin stock solution to a Coplin jar with 100 ml in a 37°C water bath and incubate the slide for 10 min in the Pepsin solution. **Pepsin**

4. Wash 1 x 5 min in 1xPBS in a Coplin jar on a shaker at room temperature.

5. Wash 5 min in PBS-$MgCl_2$.

6. Wash 10 min in 1% formaldehyde in PBS-$MgCl_2$.

7. Wash for 5 min in PBS.

8. Dehydrate in a series of ethanol (70%, 90%, 100%) for 2 min each.

9. Air dry slides.

Denaturation and hybridization

1. Cover slide with 100 µl 70% formamide/2×SSC and a coverslip and incubate it for 5 min on the back of a metal box in a 75°C water bath for denaturation of target DNA.

2. Dehydrate slide in a series of ethanol (70%, 90%, 100%) on ice for 2 min each to avoid rehybridization.

3. While slide is air drying, denaturate specific DNA probe in 10 µl hybridization mix for 5 min at 75°C and place on ice immediately.

4. Place denatured probe (10 µl) on one part of the slide, cover carefully with a 20×20 mm coverslip avoiding air bubbles and seal with rubber cement.

5. Incubate in a moist chamber at 37°C over night or at least 4 h for centromeric probes.

Posthybridization washes

For commercially available probes, follow the instructions of the manufacturers or:

1. Carefully remove rubber cement and coverslip with forceps.
2. Place slide in a Coplin jar with 1×SSC in a 75°C water bath for 2 min, agitate the slide for the first 1-3 s.
3. Wash slide in 2×SSC/0.1% NP-40 at room temperature for 5 s-1 min.

Counterstaining

1. Place 1 ml DAPI/propidium iodide working solution on each wet slide and incubate for 5 min in the dark.
2. Place slide shortly in 2×SSC/0.1% NP-40 and air dry slide in the dark.
3. Apply 20 μl of antifade in a row of dots across the slide and cover with a coverslip.

Storage

4. Store slide at -20°C in the dark.

Results

Analyze slides with a suitable filter set under a fluorescence microscope. According to Hook 1977 it is necessary to analyze at least 299 cells to exclude a 1% mosaicism with a confidence limit of 95%. However, in interphase FISH, the signal number obtained, varies artificially to a certain degree, because of several factors which are discussed in the Comments section. Therefore it is necessary to analyze a larger number of nuclei and interpret the results based on statistics from normal controls. With directly labeled probes from VYSIS Corp., the following results were achieved. For normal controls, the mean percentage of cells with a normal signal pattern on buccal smears was 99% for the repetitive centromeric probe CEP X and 98% for the single copy probes LSI 21 and 20q13. In normal controls of urinary sediment, a normal signal pattern of LSI 20q13 was found in 98.7% of cells (see Table 1). If a threshold standard deviation of 4 is considered sufficient, this method would detect trisomic cell lines constituting at least 1.4% (CEP) and 2% (LSI) of the total cell population and monosomic cell lines constituting at least 2.5% (CEP) and 4% (LSI). However, for diagnostic purposes, each lab has to set up their own normal values and a second DNA probe should be cohybridized as an internal control of hybridization efficiency.

Table 1. Signal distribution of FISH with directly labeled centromeric probe CEP X, single copy probe LSI 20q13 and 21 (VYSIS) on cells from buccal smears or urinary sediment from normal controls

	CEP X	LSI 21	LSI 20q13	LSI 20q13
Tissue	Buccal smear n=10	Buccal smear n=3	Buccal smear n=3	Urinary sediment n=9
1 signal	0.5 ± 0.5%	1.25 ± 02 %	1.23 ± 0.3%	1.1 ± 0.7%
2 signals	99.25 ± 0.6%	98.25 ± 0.2%	98.03 ± 0.1%	98.7 ± 0.7%
3 signals	0.2 ± 0.3%	0.5 ± 0.2%	0.60 ± 0.3%	0.1 ± 0.2%
4 signals	0	0	0.13 ± 0.1%	0.1 ± 0.2%
Number of analyzed nuclei per case	200	400	1000	200

Troubleshooting

No cells

If urine cells are not separated from urine within a few hours they disappear due to toxic urine components. Until fixed with fixative, buccal smear or urinary cells may be washed from the slide if shaken too heavily.

No or weak signals

Reasons for no or weak signals may be too short a denaturation or hybridization time or washing the slide after hybridization with too high stringency. Stringency decreases with increase of salt concentration, or decrease of temperature and formamide concentration. Hybridization efficiency is increased by higher dextran sulfate and probe concentrations.

Too much background

Too much background fluorescence may result from any soiling and can be decreased during hybridization or posthybridization washing by increasing the temperature or formamide concentration, or decreasing salt concentrations.

Diffuse signals and cells

Decrease denaturation time or temperature by 2°C.

Comments

FISH analysis on interphase nuclei generally suffers from systematic errors. This means, that the number of detected signals varies from the number of actual DNA binding sites. The signal number may be artificially decreased by:

- insufficient tissue penetrability, resulting in minor hybridization efficiency
- a certain number of two actual signals lying so close together that they appear as one
- analysis of damaged or overlapping cells

or increased by:

- signal splitting due to dispersed DNA
- signal splitting due to S or G2-phase cells
- signal imitating by bacteria or other background fluorescence

False negative signals from insufficient tissue penetrability may be reduced by optimal tissue pretreatment, while the fact that a certain number of signals sticking together simulating a single signal has to be addressed

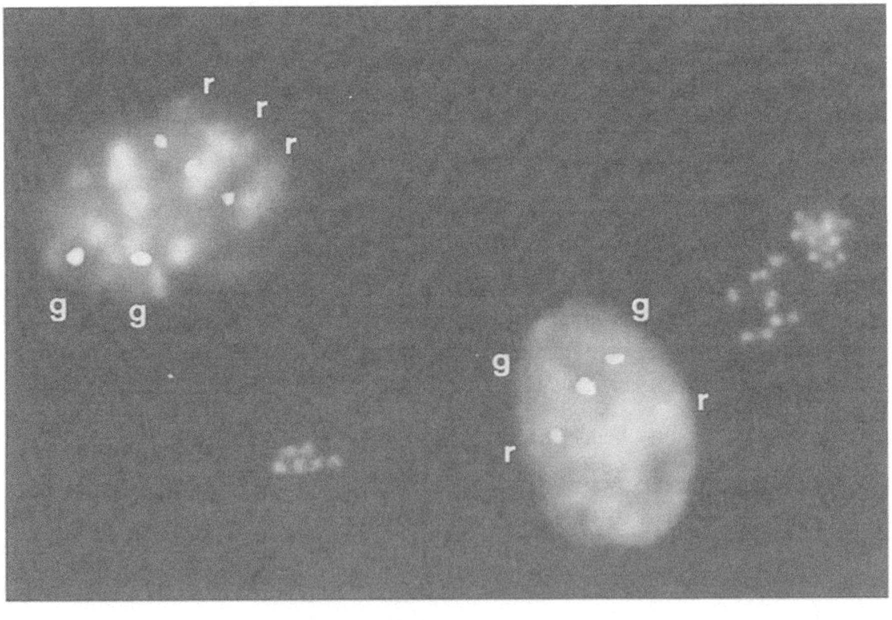

Fig. 2. FISH with probe VYSIS LSI 20q13 (red *[r]*) and control probe VYSIS CEP 18 (green *[g]*) to epithelial cells from urinary sediment of a child with mosaicism of an isochromosome 20q (cell with three red signals) in 13.7% of urinary cells

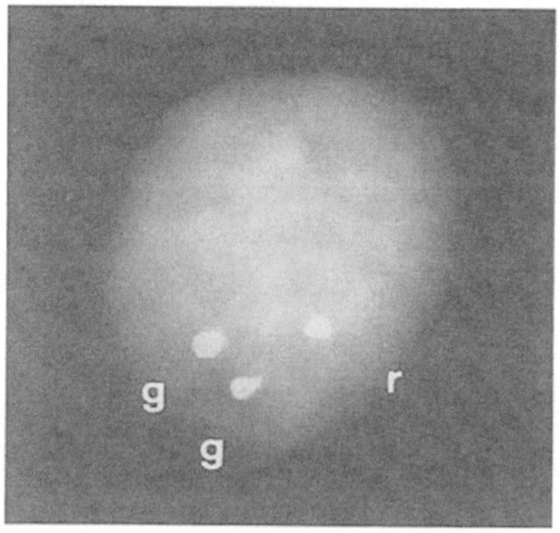

Fig. 3. FISH with VYSIS VCFS probe (red *[r]*) and control probe ARSA (green *[g]*) demonstrating monosomy 22q11.2 (VCFS) in buccal smear cells. In a total of 312 cells, 98.4% showed one VCFS signal and 1.6% showed two signals, which is in line with a nonmosaic monosomy 22q11.2. Routine karyotyping from peripheral blood lymphocytes was not possible in this patient, due to the failure of defective T-cells growth

by a statistical approach. False positive signals by background fluorescence can be reduced by working as cleanly as possible, by optimized washing procedures or by the use of directly labeled DNA probes. The described protocol takes care of all these items and results in a high accuracy of results, allowing the detection of even minor aneuploidy mosaicism.

Applications

Besides the determination of aneuploidy mosaicism (Fig. 2), interphase FISH on epithelial cells can be useful if metaphase analysis from peripheral T-lymphocytes is not possible due to a T-cell defect, which is a symptom of partial monosomy 22q11.2 (DiGeorge syndrome) (Fig. 3).

Acknowledgements. I am grateful to Dr. U. Trautmann, Mrs. S. Reichhardt, and Mrs. M. Kirsch for sharing their technical experience.

References

Hook EB (1977) Exclusion of chromosomal mosaicism: tables of 90%, 95%, and 99% confidence limits and comments on use. Am J Hum Genet 29:94-97

Körner H, Dia H, Bommer C (1999) Cells from urine sample. In: Wegner (ed) Cytogenetics labmanual. Springer, Berlin Heidelberg New York

Koskull H von, Aula P, Trejdosiewicz LK, Virtanen I (1984) Identification of cells from fetal bladder epithelium in human amniotic fluid. Hum Genet 65:262-267

Meloni AM, Peier AM, Haddad FS, Powell IJ, Block AMW, Huben RP, Todd I, Potter W, Sandberg AA (1993) A new approach in the diagnosis and follow-up of bladder cancer. FISH analysis of urine, bladder washings, and tumors. Cancer Genet Cytogenet 71:105-118

Ohashi H, Ishikiriyama S, Fukushima Y (1993) New diagnostic method for Pallister-Killian syndrome: Detection of i(12p) in interphase nuclei of buccal mucosa by fluorescence in situ hybridization. Am J Med Genet 45:123-128

Otto FJ (1988) Assessment of persisting chromosome aberrations by flow karyotyping of cloned chinese hamster cells. Z Naturforsch 43c:948-954

Pfeiffer RA, Ulmer R, Rauch A, Trautmann U, Beinder E, Rupprecht T, Mayer U, Steinkirchner B, Wüündisch F (1997) True fetal mosaicism of an isochromosome of the long arm of a chromosome 20: The dilemma persists. Prenat Diagn 17:1171-1175

FISH on Hair Root Cells

THOMAS LIEHR

Introduction

In clinical cytogenetics, patients with somatic chromosomal mosaicism are present in about 0.3 to 1% of analyzed cases. Mosaic karyotypes in connection with gonosomes are the most frequent observation, presenting e.g. with karyotypes like 45,X[10]/47,XXX[5]/46,XX[5] (unpublished case of an unfertile woman, Inst. of Human Genetics, Jena). However, autosomes can be involved, as well, as in a familial case with a supernumerary marker derived from chromosome 8, present in the peripheral blood of the father in 10% and in that of the two daughters in 97 and 98% of analyzed metaphase spreads (Rothenmund et al., 1997). Such kinds of mosaicism can easily be detected by interphase FISH using centromere-specific or locus-specific probes of the corresponding chromosomes.

As it is a well-known fact that the patterns of mosaicism can vary to a great extent in different kinds of tissues of the same patient (e.g. Pfeiffer and Schulze, 1994), the analysis of different tissue types by molecular cytogenetic methods can be of great interest. This has been demonstrated with the example of a patient with a microduplication on chromosome 17p11.2, which appeared in different patterns of mosaicism in peripheral blood [49%], in buccal mucosa [51%], in nerve tissue [74%] and in hair root cells [66%] (Liehr et al., 1996).

In the following, a modified protocol according to Lampel et al. (1993) for pretreating hair root cells before FISH is described. It may be useful for the analyses of mosaic cases as well as in cases where other types of cells are difficult to obtain.

✉ Thomas Liehr, Institut für Humangenetik, Kollegiengasse 10, Jena, 07743, Germany (*phone* +49-3641-935533; *fax* +49-3641-935502; *e-mail* i8lith@mti-n.mti.uni-jena.de)

Outline

1. Freshly extracted hairs are incubated in acetic acid and centrifuged to remove the hair root cells from the hairs themselves; then the hair root cells are incubated in fixative (= methanol/acetic acid) and spotted onto a dry clean slide.
2. For a FISH approach, these slides are treated with RNase and pepsin prior to denaturation.
3. Denaturation of the slides is performed for about 5 min at 75°C. Biotinylated and digoxigenated probes are denatured for 5 min and hybridized on the slides for 3 nights at 37°C in a humid chamber. After a postwashing series, detection of the biotinylated probe is performed with an FITC-avidin system which leads to a green signal and of the digoxigenated probe with anti-digoxigenin-rhodamine leading to red signals. After counterstaining of the nuclei with DAPI and addition of antifade solution, slides can be evaluated under a fluorescence microscope.

Procedure

Hair root cell preparation

1. Freshly extracted hairs can be used at once for the hair root cell preparation or they can be transferred into cell culture medium and stored there for up to 48 h.
2. Dip the tip of 3 to 6 hair roots into 60 µl of 50% acetic acid in a 1.5-ml tube; cut the end of the hair short enough that the tube can be closed and incubate for 10 min at room temperature (RT).
3. Centrifuge for 5 min at 4000 rpm to remove the hair root cells from the hairs themselves.
4. Remove the hairs from the tube, add 30 µl of methanol to the suspension and incubate for 30 min at RT.
5. Pipette the resulting 45 µl onto two clean dry slides and allow to air dry.

Slide pretreatment

As in a conventional FISH approach, a pretreatment of the slides with RNase and pepsin followed by a postfixation with formalin-buffer is required to reduce background (Liehr et al., 1995).

1. Slides with the hair root cells are incubated in 2×SSC for 5 min at RT (in a 100-ml Coplin jar on a shaker).
2. Remove slides from the Coplin jar, add 100 µl of RNase solution/slide and cover with a 24×50 mm coverslip (RNase solution: per slide 100 µl 2×SSC plus 1 µl of RNase stock (= 5 µg/µl)).
3. Incubate the slides in a humid chamber for 15 min at 37°C (see comment 1).
4. Put slides back into the Coplin jar with 100 ml 2×SSC (RT) and remove the coverslips with forceps. Leave slides in 2×SSC solution for 3 min with gentle agitation.
5. Discard the 2×SSC and replace it with 100 ml 1×PBS (RT) for 5 min (shaker).
6. Replace 1×PBS with 100 ml prewarmed pepsin-buffer (37°C) and incubate the slides for 10 min at 37°C, without agitation (see comment 1). (Pepsin-buffer: Add 1 ml of 1 M HCl to 99 ml of distilled water and incubate at 37°C for about 20 min; then add 50 µl of the pepsin stock solution 10% (w/v) (Serva 31855) and leave the Coplin jar at 37°C; make fresh as required).
7. Replace fluid with 100 ml 1×PBS/$MgCl_2$, incubate at RT for 5 min with gentle agitation. $MgCl_2$ will block the enzymatic activity of pepsin (1×PBS/$MgCl_2$ = 5% (v/v) 1M $MgCl_2$ in 1×PBS).
8. Postfix nuclei on the slide surfaces by replacing 1×PBS/$MgCl_2$ with 100 ml of formalin-buffer (3 ml acid free formaldehyde [37%; Roth 4979.1] in 100 ml 1xPBS) for 10 min (RT, with gentle agitation).
9. Formalin-buffer is replaced by 100 ml 1×PBS for 2 min (RT, with gentle agitation).
10. Finally, slides are dehydrated by an ethanol series (70%, 90%, 100%, 3 min each) and air dried (see comment 2).

Fluorescence in situ hybridization (FISH)

1. Add 100 µl denaturation-buffer to the slides and cover with (24×50 mm) coverslips. [Denaturation-buffer: 70% (v/v) deionized formamide, 10% (v/v) filtered double distilled water, 10% (v/v) 20×SSC, 10% (v/v) phosphate buffer; make fresh as required; phosphate buffer: prepare 0.5 M Na_2HPO_4 and 0.5 M NaH_2PO_4, mix these two solutions (1:1) to get pH 7.0, then aliquot and store at -20°C].
2. Incubate slides on a warming plate for 5 min at 75°C (see comment 3).
3. Remove the coverslips immediately with forceps and place slides in a Coplin jar filled with 70% ethanol (4°C) to conserve target DNA as single strands.
4. Dehydrate slides in ethanol (70%, 90%, 100%, 4°C, 3 min each) and air dry.
5. For each slide to be hybridized, treat the labeled YAC-, BAC- or cosmid-DNA according to the manufacturers instructions or add 1.5 µl of biotin or digoxigenin labeled chromosome-specific satellite probes (e.g. Oncor Inc., Vysis, Inc.) plus 0.6 µl of 1 µg/µl COT1-DNA (Gibco) to 20 µl of the hybridization-buffer in a 1.5-ml microtube, vortex and spin down. (Hybridization-buffer: Dissolve 2 g dextran sulfate in 10 ml 50% deionized formamide/2×SSC/50 mM phosphate buffer for 3 h at 70°C. Aliquot and store at -20°C).
6. Denature probe-solution at 75°C for 5 min and cool immediately on ice to conserve probe DNA in single strands for satellite probes or handle the probe according to manufacturers instructions (e.g. do a prehybridization step at 37°C for 20 min for YACs, BACs, cosmids, P1-clones).
7. Add 20 µl of probe-solution onto each denatured slide, put 24×50 mm coverslips on the drops and seal with rubber cement (Fixogum; Marabu).
8. Incubate slides for three nights at 37°C in a humid chamber (see comment 4).
9. Take the slides out of 37°C, remove rubber cement with forceps and the coverslips by letting them swim off in 4×SSC/0.2%Tween (RT, 100-ml Coplin jar) (see comment 5).
10. Postwash the slides 3×5 min in formamide-solution (45°C) followed by 3×5 min in 2×SSC (37°C) in a 100-ml Coplin jar, with gentle agitation.

11. Put the slides in 4×SSC/0.2% Tween (100 ml, RT), for a few seconds.
12. Add 50 μl of solution I to each slide, cover with 24×50 mm coverslips and incubate at 37°C for 30 min in a humid chamber. [Solution I: FITC-avidin (CAMON Vector Laboratories A2011)/4×SSC + 0.2% Tween + 5% BSA (1:300); make fresh as required].
13. Remove the coverslips and wash 3×3 min in 4×SSC/0.2% Tween (RT, with gentle agitation).
14. Add 50 μl of solution II to each slide, cover with 24× 50 mm coverslips and incubate at 37°C for 75 min in a humid chamber. [Solution II: Biotinylated antiavidin (CAMON Vector Laboratories BA0300)/anti-digoxigenin-rhodamine (Boehringer Mannheim, 1207750)/ 4×SSC + 0.2% Tween + 5% BSA (1:20:100); make fresh as required].
15. Repeat step 13.
16. Repeat steps 12 and 13.
17. Counterstain the slides with DAPI-solution (100 ml in a Coplin jar, RT) for 8 min. [DAPI-solution: Dissolve 5 μl of DAPI (4,6-diamidino-2-phenylindol. 2HCl stock-solution; Serva 18860) in 100 ml 4×SSC + 0.2% Tween; make fresh as required].
18. Wash slides several times in water for a few seconds and air dry.
19. Add 15 μl of antifade Vectashield (CAMON Vector Laboratories H1000), cover with coverslips and look at the results under a fluorescence microscope.

Comments

1. RNase and pepsin pretreatment conditions should be tested in each laboratory on a single slide first. Both RNase and pepsin concentrations can be too stringent, resulting in clean slides without any remaining nuclei.
2. The pretreated slides can be hybridized immediately or stored at RT for up to 3 weeks. If longer storage is necessary, slides are stable at -20°C for several months.
3. In other FISH-protocols with denaturation times of 2-3 min only, the maintenance of available metaphase chromosomes is the main aspect, which is of no significance in the actual protocol.

4. Incubation can be stopped – if necessary – after 48 or 96 h, as well. While in the first case weaker signals are possible, in the second case some cross-hybridization problems may arise.
5. During the washing steps it is important to prevent the slide surfaces drying out, otherwise background problems may arise.

Acknowledgements. Supported by the EU (ICA2-CT-2000-10012 and QLRT-1999-31590). The continuos support of the Carl Zeiss GmbH (Jena, Germany) is gratefully acknowledged.

References

Lampel, S, Steilen, H, Zang, KD, Wullich, B (1993) Sex chromatin in hair roots-25 years later: fluorescence in situ hybridization of hair root cells for detection of numerical chromosome aberrations. Cytogenet Cell Genet, 63:244-246

Liehr, T, Thoma, K, Kammler, K, Gehring, C, Ekici, A, Bathke, KD, Grehl, H, Rautenstrauss, B (1995) Direct preparation of uncultured EDTA-treated or heparinized blood for interphase FISH analysis. Appl Cytogenet, 21:185-188

Liehr, T, Rautenstrauss, B, Grehl, H, Bathke, KD, Ekici, A, Rauch, A, Rott, HD (1996) Mosaicism for the Charcot-Marie-Tooth disease type 1A duplication suggests somatic reversion. Human Genet, 98:22-28

Pfeiffer, RA, Schulze, T (1994) Mosaicism in three cases 47,XY (or XX), +i(18)(p10) detected by interphase FISH of buccal mucosa. Ann Génét 37:210-214

Rothenmund, H, Chudley, AE, Dawson, AJ (1997) Familial transmission of a small supernumerary marker chromosome 8 identified by FISH: an update. Am J Med Genet, 72:339-334

Amniotic Fluid Cells – Uncultured

RENATE ULMER

Introduction

Cytogenetic analysis of fetal cells after amniocentesis is routinely offered to women who have an increased risk of carrying a fetus with a chromosomal abnormality. Indications for fetal karyotyping include advanced maternal age, increased risk for fetal trisomy by elements of the triple screen (i.e. alphafetoprotein, human chorionic gonadotropin, and estriol), a previous pregnancy with an aneuploid fetus or potential risks due to family history.

The rate of chromosomal disorders in low-risk groups, like advanced maternal age is about 2.0%. Rates of ›20% of abnormalities are found in high risk groups with pathological findings in ultrasonography (Eiben et al.1997, Claussen et al.1994).

Conventional prenatal diagnosis requires a long-term culture of amniocytes (6-14 days) and is routinely accomplished by GTG banding of metaphase chromosomes. This technique is accurate and reliable, allowing the detection of a variety of numerical and structural aberrations. Disadvantages are the time-consuming cell-culture and the banding analysis which requires some experience.

In certain clinical situations (sonographically detected severe fetal abnormalities) a result within the shortest time possible is urgently necessary for the management of the pregnancy. In other cases, a rapid result could moderate the anxiety of the pregnant woman, especially in cases such as an advanced pregnancy, a pathological result of the triple screen, some conspicuous sonographical findings like nuchal translucency, choroid plexus cyst and singular umbilical cord artery.

Renate Ulmer, Institut für Humangenetik, Schwabachanlage 10, Erlangen, 91054, Germany (*phone* +49-9131-85-2-4134; *fax* +49-9131-209297; *e-mail* rulmer@humgenet.uni-erlangen.de)

In most of the above mentioned situations, a method which provides a rapid and accurate identification of two thirds of all chromosomal defects (Philip et al., 1994) is a useful adjunctive diagnostic test to conventional cytogenetic analysis.

There are two more situations, where a rapid testing for aneuploidy can also be useful in a prenatal diagnostic laboratory.

First: the amniocytes of a special case show no or only very poor growth in culture. With FISH on the uncultivated cells, a re-amniocentesis can be avoided in some instances.

Second: one or more aneuploid cell colonies (after in situ culture) or metaphases (after flask culture) are found by analyzing the metaphase spreads. For distinguishing between pseudomosaicism and true mosaicism, FISH on both uncultured (interphases) and cultured cells (interphases and metaphases) can be useful.

For each of the above-mentioned purposes a suspension of amniotic fluid cells is necessary. Normally, there is no amniotic fluid left after setting up the culture. When the culture-medium is changed (as a rule 4 or 5 days after setting up the cultures), many amniocytes that did not settle down are discarded. This medium from each amniotic fluid sample can be kept in the refrigerator until the case is finished. If necessary, FISH studies can be performed on this material.

FISH has become a powerful diagnostic tool, which allows a detection of the aneuploidies of chromosomes 13, 18, 21, X and Y within 24 h. As has been shown by Liehr et al. (1999), even a translocation trisomy dup(21q) can be distinguished from a free trisomy 21 by interphase-FISH. The reliability of the FISH- technique depends highly on the specificity and hybridization efficiency of the DNA-probes used.

Rapid and accurate detection of chromosomal aneuploidies has been demonstrated in a blinded prospective study which compared aneuploid detection by FISH for chromosomes 13, 18, 21, X and Y to results obtained by cytogenetics (Klinger et al. 1992). The first clinical application of this technology was published by Ward et al. (1993, 1997). They reported informative results on 89.5% of 10,342 processed uncultured amniotic fluid samples.

The breakthrough in using FISH on uncultured amniocytes was made possible by the development of new DNA-probes. Up to 1992, probes specific for chromosome 18, X and Y were commercially available but the detection of trisomy 13 and 21 in interphase cells was hampered by the fact that available probes could not distinguish between chromosome types 13 and 21 because of cross-hybridization (Philip et al., 1994). While Klinger (1992) and Ward (1993) constructed their own cosmid probes for chromosomes 13 and 21, labeled α-satellite centromere-specific DNA-probes for chromosomes 18, X and Y and labeled unique DNA probes for chromosomes 13 (hybridizing to 13q14) and 21 (hybridizing to 21 q22.13-21q22.2) are now commercially available as a kit (AneuVysion EC by Vysis GmbH, 51429 Bergisch Gladbach, Germany).

The first application of these new DNA-probes was published by Eiben et al. (1998). In this prospective comparative study no false-negative or false-positive results were obtained in 91% (798/904) successfully analyzed cases, with a minimum of 50 nuclei counted for each probe. In contrast to Bryndorf (1997), who reported 19% of cases with less than 50 scorable nuclei, this value dropped in the study of Eiben to 8%.

In our own study (Ulmer et al., 2000) in 86% of cases, 50 nuclei could be scored for each probe. However, only 12/230 cases were classified as not informative (less than 15 analyzable nuclei). More than 30 scorable nuclei were obtained for the probes X,Y and 18 in 94.7% and for the probes 13 and 21 in 90% of the cases.

According to the recipe, which is included in the AneuVysion EC-Kit of Vysis together with the experience we had with FISH in our institute, we tried to find a procedure which allows the preparations to be done in parallel to other work in a prenatal diagnostic laboratory. It should be possible with minimal technical equipment, to accomplish this in a short time, with a high success-rate and accurate results.

The procedure for FISH on uncultivated amniotic fluid cells will now be described as it has been developed in our laboratory with the equipment available there. Consequently, there are several steps which can be varied or adapted to the available conditions in other laboratories.

Outline

There are four main steps from the native amniotic fluid to the slide which is ready for fluorescence microscopy:

1. Preparation of cells: from amniotic fluid to a slide with swollen and well-fixed nuclei.
2. Slide pretreatment: in order to increase the accessibility of the probe-DNA to the target sequences. This step normally includes a protease and an RNase step. The RNase can be omitted; duration of the pepsin-treatment depends on the number of nuclei (and thus is dependent on the week of gestation in which the amniocentesis was done) on the slide.

3. Denaturation and hybridization: These two steps are combined in the procedure described here. Of course a separate denaturation of the target-sequences gives equal or even better results. Co-denaturation is time-saving. The adequate denaturation time has to be tested for each lab.

4. Post-hybridization washes and counterstain: The stringency of the washing steps has to be evaluated for each lab. In the case where the signals of probes for chromosomes 13 and 21 are too weak but signals of probes for chromosomes X,Y, and 18 are good, the stringency of washing can be reduced for the cosmid probes by using a second slide for hybridization. The slide with unique DNA-probes for 13 and 21 can be washed with 1×SSC while the α-satellite-probes for X,Y and 18 can be washed more stringently with 0.4×SSC.
By using our own DAPI counterstain and mounting medium, we had better results than with the combined mounting medium of Vysis. The ready-mixed mounting medium included in the Vysis-kit containing DAPI should be tested anyway.

5. Signal enumeration: including the points microscopical equipment, enumeration scan, criteria for interpretation and a warning concerning maternal cell contamination.

Materials

Laboratory equipment

- incubator (37°C)
- water baths (37±2°C, 74±2°C)
- microcentrifuge
- magnetic stirrer
- microliter pipettor (1-1000 μl)
- sterile tips (blue and yellow)
- pH meter
- calibrated thermometer
- timer
- forceps

- Coplin jars
- precleaned microscope slides
- glass coverslips (20×20 mm, 15×15 mm, 24×60 mm)
- polypropylene microcentrifuge tubes (0.5 or 1.5 ml)
- centrifuge tube 10-15 ml (conical)
- graduated cylinder
- humidified chamber
- rubber cement

Reagents and buffers

- Ethanol (100%)
- concentrated HCl (12 M)
- NaOH (1 M)
- purified water
- Carnoy's fixative (3:1 methanol/acetic acid)
- $MgCl_2$, sodium citrate, NaCl
- 10×PBS (phosphate buffered saline)
- 20×SSC (salt sodium citrate)
- pepsin (from hog stomach, Fluka; prepare a stock solution of 10% pepsin)
- Hyaluronidase (from bovine testes, Sigma; prepare a stock solution of ca. 50 IU per ml)
- Dabco (Triethylenediamine, by Sigma)
- Vectashield mounting medium (Vector Laboratories)
- DAPI (4,6-diamidino-2-phenylindol, prepare a stock solution of 0.1 mg/100 ml)
- NP 40 (included in the Vysis-kit)
- Paraformaldehyde (37%)

Preparing the solutions

Table 1.

ethanol series	70%, 90 %, 100 %
hypotonic solution	10 ml 1% sodium citrate 15 ml 0.33% MgCl2 1 ml hyaluronidase (150 U) alternatively: 0.28% KCl solution
1 M HCl	8.62 ml 32% HCl 91.38 ml purified water
1 M $MgCl_2$	101.65 g $MgCl_2 \times 6\ H_2O$ add distilled water to a final volume of 500 ml
20×SSC	175.3 g NaCl 88.2 g sodium citrate 800 ml distilled water adjust to pH 7.0 add distilled water to a final volume of 1000 ml
10×PBS	98.4 g PBS (powder) 1000 ml distilled water
1×PBS	40 ml 10×PBS 360 ml distilled water
1×PBS/$MgCl_2$	190 ml PBS 10 ml 1 M MgCl2
1×PBS/$MgCl_2$/ formaldehyde	100 ml PBS/$MgCl_2$ 3 ml paraformaldehyde
pepsin working solution	50 µl of 10% pepsin solution 100 ml distilled water 1 ml 1M HCl
mounting solution	Dabco + Vectashield mixed 1:1

Procedure

Slide preparation from uncultured amniotic fluid

1. Put clean slides (cleaned for example with tissue paper) into a Coplin jar with purified water. Place into the refrigerator.
2. Centrifuge 3-5 ml of whole amniotic fluid (AF) specimen for 5 min at 1000 rpm.
3. Resuspend the pellet in 2-5 ml hypotonic solution (37°C) and incubate for 20 min at 37°C. (In order to lose as few cells as possible, use a pipette with a plastic tip or a siliconized pipette for resuspending the pellets).
4. For prefixation add 0.5-1 ml cold fixative and mix gently.
5. Centrifuge the suspension for 5 min at 1000 rpm.
6. Resuspend the pellet in 1-2 ml fresh fixative, centrifuge again at 1000 rpm and resuspend the pellet again in fresh fixative. After 5 min or longer, centrifuge one last time at 1000 rpm for 5 min.

7. Before placing the cells on slides, adjust volume of cell suspension according to the size of the pellet; if the pellet is very small, reduce the suspension to a few drops (2-4) out of a blue or better a yellow tip.
8. Take the cold wet slide, let the water run off and place drops onto the slide making two hybridization areas or use two slides with one hybridization area on each for differential post-hybridization washes (as described in outline).

   Note: In the case of a small pellet which can hardly be seen, it is better to drop the suspension on a nearly dry cold slide and to make an unequal distribution: the hybridization area for the LSI 13/21 probe should contain more nuclei than the area for CEP 18/X/Y probe.
9. Dry the slide(s) for ca. 20 min at about 60°C.

Slide pretreatment

1. Place slide(s), prepared from uncultured amniocytes, in 1×PBS at room temperature (RT) for 10 min.
2. Transfer slide(s) to freshly made pepsin working solution for 3-10 min (depending on the density of nuclei on the slide and on the batch of pepsin) at 37°C.
3. Rinse the slide(s) in 1×PBS/$MgCl_2$ at RT for 5 min.
4. Place the slides in post fixation solution (1×PBS/$MgCl_2$/formaldehyde) for 10 min at RT. (After use, this solution should be poured into a special waste disposal).
5. Rinse the slides in 1×PBS at RT for 5 min.
6. Dry slides.
7. Immerse slide(s) in 70% ethanol for 5 min, then remove slides and place in the 90% ethanol for 5 min, followed by 100% ethanol for 5 min.
8. Dry the slides for about 20 min.

Fluorescence in situ hybridization procedure

Co-denaturation

1. Prewarm the 74°C water bath (with a steel-plate inside for laying the slides on it) or a slide warmer.

2. Prepare the humidified hybridization chamber: an airtight container with a piece of damp blotting paper.
3. Allow the probes to warm to room temperature for easier pipetting.
4. Spin the tubes briefly in a microcentrifuge to bring the contents to the bottom of the tube.
5. Apply 5-7 μl (depending on number of nuclei on the slide) of CEP 18/X/Y probe mix to one hybridization area and 5-7 μl of LSI 13/21 probe mix to the other hybridization area on the slide(s). Immediately, place a 15×15 mm (for 5 μl probe) or a 20×20 mm (for 7 μl probe) coverslip over the probe solution and allow the solution to spread evenly under the coverslip. Air bubbles should be avoided.
6. Seal coverslip with rubber cement.
7. Place the slide(s) on the 74°C hot plate inside the water-bath (or take slide-warmer with exactly this temperature) for 5-7 min (optimal time must be tested).
8. Examine the seal of rubber-cement around the coverslips.
9. Place the slides into the hybridization chamber and incubate at 37°C overnight.

Post-hybridization washes

1. Prewarm the 1×SSC (and 0.4×SSC for more stringent washing) solution by placing the Coplin jar in the 74°C water bath until the solution has reached 73±1°C.
2. Remove rubber cement seal and coverslips from the slides with forceps. Place the slides in the Coplin jar containing the 1×SSC at 73±1°C and incubate for 2 min.
3. Remove the slides from the wash bath and place in a jar with 2×SSC/NP40 at RT for 2 min.
4. During this time prepare the DAPI-counterstain solution: per slide mix 1 ml of the 2×SSC/NP 40 + 1 μl DAPI stock solution.
5. Add 1 ml of the counterstain onto each slide and leave for 4 min in the dark.

6. Tip off the stain solution into a bottle for special waste disposal (DAPI is a possible mutagen) and place the slides into a Coplin jar. Rinse 3 times with deionized water.
7. Let the slides dry in the dark.
8. Apply 20 μl mounting solution (Dabco + Vectashield 1:1) in little drops along the middle of the slide and apply a glass coverslip.
9. Store slides in the dark until signal analyzation with the fluorescence microscope.
10. The slides can be stored at – 20°C in the dark.

Signal enumeration

Microscope: For viewing the hybridization-signals, an epi-illumination fluorescence microscope is required.

Light source: A 100 watt mercury lamp is the recommended excitation light source for the Vysis probes.

Objectives: oil immersion fluorescence objectives with numeric apertures › 0.75; a 10× objective, a 63× and a 100× oil immersion type objective are necessary for scanning and signal analysis.

Filters: multi-bandpass filters and single bandpass fluorescence microscope filter sets.

Enumeration scan: a minimum of 50 nuclei for each DNA probe should be counted whenever possible and the percentage of aneuploid signals should be determined.

Depending on the week of gestation in which amniocentesis was done, you will find more or less nuclei showing hybridization signals. In the area hybridized with the cosmid-DNA probes (LSI13/21) there are always less nuclei showing good signals than in the area hybridized with the CEP 18/X/Y probe. Therefore it is recommended to drop more suspension with nuclei on this area of the slide.

Nuclei with diffuse and weak signals, split signals and very stringy signals will also be found. Counting these nuclei should be avoided. For the analysis, only nuclei with clear and distinguishable signals should be used.

For further instructions to "signal enumeration" and "troubleshooting" we refer to the instructions from Vysis, which are included in each AneuVysion EC kit.

Criteria for interpretation (in accordance with Eiben et al., 1998):

- in cases where <10% aneuploid signals are observed, euploidy is assumed.
- whenever >60% aneuploid signals are found, a case is classified as aneuploid.
- If 10-60% aneuploid signals are obtained, the case should be studied intensively by further cytogenetic analyses. Such a result could indicate mosaicism. In these cases more than 100 nuclei should be scored.

Maternal cell contamination (MCC): In amniotic fluid samples which are bloody or brownish, a high percentage of MCC can be expected (in bloody fluids we found up to 82 % maternal cells). False negative results are possible. We did not find clear morphological differences between maternal and fetal interphase nuclei (except in same cases). Winsor et al.(1996) found a rate of maternal cell contamination (MCC) in uncultured amniotic fluids, which is much higher than expected with cultured fluid. They identified MCC in 107/500 (21.4%) uncultured interphase nuclei using X and Y FISH probes, compared with 1/500 (0.2%) in metaphases after cell culture. Winsor mentioned that it is not always possible to differentiate between maternal and fetal cells based on nuclear morphology. Although they found it likely that most of the multi-lobulated nuclei are segmented neutrophils from maternal blood, lobulated nuclei may also be fetal.

Specimens containing blood should therefore be excluded from analysis or the result should only be given if there are signals with a male constellation.

Acknowledgements. Thanks are extended to Dr. Anita Rauch and Dr. Udo Trautmann for helpful advice, to Michaela Kirsch and Anja Kollert for their advice and technical assistance and Dr. Thomas Liehr for critically reading the manuscript.

References

Bryndorf T, Christensen B, Vad M, Parner J, Brocks V, Philip J (1997) Prenatal detection of chromosome aneuploidies by fluorescence in situ hybridization: experience with 2000 uncultured amniotic fluid samples in a prospective preclinical trial. Prenat Diagn 17:4, 333-341

Claussen U, Ulmer R, Beinder E, Voigt H-J (1994) Six years experience with rapid karyotyping in prenatal diagnosis: correlations between phenotype detected by ultrasound and fetal karyotype. Prenat Diagn 14:113-121

Eiben B, Trawicki W, Hammans W, Goebel R, Epplen J T (1998) A prospective comparative study on fluorescence in situ hybridization (FISH) of uncultured amniocytes and standard karyotype analysis. Prenat Diagn 18: 901-906

Klinger K, Landes G, Shook D, Harvey R, Lopez L, Locke P, Lerner T, Osathanondh R, Leverone B, Houseal T, Pavelka K, Dackowski W (1992) Rapid detection of chromosome aneuploidies in uncultured amniocytes by using fluorescence in situ hybridization (FISH). Am J Hum Genet 51: 55-65

Liehr T, Starke H, Beensen V, Kähler Ch, Herbich M, Brude E, Ziegler M, Claussen U (1999) Translocation trisomy dup(21q) and free trisomy 21 can be distinguished by interphase-FISH. Intern J MolekGenet 3: 11-14

Philip J, Bryndorf T, Christensen B (1994) Prenatal aneuploidy detection in interphase cells by fluorescence in situ hybridization (FISH). Prenat Diagn 14: 1203-1215

Ulmer R, Pfeiffer R-A, Kollert A, Beinder E (2000) Aneuploidiediagnostik mittels Fluoreszenz-in-situ-Hybridisierung (FISH); Stellenwert bei Schwangerschaften mit erhöhtem Risiko für Chromosomenaberrationen. Z. Geburtsh. Neonatol. 204:1-7

Ward B E, Gersen S L ,Thev R, Wright M, Carelli M P (1997) Rapid prenatal diagnosis of chromosomal aneuploidies by fluorescence in situ hybridization: clinical experience on 10,000 specimens. In: Pinkel D, Gray J W, (Eds) Clinical applications of Fluorescence In Situ Hybridization, Chichester: John Wiley

Ward B E, Gersen S L, Carelli M P, McGuire N M, Dackowski W R, Weinstein M, Sandlin C, Warren R, Klinger K W (1993) Rapid prenatal diagnosis of chromosomal aneuploidies by fluorescence in situ hybridization: clinical experience with 4,500 specimens. Am J Hum Genet 52: 854-865

Winsor E J T, Silver M P, Theve R, Wright M, Ward B (1996) Maternal cell contamination in uncultured amniotic fluid. Prenat Diagn 16: 49-54

Suppliers

Biochrom KG
Leonorenstr. 2-6
D-12247 Berlin
+49-30-779906-0
(buffers, supplements and media for cell culture)

Carl Roth GmbH
Schoemperlenstr. 1-5
D-76185 Karlsruhe
+49-721-5606-0
(laboratory equipment, chemicals)

Fluka, Sigma: Sigma-Aldrich Chemie GmbH
Grünwalder Weg 30
D-82041 Deisenhofen
+49-89-6513-0
(chemicals)

Vector Laboratories
Inc. Burlingame
CA 94010

Vysis GmbH
Technologiepark
Friedrich-Ebert Str.
D-51429 Bergisch Gladbach
+49-2204-84-3930

Human Sperm Cells

ANITA RAUCH

Introduction

At least one third of human germ cells are carrying a chromosomal abnormality resulting from errors in meiosis. These chromosomal abnormalities often result in early abortions and sometimes in the birth of a disabled child. Since 1978 it has been possible to karyotype human sperm cells by reactivating them with hamster ova (Rudak et al. 1978). However, this technique is difficult, time-consuming, expensive, and yields only a small number of karyotypes, whereas interphase FISH is a rapid and relatively easy alternative technique to study aneuploidy in human sperm cells. As sperms are extremely condensed due to extensive intermolecular disulfide cross-links, they must be decondensed prior to the FISH procedure to allow the DNA probes access to the chromatin (Fig. 1). As dithioerythritol quantitatively reduces disulfide groups by forming cyclic disulfide (Cleland 1963), it was already used in 1977 to prepare mouse sperm for protein analyses (Balhorn et al. 1977).

The described decondensation-solution was used by Otto to investigate Chinese hamster cell lines by flow karyotyping (Otto 1988). In order to achieve a high hybridization efficiency we recommend pretreatment of decondensed cells with RNase and pepsin. In addition, application of directly labeled probes reduces background fluorescence.

Anita Rauch, Institut für Humangenetik, Schwabachanlage 10, Erlangen, 91054, Germany (*phone* +49-9131-8522318; *fax* +49-9131-209297; *e-mail* arauch@humgenet.uni-erlangen.de)

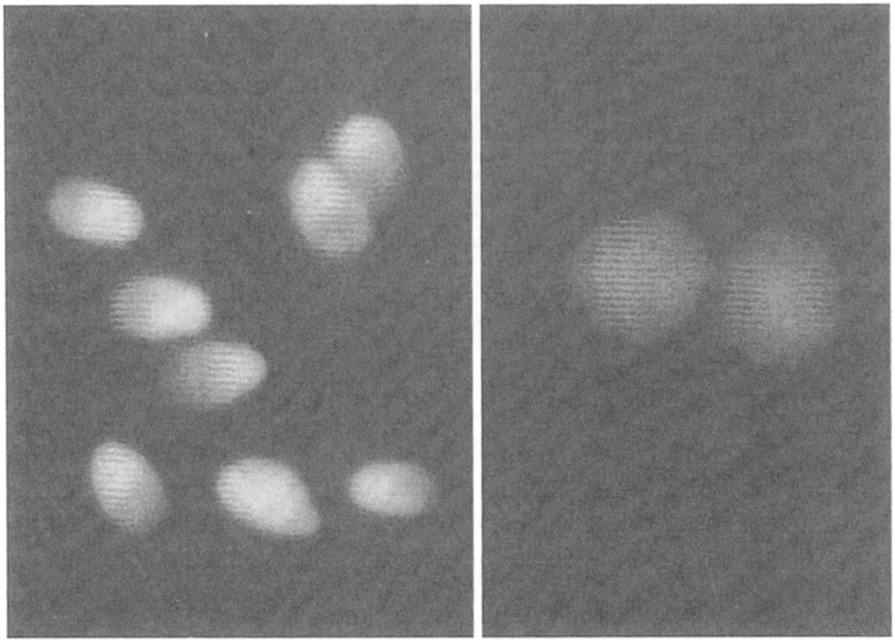

Fig. 1. Sperm cells without (*left*) and with (*right*) decondensation treatment with Otto's solutions

Outline

The procedure of interphase FISH on human sperm cells is outlined in Fig. 2.

Materials

Slide preparation

- Quadriperm trays (Invitro Systems & Services)
- Otto's solution:
  - 100 ml Tris-HCl buffer 10 mmol/l, pH 7.4: 1.21 g Tris-aminomethan and 1 l aqua bidest., adjust with HCl to pH 7.4
  - 220 mg KCl (30 mmol/l)
  - 100 mg $MgCl_2 \times 6H_2O$ (5 mmol/l)
  - 46 mg dithioerythritol (3 mmol/l)
- Fixative: Acetic acid/methanol 1:4 (e.g. 50 ml acetic acid plus 150 ml methanol, prepare freshly)

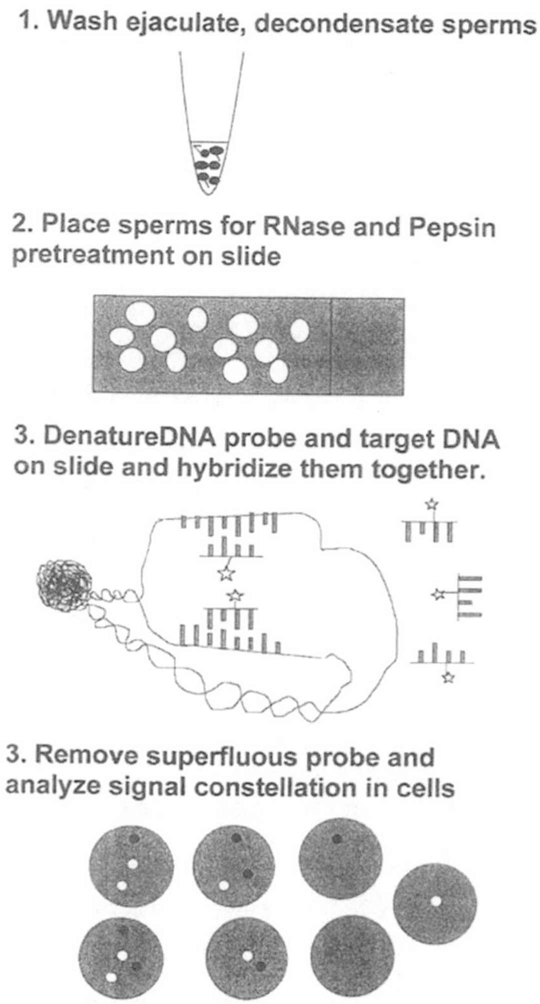

Fig. 2. Outline of interphase FISH from human sperm cells

- RNase working solution (100 ng/μl): 1 μl stock solution plus 99 μl 2×SSC
  Stock solution: dissolve 10 mg RNase A (Roche) in 1 ml 2×SSC, boil for 10 min and store in portions of 100 μl in -20°C
- Pepsin stock solution 10%: Dissolve 1 g pepsin (Fluka) in 10 ml aqua bidest and store in portions of 50 μl at -20°C
- 2×SSC: 50 ml 20×SSC plus 450 ml aqua bidest

- $20\times SSC$: 87.65 g NaCl plus 44.1 g Na-citrate plus 500 ml aqua bidest, adjust pH to 7.0 with HCl 32%, store at room temperature
- PBS: 50 ml $10\times PBS$ plus 450 ml aqua bidest
- * $10\times PBS$: 80 g NaCl plus 2 g KCl plus 11.5 g Na_2HPO_4 plus 2 g KH_2PO_4 add 1 l aqua bidest, adjust pH to 7.0 with NaOH, store at room temperature
- 1 M $MgCl_2$: dissolve 20.3 g $MgCl_2$ in 100 ml aqua bidest, store at room temperature
- PBS-$MgCl_2$: 10 ml 1M $MgCl_2$ plus 190 ml PBS
- 1% Formaldehyde: 2.7 ml formaldehyde (37%, Merck) plus 97.3 ml PBS-$MgCl_2$

Denaturation

- 70% formamide/$2\times SSC$:× per slide: 70 µl deionized formamide plus 10 µl phosphate buffer plus 10 µl $20\times SSC$ plus 10 µl aqua bidest (may be stored at -20°C)
- Phosphate buffer: dissolve 0.7 g sodium dihydrogen phosphate monohydrate in 10 ml aqua bidest, dissolve 0.89 g sodium dihydrogen phosphate dihydrate in 10 ml aqua bidest and mix both, store at -20°C in portions of 1 ml.
- * Deionized formamide: add 2.5 g ion exchanger (mixed bed resin) to 50 ml formamide, stir for 2 h, let the ion exchanger settle down and aliquot formamide in 1 ml portions, store at -20°C.

Hybridization mix

- Commercially available direct labeled probes are used with the hybridization mix provided, following supplier's instructions.
- Self-labeled probes (e.g. with Roche Corp. Nick Translation Mix according to supplier's instructions) use with 20% dextran sulfate: dissolve 2 g dextran sulphate in 5 ml deionized formamide plus 1 ml phosphate buffer plus 1 ml $20\times SSC$ plus 3 ml aqua bidest, store at -20°C.

Washing solutions

- $1\times SSC$: add 25 ml $20\times SSC$ to 475 ml aqua bidest, adjust pH to 7.0-7.5 with 5 drops of 1 M HCl. Store at room temperature.
- $2\times SSC$/0.1% NP-40: add 50 ml $20\times SSC$ to 450 ml aqua bidest, add 0.5 ml NP-40, adjust pH to 7.0-7.5 with 1 drop of HCl 32%

Counterstaining

- Dapi-solution: per slide, mix 1 ml 2×SSC/0.1% NP-40 with 1 μl Dapi stock solution (plus 1 μl propidium iodide stock solution if a red fluorochrome is not being used to detect DNA probe)
- Dapi stock solution: dissolve 10 mg DAPI (Serra) in 10 ml aqua bidest and store aliquots of 50 μl at -20°C (the aliquot in use can be stored at 4°C)
- Propidium iodide stock solution: dissolve 10 mg propidium iodide (Sigma) in 10 ml aqua bidest and store aliquots of 50 μl at -20°C (the aliquot in use can be stored at 4°C)
- Antifade: Mounting medium for fluorescence (Vectashield)

Procedure

Preparation of sperm cells

1. Suspend ejaculate in 8 ml PBS.
2. Centrifuge at 1250 rpm (280 g) for 10 min.
3. Discard supernatant and resuspend in 8 ml PBS.
4. Centrifuge at 1250 rpm (280 g) for 10 min.
5. Discard supernatant and resuspend in 8 ml PBS.
6. Centrifuge at 1250 rpm (280 g) for 10 min.
7. Discard supernatant.
8. While vortexing semen sediment add 20 drops of Otto's solution drop by drop, then suspend in 8 ml Otto's solution and incubate at 37°C for 30 min.
9. Centrifuge at 1250 rpm (280 g) for 10 min.
10. Discard supernatant.
11. While vortexing pellet add 20 drops ice-cold fixative drop by drop, then add 8 ml fixative.
12. Centrifuge at 1250 rpm (280 g) for 10 min.
13. If there is a little pellet, drop it onto a slide immediately, if there is a very visible pellet, wash 3 times with 4 ml fixative and then drop onto a slide.

14. Let the slide air dry.
15. If not fixed at step 13, transfer the slide into a Quadriperm tray.
16. Overlay the slide with fixative, incubate at room temperature for 5 min.
17. Remove fixative and incubate another 5 min at room temperature.
18. Remove fixative once again and incubate another 5 min at room temperature.
19. Remove fixative and let the slide air dry.
20. If the slide is not being used within a few days store it at -20°C.

RNase/pepsin pretreatment

RNase

1. Cover slide with 100 µl RNase working solution and a coverslip and incubate in a metal tray in a 37°C water bath for 3 min.
2. Remove coverslip and wash 3×3 min in 2×SSC at room temperature in a Coplin jar on a shaker.

Pepsin

3. Add 50 µl of pepsin stock solution to a Coplin jar with 100 ml in a 37°C water bath and incubate the slide for 8 min in the pepsin solution.
4. Wash 1 x 5 min in 1×PBS in a Coplin jar on a shaker at room temperature.
5. Wash 5 min in PBS-$MgCl_2$.
6. Wash 10 min in 1% formaldehyde in PBS-$MgCl_2$.
7. Wash for 5 min in PBS.
8. Dehydrate in a series of ethanol (70%, 90%, 100%) for 2 min each.
9. Air dry slides.

Denaturation and hybridization

1. Cover slide with 100 µl 70% formamide/2×SSC and a coverslip and incubate it for 6 min on the back of a metal box in an 86°C water bath for denaturation of the target DNA.
2. Dehydrate slide in a series of ethanol (70%, 90%, 100%) on ice for 2 min each to avoid rehybridization.

3. While slide is air drying, denature specific DNA probe in 10 µl hybridization mix for 5 min at 75°C and place on ice immediately.
4. Place denatured probe (10 µl) on one part of the slide, cover carefully with a 20×20 mm coverslip avoiding air bubbles and seal with rubber cement.
5. Incubate in a moist chamber at 37°C over night or at least 4 h for centromeric probes.

Posthybridization washes

For commercially available probes, follow the instructions of the manufacturers or:

1. Carefully remove rubber cement and coverslip with forceps.
2. Place slide in a Coplin jar with 1×SSC in a 75°C water bath for 2 min, agitate the slide for the first 1-3 s.
3. Wash slide in 2×SSC/0.1% NP-40 at room temperature for 5 s-1 min.

Counterstaining

1. Place 1 ml DAPI/propidium iodide working solution on each wet slide and incubate for 5 min in the dark.
2. Place slide shortly in 2×SSC/0.1% NP-40 and air dry slide in the dark.
3. Apply 20 µl of antifade in a row of dots across the slide and cover with a coverslip.
4. Store slide at -20°C in dark. **Storage**

Results

Analyze slides with a suitable filter set with a fluorescence microscope. Be aware that normal sperms are haploid (Fig. 3)! According to several studies, most autosomes have a disomy frequency in sperms of approximately 1 per 1000. Therefore depending on the study purpose, usually about 10,000 sperms per donor have to be analyzed. To distinguish disomy

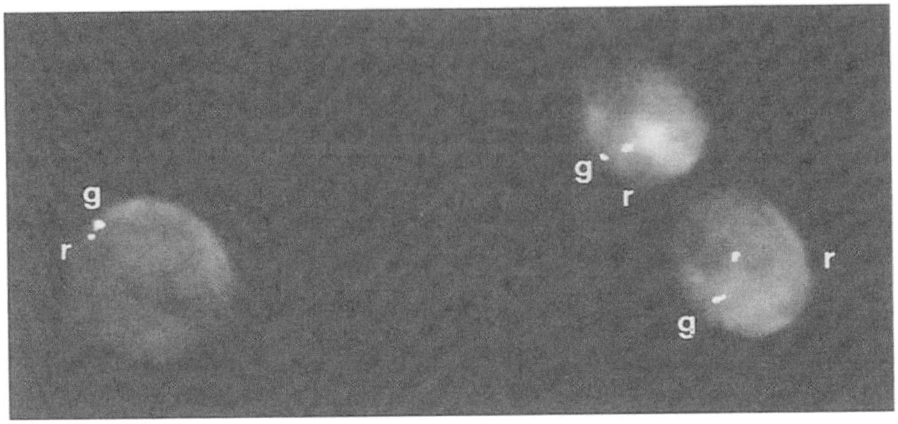

Fig. 3. FISH on human sperm cells with two single copy probes from chromosome 11 labeled in red (*r*) and green (*g*) indicating euploidy for chromosome 11

from diploidy, two color FISH and the simultaneous scoring of two different chromosomes are recommended. Moreover, in interphase, the FISH obtained signal number varies artificially to a certain degree, because of several factors discussed in the Comments section.

Troubleshooting

No or weak signals Reasons for no or weak signals may be too short a denaturation or hybridization time or washing slide after hybridization with too high stringency. Stringency decreases with increase in the salt concentration or decrease in temperature and formamide concentration. Hybridization efficiency is increased by higher dextran sulfate and probe concentrations.

Too much background Too much background fluorescence may result from any soiling and can be decreased during hybridization or posthybridization washing by increasing the temperature or formamide concentration, or decreasing salt concentrations.

Diffuse signals and cells Decrease denaturation time or temperature by 2°C or shorten decondensation time.

Comments

FISH analysis on interphase nuclei generally suffers from systematic errors. This means, that the number of detected signals varies from the number of actual DNA binding sites. The signal number may be artificially decreased by:

- insufficient tissue penetrability resulting in minor hybridization efficiency,
- a certain amount of two actual signals lying so close together that they appear as one,
- analysis of damaged or overlapping cells,

or increased by:

- signal splitting due to dispersed DNA
- signal imitating by bacteria or other background fluorescence.

False negative signals from insufficient tissue penetrability may be reduced by optimal tissue pretreatment, while the fact that a certain number of signals stick together, simulating a single signal has to be addressed by a statistical approach. False positive signals by background fluorescence can be reduced by working as cleanly as possible, by optimized washing procedures or by the use of direct labeled DNA probes.

Applications

FISH analyses of human sperm cells have been applied for basic studies of aneuploidy (Martin et al. 1995, Hu et al. 1996, Martin et al. 1996, Spriggs et al. 1996, Rademaker et al. 1997, Mennicke et al. 1997), to study chromosomal aberrations in infertile men (Moosani et al. 1995, Martin 1996, Lahdetie et al. 1997) and men exposed to potential mutagens (Martin and Barclay 1996, Martin 1997, Martin et al. 1997, Robbins et al. 1997, Robbins et al. 1997, Harkonen et al. 1999, Sram et al. 1999) as well as to analyze the frequency of structural abnormal chromosomes (Vanhummelen et al. 1996, Pfeiffer et al. 1998, Martini et al. 1998, Colls et al. 1998, Baumgartner et al. 1999, Durak et al. 1999).

Acknowledgements. I am grateful to Dr. U. Trautmann and Dr. P. Binder-Wirth for sharing their technical experience.

References

Balhorn R, Gledhill BL, Wyrobek AJ (1977) Mouse sperm chromatin proteins: quantitative isolation and partial characterization. Biochemistry 16:4074-4080

Baumgartner A, Van Hummelen P, Lowe XR, Adler ID, Wyrobek AJ (1999) Numerical and structural chromosomal abnormalities detected in human sperm with a combination of multicolor FISH assays

Cleland WW (1963) Dithiothreitol, a new protective reagent for SH groups. Biochemistry 3:480-482

Colls P, Martinez-Pasarell O, Perez MM, Egozcue J, Templado C (1998) Cytogenetic analysis of spermatozoa in the father of a child with a de-novo reciprocal translocation t(7;9)(q22;p23). Mol Hum Reprod 4:1145-1149

Durak B, Ozon YH, Ozdemir M, Artan S, Basaran N, Basaran S, Ozkinay C (1999) FISH analysis with locus-specific probes in sperm from two translocation carrier men. Clin Genet 56:129-135

Harkonen K, Viitanen T, Larsen SB, Bonde JP, Lahdetie J (1999) Aneuploidy in sperm and exposure to fungicides and lifestyle factors. Environmental & Molecular Mutagenesis 34:39-46

Hu H, Miharu N, Mizunoe T, Nakaoka Y, Okamoto E, Ohama K (1996) Detection of aneuploidy in human spermatozoa using fluorescence in situ hybridization (FISH). Jpn J Hum Genet 41:381-389

Lahdetie J, Saari N, Ajosenpaasaari M, Mykkanen J (1997) Incidence of aneuploid spermatozoa among infertile men studied by multicolor fluorescence in situ hybridization. Am J Med Genet 71:115-121

Martin RH (1996) The risk of chromosomal abnormalities following ICSI. Hum Reprod 11:924-925

Martin RH (1997) Sperm chromosomal abnormalities in a Familial Mediterranean Fever patient treated with colchicine. Middle Eastern Fertility Society Journal 2:82-83

Martin RH, Barclay L (1996) Sperm aneuploidy and diploidy frequencies after treatment of psoriatic arthritis with methotrexate. Ind J Hum Genet 2:65-67

Martin RH, Ernst S, Rademaker A, Barclay L, Ko E, Summers N (1997) Chromosomal abnormalities in sperm from testicular cancer patients before and after chemotherapy. Hum Genet 99:214-218

Martin RH, Spriggs E, Ko E, Rademaker AW (1995) The relationship between paternal age, sex ratios and aneuploidy frequencies in human sperm as assessed by multicolour fluorescence in situ hybridization. Am J Hum Genet 57:1395-1399

Martin RH, Spriggs E, Rademaker AW (1996) Multicolour fluorescence in situ hybridization analysis of aneuploidy and diploidy frequencies in 225,846 sperm from ten normal men. Biol Reproduction 54:394-398

Martini E, Vonbergh ARM, Coonen E, Dediesmulders CEM, Hopman AHN, Ramaekers FCS, Geraedts JPM (1998) Detection of structural abnormalities in spermatozoa of a translocation carrier t(3;11)(q27.3q24.3) by triple FISH. Hum Genet 102:157-165

Mennicke K, Diercks P, Schlieker H, Balspratsch M, Alhasani S, Diedrich K, Schwinger E (1997) Molecular cytogenetic diagnostics in sperm. Int J Andrology 20 (Suppl. 3):11-19

Moosani N, Pattinson HA, Carter MD, Cox DM, Rademaker AW, Martin RH (1995) Chromosomal analysis of sperm from men with idiopathic infertility using sperm karyotyping and fluorescence in situ hybridization. Fertility and Sterility 64(4):811-817

Otto FJ (1988) Assessment of persisting chromosome aberrations by flow karyotyping of cloned Chinese hamster cells. Z Naturforsch 43c:948-954

Pfeiffer RA, Rauch A, Ulmer R, Beinder E, Trautmann U (1998) Interstitial deletion del(3)(p12p21) in a malformed child subsequent to paternal paracentric insertion (or intraarm shift) 46,XY,ins(3)(p24.1p12.1p21.31). Ann Génét 41:17-21

Rademaker A, Spriggs E, Ko E, Martin RH (1997) Reliability estimates of diploid human sperm using multicolour fluorescence in situ hybridization. Human Reproduction 12:77-79

Robbins WA, Meistrich ML, Moore D, Hagemeister FB, Weier HU, Cassel MJ, Wilson G, Eskenazi B, Wyrobek AJ (1997) Chemotherapy induces transient sex chromosomal and autosomal aneuploidy in human sperm. Nat Genet 16:74-78

Robbins WA, Vine MF, Truong KY, Everson RB (1997) Use of fluorescence in situ hybridization (FISH) to assess effects of smoking, Caffeine, and alcohol on aneuploid load in sperm of healthy men. Environmental & Molecular Mutagenesis 30:175-183

Rudak E, Jacobs PA, Yanagimachi R (1978) Direct analysis of the chromosome constitution of human spermatozoa. Nature 274:911-913

Spriggs EL, Rademaker AW, Martin RH (1996) Aneuploidy in human sperm: the use of multicolor fluorescence in situ hybridization (FISH) to test various theories of nondisjunction. AM J Hum Genetics. 58:356-362

Sram RJ, Binkova B, Rossner P, Rubes J, Topinka J, Dejmek J (1999) Adverse reproductive outcomes from exposure to environmental mutagens. Mutation Research 428:203-215

Van Hummelen P, Lowe XR, Wyrobek AJ (1996) Simultaneous detection of structural and numerical chromosome abnormalities in sperm of healthy men by multicolor fluorescence in situ hybridization. Hum Genet 98:608-615

Part C
FISH on Archival, Human Tissues

Archival Cytogenetic Slides

ERICH GEBHART

Introduction

A great number of cytogenetic slides stored worldwide are available for FISH studies by the use of a recovery technique which would allow the use of, e.g., archived tumor material. This permits the application of various FISH techniques – particularly interphase FISH – to reevaluate chromosomal anomalies for summary studies of the diagnosis, prognosis and therapeutic response (Schad, Dewald 1995).

A rather uncomplicated technique will be reported which reactivates slides made permanent by embedding in Eukitt (storage age up to 15 years). It is based on the experience obtained from examinations on archival cytogenetic slides prepared from 33 breast carcinomas, from 23 cancerous effusions, from 100 rectal tumors (Gebhart et al 1996), and from 65 bone marrow samples of human leukemias.

Outline

This technique consists of :
A coverslip removal procedure; a pretreatment step for exclusion of vestiges of RNA or proteins, and a modified FISH protocol. The technique can be applied to slides made from the bone marrow of leukemic patients as well as from slides prepared from cell suspensions of human solid tumors and leukemias.

Erich Gebhart, Universität Erlangen/Nürnberg, Institut für Humangenetik, Schwabachanlage 10, Erlangen, 91054, Germany (*phone* +49-9131-8522351; *fax* +49-9131-209297; *e-mail* egebhart@ humgenet.uni-erlangen.de)

Materials

Equipment

The equipment for FISH and its microscopic evaluation have been presented in detail in the previous chapters.

For the procedures described here in detail, nothing special is required, except for the usual laboratory equipment such as a heating plate, water bath, Coplin jars, etc.

Buffers and solutions

- Xylene (Merck)
- RNase stem solution 10 μg RNase A/μl 2×SCC
- SSC standard saline citrate (0.3 M NaCl +0.03 M Na citrate)
- PBS phosphate buffered saline (8 g NaCl + 0.2 g KCl + 1.15 g $Na_2H_2PO_4$ + 0.2 g KH_2PO_4)
- Pepsin stock 10% pepsin in a.d.
- Formaldehyde stock: 37% paraformaldehyde in 1×PBS

Procedure

Removal of the embedding medium

The first critical step of the procedure is the removal of the coverslips from the archived slides. This will depend on the embedding medium used. The following description refers to embedding media (e.g. "Eukitt") which are soluble in xylene. Slides embedded, for instance, in "Canada balsam" can be recovered much more efficiently than those embedded in "Eukitt" (G. Thiel, personal communication).

Slides are put into a Coplin jar which is filled with xylene at 37°C and remain therein until the coverslip floats off spontaneously, i. e. 2 to 5 days (depending on the amount of embedding medium to be dissolved). The xylene can be removed daily and replaced with fresh xylene. This daily change of xylene, however, apparently does not generally improve the result, unless the embedding medium has been coated in a rather thick layer. In an example of a study on leukemia slides, it could be shown that there was a considerable difference in the percentage of evaluable slides between a xylene exposure up to 48 h (40%) and slides exposed 3 to 4 days (10%).

The removal of the coverslips must be controlled several times a day as a longer direct action of xylene on the nuclei may destroy the latter. The removal can be supported by very cautious use of forceps, but nuclei can also be lost or destroyed if this procedure is not performed cautiously enough.

Immediately after the removal of the coverslip, the slides must be dried at room temperature and afterwards their quality has to be controlled under the microscope to detect any unremoved embedding medium, on the one hand, and to estimate the relation of undamaged versus damaged nuclei, on the other. If visible amounts of embedding medium are still left behind on the slide, it has to be placed in xylene once more (however, at room temperature now) from 20 min up to 3 days. This removal of the rest of the embedding medium is a highly critical step and must be steadily controlled using a phase contrast microscope. An excess of remaining embedding medium will disturb the FISH reactions, and if the xylene is left on too long, it will destroy the interphase nuclei.

As soon as this procedure is finished, the slides are dried at room temperature for about 30 min and afterwards destained in methanol. After another drying step, the slides can be stored now at room temperature until they are subjected to the pre-treatment procedure.

Pre-treatment

Before the FISH procedures, the slides have to be subjected to a pretreatment which removes any traces of cytoplasm and nuclear proteins as well as RNA, which would disturb the hybridization.

The slides are incubated for 30 min at 37°C in RNase A solution (5 μl RNase A stock + 95 μl 2×SSC per slide) under a coverslip. After the removal of the latter, they are rinsed three times (4 min each) in fresh 2×SSC. After 4 min in 1×PBS, hydrolysis removal of nuclear proteins is performed by placing the slides at 37°C in a Coplin jar filled with 1 ml HCl and 99 ml aqua bidest (adjusted to pH 2.0), 50 μl of a pepsin solution is added. After 10 min, slides are washed for 4 min in a solution made up of 5 ml 1 M $MgCl_2$ plus 95 ml 1xPBS.

Re-fixation is then performed for 10 min in 1% formaldehyde (5 ml 1 M $MgCl_2$ plus 95 ml 1×PBS plus 2.7 ml formaldehyde stock) and slides are washed afterwards in a Coplin jar with 100 ml 1×PBS. Dehydration is attained by placing the slides for 4 min in increasing strengths of ethanol at 4°C (70% – 90% – 100%), followed by air drying.

Fluorescence in situ hybridization

The recovered slides, which have been pretreated in the way described above, are now ready for being subjected to FISH. In particular, interphase FISH is a readily applicable technique, while chromosome painting proved to be difficult, although not impossible, on archival slides.

Results

The efficacy of the described procedure in our hands was sufficient to obtain I-FISH data not only from 32 out of 43 breast tumors (74%), from 29 out of 100 colorectal tumors (29%), from 20 out of 35 cancerous effusions (57%), from 5 out of 5 non-cancerous fibroblast controls (100%), and from 12 out of 65 leukemia samples (18.5%), but also from all archival slides of normal bone marrow and of peripheral lymphocytes (100%). In the meantime, with increased experience, however, we could also increase the success rates in leukemia slides considerably. The attained quality of signals in interphase nuclei from archival material in a breast tumor case is shown in Fig. 1.

I-FISH results of our first examinations using the described method are summarized in Table 1. They document that cell clones could be detected. These results helped to characterize the genomic DNA changes as specific gains and losses. Using a conservative estimate of the percentage of nuclei showing either gains or losses of signals allowed guidelines to be established for determining the presence of significant clonality (Table 1). The percentage of nuclei used to define clonality will probably decline as the growth of the data base is increased, as, for instance shown for hematopoietic neoplasms (Gebhart et al 1995).

Troubleshooting

The percentage of slides accessible to interphase FISH, from our experience, depends on the tissue type, but not on the age of the slides. The most critical step of the procedure is the removal of the embedding medium without destroying too many of the nuclei. If the coverslip and embedding medium can be removed totally, 100% of the slides will show reliable FISH signals.

In colorectal tissue, the interphase nuclei proved to be extremely sensitive to washing in xylene: During this procedure, most of the nuclei expanded and were lost to subsequent analysis.

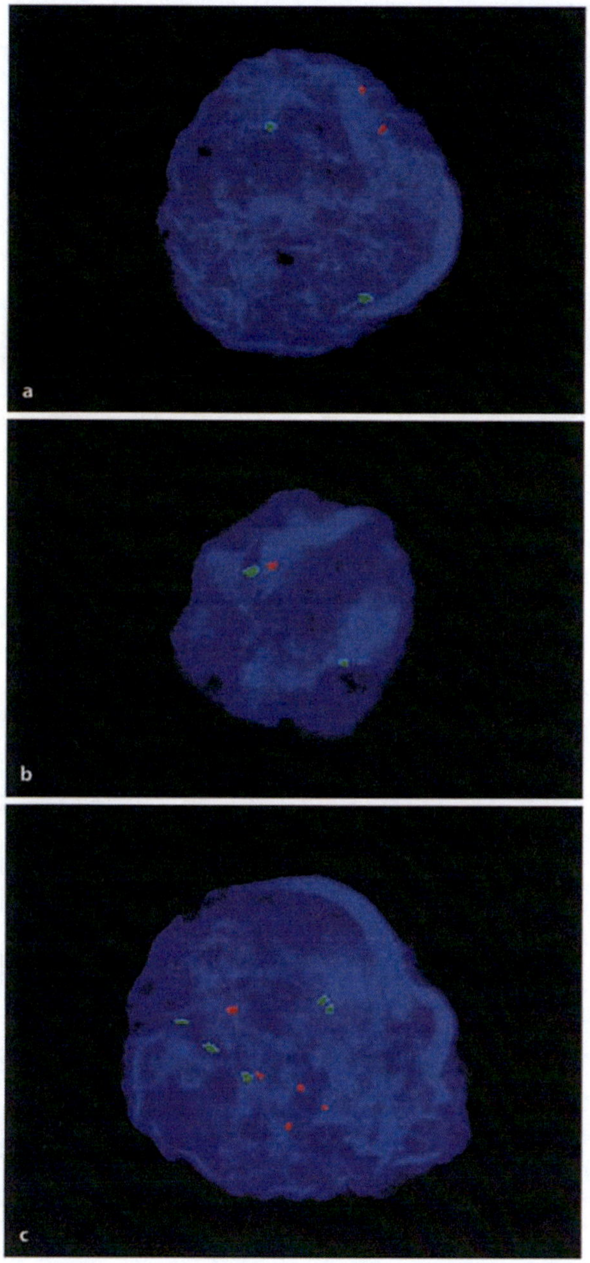

Fig. 1. Three interphase nuclei of a breast tumor are shown. Preparation was counterstained with diamonophenylindol (DAPI) and hybridized to alphoid DNA probes of chromosome 17 (green spots) and chromosome 7 (red spots). **a** One nucleus with two signals each; **b** one with only one signal of chromosome 7; **c** one with five signals for each of the two labeled chromosomes are presented. Pictures were taken using a ZEISS Axioplan microscope equipped with a CCD camera Imac S30 (Compulog) and ISIS 3 software (Metasystems, Altlussheim)

Table 1. Gains and losses of chromosomes detected by I-FISH in archived slides (extended from Gebhart et al. 1996)

Tumor type	Alteration	Chromosome no.					
		1	3	7	8	17	18
Breast ca.	Gain	nd	nd	3/14	4/31	2/13	2/31
	Loss	nd	nd	4/14	13/31	7/13	15/31
Colorectal ca.	Gain	3/12	2/12	9/28	nd	10/29	nd
	Loss	1/12	0/12	8/28	nd	6/29	nd
Cancerous effusions	Gain	0/2	0/6	4/8	1/16	nd	3/16
	Loss	2/2	4/6	1/8	7/16	nd	5/16
Controls I	Gain	nd	nd	0/2	0/5	nd	0/5
	Loss	nd	nd	0/2	0/5	nd	0/5
Leukemias	Gain	nd	1/2	0/5	2/5	0/5	nd
	Loss	nd	1/2	3/5	3/6	2/2	nd
Controls II	Gain	nd	nd	0/8	0/8	nd	nd
	Loss	nd	nd	0/8	0/8	nd	nd

Number of cases (out of total number studied) with significant gains (i.e. more than 15% of interphase nuclei showing more than two signals) and losses (i.e. more than 30% of interphase nuclei showing less than two signals) of chromosomes as found in archived cytogenetic slides of various neoplasias (nd = not done)

Comments

The presented data obtained from archived chromosome preparations of primary tumors, tumor cell cultures, and leukemia preparations document the suitability of this technique for retrospective cytogenetic analysis. Much smaller DNA probes (down to less than 10 kb) for archival paraffin-embedded material (Liehr et al 1995) may be applicable to this recovery procedure, in addition to the centromeric probes presented above. Also, whole chromosome libraries have just been applied for chromosome painting on archived slides in exemplary cases.

Acknowledgements. The technique described here and its application on slides of various neoplastic samples were originally gained from the work of the following coworkers: Peter Hofbeck, Yvonne Hofmann, Rita Lerch, Christian Schlüter (during their MD theses), Thomas Liehr, and Gabriele Schmitt.

References

Gebhart E, Liehr T, Harrer P, Reichardt S, Schmitt G, Thoma K, Gramatzki M, Trautmann U (1995) Determination by interphase FISH of the clonality of aberrant karyotypes in human hematopoietic neoplasia. Leukemia & Lymphoma 17: 295-302.

Gebhart E, Hofbeck P, Hofmann Y, Lerch R, Schmitt G, Liehr T (1996) Recovering of archival tumor cytogenetic slides for two-colour-interphase FISH. Appl. Cytogenet 22: 146-148.

Liehr T, Grehl H, Rautenstrauß B (1995) FISH analysis of interphase nuclei extracted from paraffin-embedded tissue., Trends in Genetics 11: 377-378.

Schad CR, Dewald GW (1995) Building a new clinical test for FISH. Appl Cytogenet 21: 1-14.

Formalin-Fixed and Paraffin-Embedded Tissue Sections

HARALD RIEDER, CHRISTINE SOMMERLAD, YASMIN MEHRAEIN, MARINA GIERSBERG, KERSTEN MARBEN, and HELGA REHDER

Introduction

In situ hybridization in tissue sections has primarily been associated with RNA analysis in gene expression studies. More recently, in situ DNA hybridization in tissue sections has also gained increasing importance. The tissue context allows the correlation of specific genetic and morphological characteristics which is not possible by genetic analysis on a single cell level, e.g. in metaphase preparations or isolated interphase nuclei. Different cell types may be distinguished and malignant cells may be differentiated from non-malignant cells by genetic as well as morphological criteria.

The maintenance of the original tissue context may also be of fundamental importance when tissue areas carry different genetic markers – a situation often found in tumors.

A major advantage of this technique is the use of formalin-fixed and paraffin-embedded archival material, allowing the investigation of cases in which vital cells for conventional cell culture and metaphase chromosome

✉ Harald Rieder, Klinikum der Philipps-Universität, Institut für Klinische Genetik, Zentrum für Humangenetik, Bahnhofstraße 7, Marburg, 35033, Germany
(*phone* +49-6421-28-62986; *fax* +49-6421-28-63984; *e-mail* riederh@mailer.uni-marburg.de)
Christine Sommerlad, Klinikum der Philipps-Universität, Institut für Klinische Genetik, Zentrum für Humangenetik, Bahnhofstraße 7, Marburg, 35033, Germany
Yasmin Mehraein, Klinikum der Philipps-Universität, Institut für Klinische Genetik, Zentrum für Humangenetik, Bahnhofstraße 7, Marburg, 35033, Germany
Marina Giersberg, Klinikum der Philipps-Universität, Institut für Klinische Genetik, Zentrum für Humangenetik, Bahnhofstraße 7, Marburg, 35033, Germany
Kersten Marben, Klinikum der Philipps-Universität, Institut für Klinische Genetik, Zentrum für Humangenetik, Bahnhofstraße 7, Marburg, 35033, Germany
Helga Rehder, Klinikum der Philipps-Universität, Institut für Klinische Genetik, Zentrum für Humangenetik, Bahnhofstraße 7, Marburg, 35033, Germany

analysis are not, or no more available, thus giving access to a wide pool of material hitherto not considered for genetic studies.

Due to distinctive features of the archival material, in situ hybridization in tissue sections is complicated by a variety of problems which arise either from characteristics of the tissue itself or from preparation procedures. Major difficulties include reduced probe penetration, high level of tissue autofluorescence, cell overlap or truncated nuclei, and last but not least, low hybridization efficiency caused by a long-formalin fixation.

To overcome these problems, different pretreatment procedures have been developed for tissue sections, involving proteolytic digestion (e.g. treatment with pepsin, trypsin, proteinase K), microwave treatment, preincubation in formamide solutions, acid treatment or the so called chaotropic treatment with NaSCN, first described by Hopman and colleagues in 1991.

Principles and applications

Although light microscopy is well established in histology, in situ hybridization with fluorescence detection has the major advantages (compared to bright-field microscopic detection), of a higher sensitivity and the possibility of using various fluorescent dyes for multi-target detection. The application of direct fluorochrome labeled probes greatly facilitates the detection procedure.

Main diagnostic applications of FISH in tissue sections relate to questions concerning:

- abortion material where fetal and maternal cells may be distinguished, as in the placenta or where the differentiation between tissue types allows correlation between genetic markers and morphological features in cases of genetic mosaicism,
- tumor material where tumor cells can be distinguished from normal cells, and seemingly normal cells can be unmasked to show the presence of genetic abnormalities. Cell clones with different genetic markers can be separated from each other and, because the tissue maintains its original cell formation, can be related directly to the tissue regions in haemotoxylin and eosin(HE)-stained serial sections.

In the following, we outline a protocol which has proved highly efficient for hybridization with direct fluorochrome labeled probes on formalin-fixed paraffin-embedded tissue sections.

Outline

Essential steps working with FISH on formalin-fixed paraffin-embedded tissue sections are shown in the following scheme:

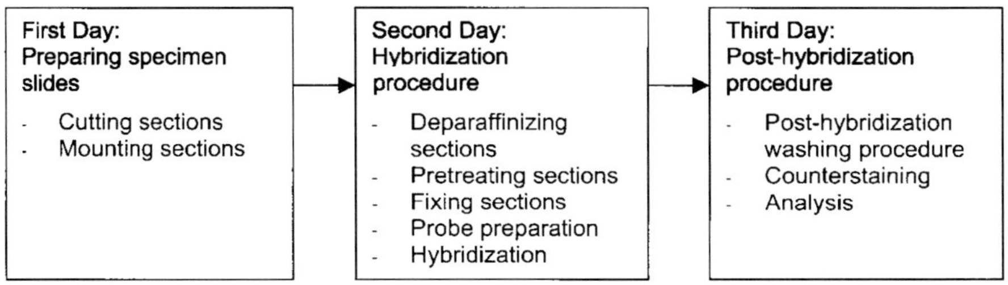

Materials

First day: preparing specimen slides

Equipment

- Microtome (e.g. Leitz Type 530-492)
- Water bath (e.g. GFL Type 1052)

Supplies

- Positively charged slides (e.g. Super Frost/Plus Menzel) or coated slides (poly-l-lysine, silane)

Specimens

- Paraffin blocks of conventionally embedded tissue

Second day: hybridization procedure

Equipment

- Two water bath circulators (for simultaneous use), one heating up to 100°C (e.g. RM 6 Lauda GFL 20 l)
- Heating plate, heating up to 50°C
- Shaker (e.g. Vibrax, IKA Labortechnik)
- Microwave with temperature sensor (can be used facultatively for alternative wash procedure)

Supplies

- Coverslips, 22×22 mm and 22×60 mm
- Rubber Cement (e.g. Fixogum, Marabu)
- Heat resistant Coplin jars

Reagents and solutions

- Xylol 100%
- Ethanol 100%, 85% and 70%, store at 4°C
- 1 M NaSCN, autoclaved
- 0.2 M HCl
- Pepsin, lyophilized 2500-3500 units/mg
- 0.9% NaCl, pH 2.0
- 2×SSC, pH 7.0 (prepare from 20×SSC)
- 70% Formamide in 2×SSC: Deionise formamide using mixed bed resins (e.g. Dowex, mix 70 ml deionised formamide with 10 ml 20×SSC, add distilled water to 100 ml volume, adjust to pH 7.0
- Formalin 3.7% in phosphate buffered saline

Third day: post-hybridization procedure

Reagents and solutions

- 2×SSC/0.3% Nonidet p40
- DAPI staining solution: 0.2 µg DAPI/ml 2×SSC, prepare from stock solution (200 µg DAPI/ml distilled water)
- Mounting medium for fluorescence microscopy (e.g. Vectashield)

Procedure

First day: preparing specimen slides

1. Cut 5 µm thick paraffin sections using a microtome.
2. Float the section on a protein free water bath at 40°C.
3. Mount the section on slides.
4. Air dry the slides.
5. Bake the slides overnight at 56°C.

Second day: hybridization procedure

- Prewarm Coplin jar with NaSCN to 80°C in a water bath
- Prewarm a bottle with 0.9% NaCl and an empty Coplin jar in a 37°C water bath
- Adjust the heating plate to approx. 47°C

It is advisable to place the Coplin jars onto a gently moving shaker during incubation to improve the tissue-solution contact.

Deparaffinization

1. Immerse slides in Xylol: 3×10 min.
2. Dehydrate slides in 100% ethanol: 2×5 min.
3. Dry slides on a heating plate: 2-3 min.

Acid treatment

Treatment with acid (e.g. HCl) pulls the basic histone proteins away from DNA; especially in formalin fixed material, it helps to deproteinize the DNA if signals are weak.

4. Immerse slides in 0.2 N HCl at room temperature: up to 20 min.
5. Rinse in distilled water: 3 min.
6. Rinse in 2×SSC: 3 min.

Chaotrope treatment

Chaotrope treatment by a solvent disrupting molecular structures – which is a compound or group of secondary structures that interfere with the mechanism which stabilizes the complex proteins, nucleic acids and polysaccharides. These agents denature macromolecules primarily through their disruption of the water lattice.

7. Incubate slides in 1 M NaSCN, 80°C: 30 min.
8. Rinse in distilled water: 1 min.
9. Immerse slides in 2×SSC at room temperature (solutions can be kept for step 13): 2×5 min.
10. Turn down water bath to 73°C and place Coplin jar containing formamide into it.

    Note: Caution: glass jars may crack if not prewarmed!

Protease treatment

11. Add lyophilized pepsin to the bottle of prewarmed 0.9% NaCl, pH 2.0, to a final concentration of about 1250-1750 units/ml, invert several times to mix and pour into the prewarmed Coplin jar.
12. Remove slides from 2×SSC, blot off excess liquid and incubate in pepsin solution at 37°C: 10 min.

    Note: Make sure that temperature has reached 37°C after adding all slides (at most 6/jar) and check temperature during the whole procedure.
13. Rinse in 2×SSC at room temperature: 2×5 min.
14. Place on the slide warmer to dry: 2 min.

Postfixation

15. Immerse slides in 3.7% buffered formalin solution at room temperature: 10 min.

    Note: Do not shake Coplin jars during this step.
16. Rinse in 2×SSC at room temperature: 2 x 5 min.
17. Place on the heating plate to dry: 2 min.

Denaturation

18. Place slides into 70% formamide/2×SSC at 73°C: 5 min.

    Note: Make sure that temperature has reached 71-73°C after adding all slides (at most 6/jar) and check temperature during the whole procedure.
19. Dehydrate slides in icecold 70%, 85% and 100% ethanol: 1-2 min each.
20. Dry slides at room temperature.

    Note: If working with direct fluorochrome labeled probes, all following steps should be performed protected from the light (in subdued light or in a dark room if possible).

Hybridization

21. Prepare 10-15 µl hybridization mixture containing labeled probe, blocking DNA, 50% formamide and 10% dextran sulfate. If commercially available probes are used, prepare according to the manufacturers recommendations (e.g. in the cases demonstrated in Figs. 3A,B a mixture of centromere and locus specific probes was used.
22. Cap tightly and denature at 73°C in a water bath: 5 min.
23. Apply denaturated hybridization mixture onto the slide.

24. Place a 22×22 mm coverslip on the hybridization area, seal with rubber cement, dry on heating plate .
25. Place slides in a humidified box and incubate in a 37°C incubator overnight (at least 14-18 h).

Third day: post-hybridization procedure

Prewarm Coplin jar containing 2×SSC/0.3% NP 40 in a 73°C water bath.

Washing

1. Carefully remove rubber cement with forceps.
2. Immerse slides in 2×SSC/0.3% NP 40 at room temperature and float off coverslips.
3. Place slides into 2×SSC/0.3% NP 40 at 73°C: 2 min.

   Note: Make sure that the temperature has reached 71-73°C after adding all slides, at most 6 slides/jar.

More rapidly, the solution can be prewarmed to 73°C by a microwave if the temperature in the Coplin jar is controlled by a sensor. When proceeding immediately, the washing procedure may even be performed without using a water bath because the decrease in the temperature during incubation is negligible.

4. Place slides into 2×SSC/0.3% NP40 at room temperature.
5. Immerse slides in 2×SSC: 5 min.
6. Immerse slides in DAPI/2×SSC: 5 min.
7. Immerse slides in 2×SSC: 5 min.
8. Remove excess liquid by wicking off the edge of the slide, add one drop of mounting medium (e.g.Vectashield)/slide and cover with a 22×60 mm coverslip.
9. Analyze hybridization signals using an epifluorescence microscope equipped with filter combinations adequate for the fluorochromes used.

Analysis (basic remarks for signal evaluation in tissue sections): To provide reliable results, the counting of overlapping and ruptured nuclei should be avoided. Out of focus signals should be considered and focussing the whole depth of the section is therefore necessary. It has to be kept in mind that part of the nuclei is truncated by the preceding tissue sectioning, leading to a loss of genetic material and reduced numbers of hybridization signals. Thus diagnostic results in tissue sections, although based on information from a single cell, require the analysis of large numbers of nuclei and statistical evaluation.

Results

We performed this method on tissue sections of various thicknesses (2, 5, 10, 15 and 20 μm) of archived, formalin-fixed and paraffin-embedded tissue specimens derived from placenta, fetal organs, lymphomas and breast cancers. Centromeric probes as well as single copy probes were successfully employed. Mainly centromeric probes used in a large variety of samples of different tissue origin were the basis for our statistical evaluation of FISH results.

The hybridization efficiency was dependent on the tissue type, the conservation of the tissues with respect to autolysis as well as necrosis, and the period of formalin fixation rather than of paraffin embedding. Regarding centromeric probes, diagnostic results could be achieved in approximately 60% of placental tissues, 20-90% of fetal tissues (depending on the different tissue types), 90% of lymphoma and 60-70% of breast cancer samples.

Characteristic signal distribution patterns were found after successfully performed hybridization. They clearly differentiated between chromosome and/or gene copy numbers, but not between different levels of ploidies. Exclusion of polyploidy required the application of several chromosome probes in single, double or multiple color FISH. The combined use of differently labeled gene specific and centromeric probes of the same chromosomes helped to distinguish between changes in the gene copy number, due to complete chromosome gains or losses and due to structural chromosome rearrangements (Figs. 1, 2). Double in situ hybridization using human DNA and viral RNA probes was performed for the identification of Ebstein Barr virus positive tumor cells. By the combination of using chromosome-specific FISH with immunhistochemistry to identify tumor cell-specific antigens, a correlation of cytogenetic, immunological and morphological criteria was possible.

The section thickness as well as the chromosome copy numbers directly influenced the FISH results with respect to the detection of the true karyotype. The highest number of spots per nucleus present in at least

5% of all nuclei corresponded well with the true copy number of the respective chromosome and was independent of section thickness and also of chromosome count. The fraction of the nuclei which represented the true chromosome number decreased with the copy number of this chromosome and increased with the section thickness. In placental tissues, for instance, it was highest in monosomy and lowest in tetrasomy at a section thickness of 20 µm. Except for monosomies, the detection of low grade mosaicism usually required sections thicker than 5 µm and the evaluation of an increased number of cells.

The average number of spots per nucleus was also used to identify monosomies, disomies and trisomies in thin tissue sections, while detection of tetrasomies required sections of more than 15 µm. The influence of section thickness on signal distribution patterns was dependent on the density of nuclei within the tissue formation. The placenta, for instance,

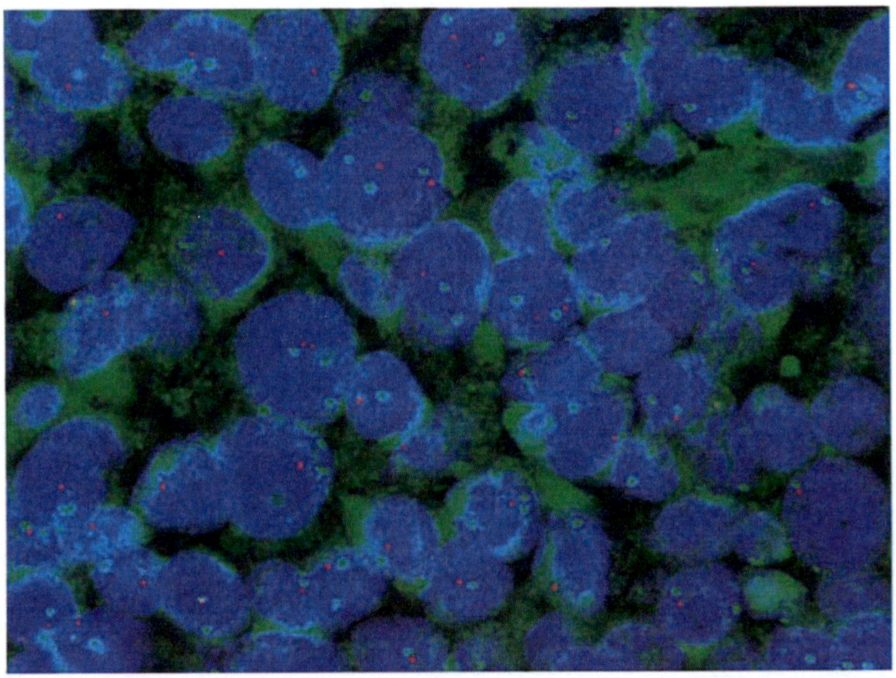

Fig. 1. Dual color FISH with differently labeled DNA probes for the p53 gene (red), localized on the short arm of chromosome 17, and for centromeric alpha satellite DNA (green) of chromosome 17, performed on sections of formalin-fixed archival breast cancer tissue. Most nuclei show two chromosome 17 centromeric together with only one p53 signal pointing to a loss of p53

represents a tissue with a low density of nuclei. In tumors which were characterized by a high tumor cell density, e.g. lymphomas or ductal breast carcinomas, 5 μm sections were sufficient for the evaluation of the signal numbers. The investigations aimed at identifying high level gene amplifications such as the proto-oncogenes HER-2/neu or c-myc in ductal breast cancer were even hampered by thick sections, so the use of 5 μm sections was preferred.

Troubleshooting

Hybridization efficiency and tissue morphology are usually negatively correlated. Aggressive pretreatment procedures improving hybridization efficiency (e.g. long incubation times of NaSCN or proteolytic digestion)

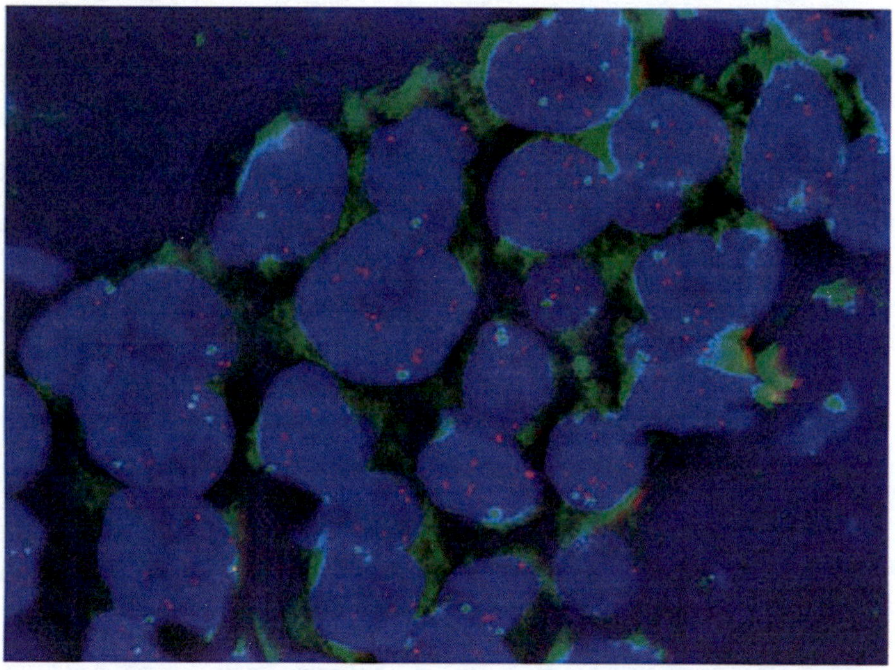

Fig. 2. Dual color FISH with differently labeled DNA probes for the c-myc gene (red), localized on the long arm of chromosome 8, and for centromeric alpha satellite DNA (green) of chromosome 8, performed on sections of formalin-fixed archival breast cancer tissue. Most nuclei show two chromosome 8 centromeric together with 8-10 c-myc signals indicating an intermediate c-myc amplification

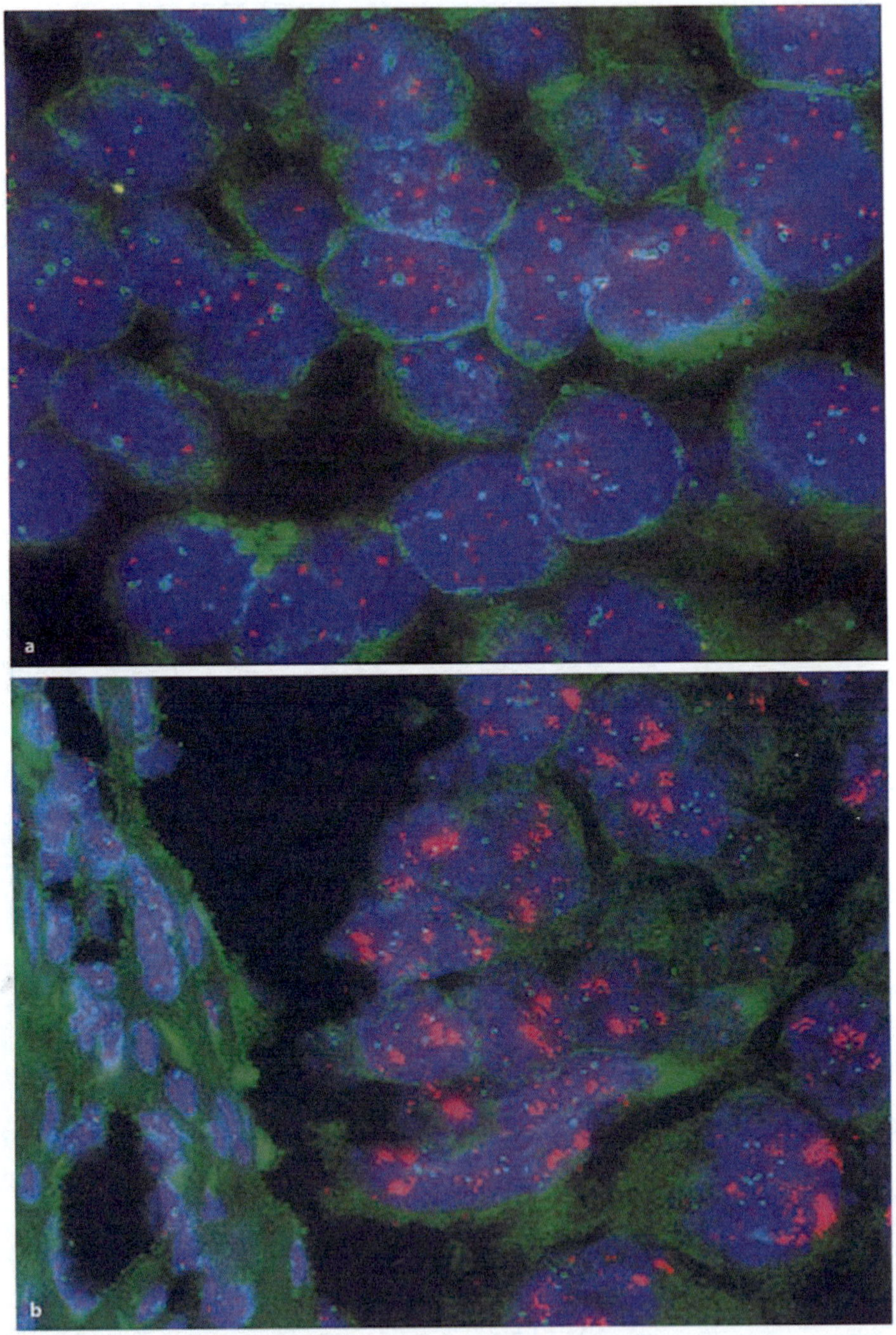
a
b

tend to destroy cell structure and tissue morphology. However, mild pretreatment saving tissue structures, may not be sufficient for probe penetration and successful FISH results. Especially locus specific probes with a single copy target may show dim or a reduced number of signals.

Generally: If the experiment did not work – before changing any of the parameters, try it again!

Problem	Presumed Cause	Solution
Porous paraffin block	– Prolonged storage	– Embed anew
Tissue loss, complete or partial	– Insufficient slide coating – Tissue gets lost during the NaSCN procedure – The tissue was not fixed thoroughly	– Use positively charged (e.g. Super Frost Menzel) or coated slides (e.g. poly-l-lysine, silane) and bake the slides at 56 °C overnight – Check temperature of NaSCN buffer (should not exceed 80 °C) – Decrease the incubation time of NaSCN – Increase the incubation time of buffered formalin (some specimens do not stick to slides at all!)
Poor tissue morphology	– The tissue was digested too extensively – The tissue was extensively fixed (specimen probably exceeded fixation times of 24–48 h) – The tissue was insufficiently fixed or autolytic	– Reduce the incubation time in protease solution – Check temperature of the protease solution – Eliminate or reduce the fixation step with formalin – Choose new specimen – Choose new specimen
High tissue autofluorescence		– Choose new specimen, problem cannot be eliminated

Fig. 3. Dual color FISH with differently labeled DNA probes for the HER-2/neu gene (red), localized on the long arm of chromosome 17, and for centromeric alpha satellite DNA (green) of chromosome 17, performed on sections of formalin-fixed archival breast cancer tissue. **a** Multiple distinct signals of HER-2/neu as well as of centromeric DNA of chromosome 17 point to polysomies of chromosome 17. **b** On the *right*, red signals appear as clusters demonstrating high amplifications of HER-2/neu in tumor cells. Distinct green signals indicate multiple copies of chromosome 17. The HER-2/neu amplifications are estimated at 10-40 signals per nucleus. On the *left*, normal connective tissue with vascular structures is visible showing normal signal patterns.

Problem	Presumed Cause	Solution
Dim or no signal	- The tissue was poorly digested (the target is not sufficiently accessible) - The probe concentration was too low - The hybridization conditions were poor - Loss of DNA (may be indicated by weak DAPI staining)	- Prepare new protease solution directly before use - Check temperature of the water bath (exactly 37+/-1 °C) - Check pH of the protease buffer (exactly 2.0) - Ensure the correct amount of probe. - If working with dual color FISH, only increase the concentration of the probe which shows weaker signals. - Check probe hybridization time (at least 14–16 h) - Check temperature of the hybridization incubator (37+/-1 °C) - Increase fixation time
Probe signal intensity varies within the tissue	- Regional loss of DNA - The hybridization mixture was unevenly distributed on the section	- Do not score the slide if DAPI staining is weak, choose new section - Score the slide if DAPI staining is strong - Ensure that there are no air bubbles in the hybridization mixture during incubation
Signals are speckled over the whole slide	- The stringency conditions were too low	- Check the concentration of the wash buffers - Check temperature of the washing buffer (73+/-1 °C) - Gently agitate the Coplin jar containing the slides to improve the washing - Increase the washing time (5 min)

References

Hopman AHN, van Hooren E, van de Kaa ChA, Vooijs PGP, Ramaekers FCS (1991). Detection of numerical chromosome aberrations using in situ hybridization in paraffin sections of routinely processed bladder cancers. Modern Path. 4, 503-513

Hopman AHN, Classen S, Speel EM (1997). Multi-color brightfield in situ hybridization on tissue sections. Histochemistry Cell Biology. 108, 291-298

Hyytinen E, Visakorpi T, Kallioniemi A, Kallioniemi OP, Isola JJ (1994). Improved technique for analysis of formalin fixed, paraffin embedded tumors by fluorescence in situ hybridization. Cytometry 16: 93-9

Kopf I, Hanson C, Delle U, Verbiene I, Weimark A (1996). A rapid and simplified technique for analysis of archival formalin fixed, paraffin embedded tissue by fluorescence in situ hybridization (FISH). Anticancer-Res. 16: 2533-6

Marben K, Mehraein Y, Jensen K, Rehder H (1997). FISH-derived diagnostic criteria for aneuploidies in paraffin embedded tissue sections. Medizinische Genetik 1/1997.

Marben K, Mehraein Y, Jensen K, Fritz B, Rehder H (submitted) Assessment of aneuploidies in paraffin-embedded placental tissue sections by FISH: Statistical evaluation of characteristic signal distribution patterns

Suppliers

Gesellschaft für Labortechnik mbH, Burgwedel, Germany

IKA Labortechnik, Staufen, Germany

Lauda-Königshofen, Münster, Germany

Leitz Leica, Wetzlar, Germany

Menzel, Braunschweig, Germany

Merck, Darmstadt, Germany

Philips, Schweden

Roth, Karlsruhe, Germany

Sigma Chemie (DAPI), Deisenhofen, Germany

Sigma Chemical Co (Nonidet p 40), St. Louis, MO, USA

Vector Laboratories INC (Vectashield), Burlingame, CA, USA

VYSIS Inc., 3100 Woodcreek Drive, Downers Grove, IL, USA

Zeiss, Oberkochen, Germany

Nucleus Extraction from Formalin Fixed/paraffin Embedded Tissue

THOMAS LIEHR

Introduction

Interphase cytogenetics using formalin fixed/paraffin embedded tissue is now a well-established technique, which renders it possible to obtain "cytogenetic information" from interphase nuclei, especially from solid tumors (Hackel and VarellaGarcia, 1997, Wolfe and Herrington, 1997). It is nearly the only tool to investigate specific numerical chromosomal aberrations, amplification of oncogenes (see Figure 1), deletion of tumor suppressor genes or chromosomal translocations in formalin-fixed/paraffin-embedded tumor tissue samples on a single cell level.

Interphase FISH-studies on formalin-fixed/paraffin-embedded tissue can be done, on the one hand, directly on sectioned and mounted material (e.g. Neubauer et al. 1994) and on the other hand, on extracted interphase nuclei, a technique first described by Hedley et al. (1983). As concluded by Qian et al. (1996), both methods are comparable and reliable for the detection of chromosomal changes in archival tissue, however, each of them has advantages and disadvantages. The first approach is recommended, when the tissue architecture must be studied intact, e.g. in the case of small and/or invasive tumors, while the second technique can be applied successfully when more or less homogenous (tumor)-tissue samples are under study (Köpf et al. 1996). During evaluation of tissue sections, the problem often arises that overlapping nuclei, due to the presence of several cell layers, one on top of the other; and that cut nuclei, leading to an artificial signal loss in interphase cytogenetic studies (Dinghra et al. 1992, Liehr et al. 1994) cannot be evaluated reliably. These problems are not present in nuclear extraction techniques (Köpf et al. 1996, Liehr et al. 1995a, Qian et al. 1996). In the following, the nucleus extraction from un-

✉ Thomas Liehr, Institut für Humangenetik, Kollegiengasse 10, Jena, 07743, Germany (*phone* +49-3641-935533; *fax* +49-3641-935502; *e-mail* i8lith@mti-n.mti.uni-jena.de)

mounted (Liehr et al. 1995a, 1997) and from mounted tissue sections (Liehr et al. 1999) is described.

Outline

1. Formalin-fixed/paraffin-embedded tissue is cut into sections, which are deparaffinized, rehydrated and treated with proteinase K to digest the cytoplasm and set the nuclei free. This can be performed with mounted and unmounted sections. Interphase nuclei are separated from debris by passing through a nylon mesh. Nuclei are washed in PBS and placed on uncoated, clean slides and fixed. Small pieces of tissue (5×2×0.5 mm) or one mounted section 1 cm in diameter are sufficient to produce two to five slides with extracted nuclei, suitable for interphase FISH analyses.
2. For a FISH approach, these slides are treated with RNase and pepsin prior to denaturation.
3. Denaturation of the slides is performed for about 12 min at 75°C. Biotinylated and digoxigenated probes are denatured for 5 min and hybridized on the slides for 3 nights at 37°C in a humid chamber. After a

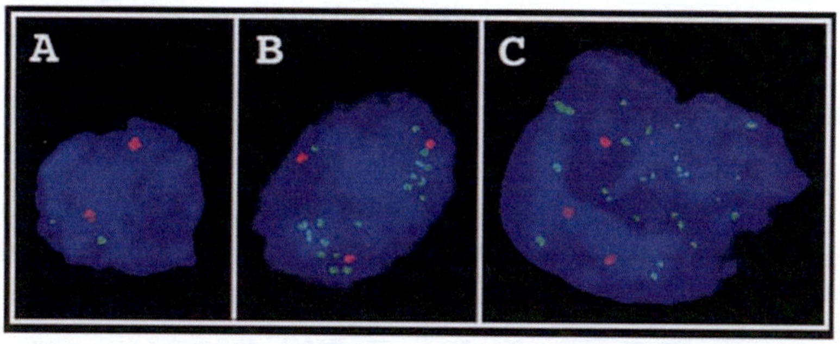

Fig. 1. Fluorescence in situ hybridization (FISH) on interphase nuclei extracted from paraffin-embedded tissue of different head and neck squamous cell carcinoma (HNSCC) specimens. The probe for INT2/FGF3 (Oncor) is labeled in green and the one for the centromeric region of chromosome 11 in red. In **A** a nucleus with two signals per probe ("normal") is shown. In **B** and **C** two nuclei with three centromeric signals for chromosome 11 and amplification of INT2/FGF3 are visible. Two different kinds of the cytogenetic equivalents of gene amplification are shown: in **B** two homogeneously stained regions (HSRs) and in **C** double minutes (DMs). Images were captured on a Zeiss Axioplan microscope using the ISIS digital FISH imaging system (MetaSystems, Altlussheim, Germany) using a XC77 CCD camera with on-chip integration (Sony)

postwashing series, detection of the biotinylated probe is performed with an FITC-avidin system which leads to green signals and of the digoxigenated probe with anti-digoxigenin-rhodamine leading to red signals. After counterstaining of the nuclei with DAPI and addition of antifade solution, slides can be evaluated under a fluorescence microscope.

Procedure

Nuclear extraction

Nuclear extraction from unmounted formalin-fixed/paraffin-embedded tissues

1. According to the diameter of the piece of tissue studied, two to twenty 20 μm sections of paraffin-embedded tissue are produced on a microtome and collected in a glass tube.
2. Dewax the tissue pieces by adding 10 ml of xylene (100%), for 10 min at room temperature (=RT).
3. Tissue is sedimented by centrifugation (1000 rpm, 3 min) and the supernatant is discarded (1-2 ml of supernatant are left in the tube to avoid loss of small tissue pieces).
4. Repeat step 2 and 3.
5. Remove the xylene rest by adding 10 ml of ethanol (100%) and incubate for 10 min at RT.
6. Repeat step 3.
7. Repeat step 5 and 6.
8. Rehydrate the tissue by adding 10 ml of 90% ethanol (5 min at RT), 10 ml ethanol 70% (5 min at RT), 10 ml ethanol 50% (5 min at RT) and 10 ml 0.9% NaCl (5 min at RT). Remove the corresponding supernatants by repeating step 3.
9. Wash the rehydrated tissue in 10 ml 0.9% NaCl (2 min at RT) and repeat step 3.
10. Put the tissue, together with appr. 1 ml of the NaCl solution from the glass tube into a 1.5-ml microtube. This can easily be done using a 1-ml Eppendorf pipette with a cut blue tip, thus enhancing the diameter of the tip.

11. Remove the NaCl solution from the microtube using a 200-μl Eppendorf pipette, add – depending on the amount of tissue – 0.2-1 ml of PK-solution and vortex the microtube (PK-solution: 5 mg proteinase K (Boehringer 745723), 50 μl 1 M Tris-HCl (pH 7.5), 20 μl 0.5 M EDTA (pH 7.0), 2 μl 5 M NaCl, make up to 1 ml with filtered double distilled water; make fresh as required).
12. Incubate the microtube at 37°C for 30 min. During this time, vortex the microtube every 5 min to promote tissue disaggregation.
13. Go on with the protocol "Purification of the released nuclei".

Nuclear extraction from mounted formalin fixed/paraffin embedded tissues

1. A 5 μm section of paraffin-embedded tissue is mounted on a glass slide. Uncoated or coated slides (e.g. with aminotriethoxysilane or polylysine) can be used.
2. The mounted section is deparaffinized in 100 ml xylene in a Coplin jar (two times 5 min) and rehydrated in an ethanol series (100%, 90%, 70%, 50%; 3 min each) and 0.9% NaCl solution (two times 2 min).
3. If necessary, undesirable parts of the tissue can be removed at this point by scratching it from the slide with a scalpel. This may be helpful when analyzing tumors infiltrating normal tissue.
4. After that, the tissue is covered with proteinase K solution (PK-solution: 5 mg proteinase K (Boehringer 745723), 50 μl 1M Tris-HCl (pH 7.5), 20 μl 0.5 M EDTA (pH 7.0), 2 μl 5 M NaCl, make up to 1 ml with filtered double distilled water; make fresh as required) and incubated at 37°C for approx. 1 h in a moist chamber. The use of a coverslip should be avoided.
5. Thereafter, the proteinase K solution together with the digested tissue and released nuclei is collected with a pipette and passed through a 55 μm nylon mesh (go on with the following protocol).

Purification of the released nuclei

1. Transfer the fluid onto a 55 μm nylon mesh (Nytal 55 (SEFAR-AG, Thal, Switzerland; see comment 1). Fluid and nuclei will pass through the mesh by force of gravity and are collected in a 15-ml plastic tube. Nuclei remaining in the mesh are washed out by 4 ml 1×PBS, passed through the mesh and collected in the 15 ml plastic tube.
2. Pellet the extracted nuclei by centrifugation (850 g, 8 min); remove the supernatant leaving about 300 μl.
3. Resuspend the pellet in 4 ml 1×PBS and repeat step 2.

4. Resuspend the remaining 300 μl of 1×PBS.
5. Distribute the suspension on 2-6 clean dry slides by pipetting the fluid onto the slide surfaces. Place one drop (30-90 μl) of about 1 cm in diameter/slide (see comment 2).
6. Allow drops to dry out on a 40°C warming plate and afterwards incubate at RT overnight.
7. Fix slides in 100 ml formalin buffer in a Coplin jar for 10 min (RT).
8. Replace formalin-buffer in the Coplin jar with 1×PBS. After 5 min of incubation at RT, 1×PBS is replaced by distilled water.
9. Remove the water after 1 min, perform an ethanol series (70%, 90%, 100%, 3 min each) to dehydrate slides and air dry. An evaluation of the success of the nuclear extraction can be performed by phase contrast light microscopy (see comment 3).

Slide pretreatment

As in a conventional FISH approach, a pretreatment of the slides with RNase and pepsin followed by a postfixation with formalin-buffer is required to reduce the background (Liehr et al. 1995b).

1. Slides with the extracted nuclei are incubated in 2×SSC for 5 min at RT (in a 100-ml Coplin jar on a shaker).
2. Remove slides from the Coplin jar, add 100 μl of RNase solution/slide and cover with a 24×50 mm coverslip (RNase solution: per slide 100 μl 2×SSC plus 1 μl of RNase stock (= 5 μg/μl)).
3. Incubate the slides in a humid chamber for 15 min at 37°C (see comment 4).
4. Put slides back into the Coplin jar with 100 ml 2×SSC (RT) and remove the coverslips with forceps. Leave slides in 2×SSC solution for 3 min with gentle agitation.
5. Discard 2×SSC and replace it with 100 ml 1×PBS (RT) for 5 min (shaker).
6. Replace 1×PBS with 100 ml prewarmed pepsin-buffer (37°C) and incubate the slides for 10 min at 37°C, without agitation (see comment 4). (Pepsin-buffer: Add 1 ml of 1 M HCl to 99 ml of distilled water and

incubate at 37°C for about 20 min; then add 50 μl of the pepsin stock solution 10% (w/v) (Serva 31855) and leave the Coplin jar at 37°C; make fresh as required).

7. Replace fluid with 100 ml 1×PBS/$MgCl_2$, incubate at RT for 5 min with gentle agitation. $MgCl_2$ will block the enzymatic activity of pepsin (1×PBS/ $MgCl_2$ =5% (v/v) 1M $MgCl_2$ in 1×PBS).

8. Postfix nuclei on the slide surfaces by replacing 1×PBS/$MgCl_2$ with 100 ml of formalin-buffer (3 ml acid free formaldehyde [37%; Roth 4979.1] in 100 ml 1xPBS) for 10 min (RT, with gentle agitation)..

9. Formalin-buffer is replaced by 100 ml 1×PBS for 2 min (RT, with gentle agitation).

10. Finally slides are dehydrated by an ethanol series (70%, 90%, 100%, 3 min each) and air dried (see comment 5).

Fluorescence in situ hybridization (FISH)

1. Add 100 μl denaturation-buffer to the slides and cover with (24×50 mm) coverslips. (Denaturation-buffer: 70% (v/v) deionized formamide, 10% (v/v) filtered double distilled water, 10% (v/v) 20×SSC, 10% (v/v) phosphate buffer; make fresh as required; phosphate buffer: prepare 0.5 M Na_2HPO_4 and 0.5 M NaH_2PO_4, mix these two solutions (1:1) to get pH 7.0, then aliquot and store at -20°C).

2. Incubate slides on a warming plate for 12 min at 75°C (see comment 6).

3. Remove the coverslips immediately with forceps and place slides in a Coplin jar filled with 70% ethanol (4°C) to conserve target DNA as single strands.

4. Dehydrate slides in ethanol (70%, 90%, 100%, 4°C, 3 min each) and air dry.

5. For each slide to be hybridized, add 200 ng of labeled YAC-, BAC- or cosmid-DNA or 1.5 μl of biotin or digoxigenin-labeled chromosome-specific satellite probes plus 3 μl of 1 μg/μl COT1-DNA (Gibco) to 20 μl of the hybridization-buffer in a 1.5-ml microtube, vortex and spin down. (Hybridization-buffer: Dissolve 2 g dextran sulfate in 10 ml 50% deionized formamide/2×SSC/50 mM phosphate buffer for 3 h at 70°C. Aliquot and store at -20°C.)

6. Denature probe-solution at 75°C for 5 min and cool immediately on ice to conserve probe DNA in single strands for satellite probes, or do a prehybridization step at 37°C for 30 min for the other probes (YACs, BACs, cosmids).

7. Add 20 µl of probe-solution to each denatured slide, put 24×50 mm coverslips on the drops and seal with rubber cement (Fixogum; Marabu).

8. Incubate slides for three nights at 37°C in a humid chamber (see comment 7).

9. Take the slides out of 37°C, remove rubber cement with forceps and coverslips by letting them swim off in 4×SSC/0.2%Tween (RT, 100-ml Coplin jar) (see comment 8).

10. Postwash the slides 3×5 min in formamide-solution (45°C) followed by 35 min in 2×SSC (37°C) in a 100-ml Coplin jar, with gentle agitation.

11. Put the slides in 4×SSC/0.2% Tween (100 ml, RT), for a few seconds.

12. Add 50 µl of solution I to each slide, cover with 24×50 mm coverslips and incubate at 37°C for 30 min in a humid chamber. [Solution I: FITC-avidin (CAMON Vector Laboratories A2011)/4xSSC + 0.2% Tween + 5% BSA (1:300); make fresh as required].

13. Remove the coverslips and wash 3 x 3 min in 4×SSC/0.2% Tween (RT, with gentle agitation).

14. Add 50 µl of solution II to each slide, cover with 24×50 mm coverslips and incubate at 37°C for 75 min in a humid chamber. [Solution II: Biotinylated antiavidin (CAMON Vector Laboratories BA0300)/anti-digoxinenin-rhodamine (Boehringer Mannheim, 1207750)/ 4×SSC + 0.2% Tween + 5% BSA (1:20:100); make fresh as required].

15. Repeat step 13.

16. Repeat steps 12 and 13.

17. Counterstain the slides with DAPI-solution (100 ml in a Coplin jar, RT) for 8 min. [DAPI-solution: Dissolve 5 µl of DAPI (4,6-diamidino-2-phenylindol.2HCl stock-solution; Serva 18860) in 100 ml 4×SSC + 0.2% Tween; make fresh as required].

18. Wash slides several times in water for a few seconds and air dry.

19. Add 15 µl of antifade Vectashield (CAMON Vector Laboratories H1000), cover with coverslips and look at the results under a fluorescence microscope.

Comments

1. Nytal 55 – the nylon mesh, has to be cut in 5×5 cm squares. Such a square can be formed into a funnel, pinned with a stapler and used as the required filter.
2. The resulting suspension has a certain turbidity reflecting the number of extracted nuclei. With experience, an assessment of the number of slides on which the suspension should be distributed, will be possible. For the first-time user, the distribution of the suspension onto two slides is suggested.
3. An evaluation of quantity and quality of extracted nuclei was not possible before this step of the protocol, due to crystallization of PBS salts on the slide surface. Now they have been washed away during the steps 6-9.
4. RNase and pepsin pretreatment conditions should be tested in each laboratory on a single slide first. Both RNase and pepsin concentrations can be too stringent, resulting in clean slides without any remaining nuclei.
5. The pretreated slides can be hybridized immediately or stored at RT for up to 3 weeks. If longer storage is necessary, slides are stable at -20°C for several months.
6. Due to the fact that DNA in archival tissues has undergone some fixation steps and has been stored up to several years, a prolonged denaturation time seems useful. Moreover, in other FISH-protocols with denaturation times of 2-5 min only, the maintenance of available metaphase chromosomes is the main aspect, which is of no significance in the actual protocol.
7. Incubation can be stopped – if necessary – after 48 or 96 h, as well. While in the first case, weaker signals are possible, in the second case some cross-hybridization problems may arise.
8. During the washing steps it is important to prevent the slide surfaces drying out, otherwise background problems may arise.

Acknowledgements. Supported by the Wilhelm Sander-Stiftung (99.105.1) and the EU (ICA2-CT-2000- 10012 and QLRT-1999-31590). The continuos support of the Carl Zeiss GmbH (Jena, Germany) is gratefully acknowledged.

References

Dhingra, K, Sahin, A, Supak, J, Kim, SY, Hortobagyi, G, Hittelman, WN (1992) Chromosome in situ hybridization on formalin-fixed mammary tissue using non-isotopic, non-fluorescent probes: technical considerations and biological implications. Breast Cancer Res Treat, 23:201-210

Hackel, C, VarellaGarcia, M (1997) Interphase cytogenetics using fluorescence in situ hybridization: An overview of its application to diffuse and solid tissue. Braz J Genet, 20:97-106

Hedley, DW, Friedlander, ML, Taylor, IW, Rugg, CA, Musgrove, EA (1983) Method for analysis of cellular DNA content of paraffin-embedded pathological material using flow cytometry. J Histochem Cytochem, 31:1333-1335

Köpf, I, Hanson, C, Delle, U, Verbiené, I, Weimarck, A (1996) A rapid and simplified technique for analysis of archival formalin-fixed, paraffin-embedded tissue by fluorescence in situ hybridization (FISH). Anticancer Res, 16:2533-2536

Liehr, T, Stübinger, A, Thoma, K, Tulusan, HA, Gebhart, E (1994) Comparative interphase cytogenetics using FISH on human ovarian carcinomas. Anticancer Res, 14:183-188

Liehr, T, Grehl, H, Rautenstrauss, B (1995a) FISH analysis of interphase nuclei extracted from paraffin-embedded tissue. Trends in Genetics, 11:377-378

Liehr, T, Thoma, K, Kammler, K, Gehring, C, Ekici, A, Bathke, KD, Grehl, H, Rautenstrauss, B (1995b) Direct preparation of uncultured EDTA-treated or heparinized blood for interphase FISH analysis. Appl Cytogenet, 21:185-188

Liehr, T, Grehl, H, Rautenstrauss, B (1997) Molecular diagnosis of PMP22-associated neuropathies using fluorescence in situ hybridization (FISH) on archival peripheral nerve tissue preparations. Acta Neuropathol, 94:266-271

Liehr, T, Claussen, U, Gebhart, E (1999) Nucleus extraction from single mounted tissue sections. Genet Anal-Biomol Engineering, in press.

Neubauer, S, Liehr, T, Tulusan, HA, Gebhart, E (1994) Interphase cytogenetics by FISH on archival paraffin material and cultured cells of human ovarian tumors. Int J Oncol, 4:317-321

Qian, JQ, Bostwick, DG, Takahashi, S, Borell, TJ, Brown, JA, Lieber, MM, Jenkins, RB (1996) Comparison of fluorescence in situ hybridization analysis of isolated nuclei and routine histological sections from paraffin-embedded prostatic adenocarcinoma specimens. Am J Pathol, 149:1193-1199

Wolfe, KQ, Herrington, CS (1997) Interphase cytogenetics and pathology: A tool for diagnosis and research. J Pathol, 181:359-361

Combined in Situ Nick Translation and mRNA FISH for Simultaneous Detection of Apoptosis and Expression of Apoptosis-Related Genes in Sections from Paraffin-Embedded Tissue

NORBERT HECKMANN

Introduction

Simultaneous demonstration of apoptotic cells and apoptosis-related gene expression on the same tissue section correlated to histomorphological features, appears to be of great interest.

Apoptosis or programmed cell death is an important factor during the evolution of various types of cancer. The proto-oncogene bcl-2 is known to act as an inhibitor of apoptosis. P53 has an apoptosis-inducing function usually demonstrated at the protein level, but due to the short half-life, an immunohistochemical detection of wild-type protein is not possible in most benign tumors.

It is possible to correlate histomorphological architecture of colonic mucosa with fluorescence staining. Published immunohistochemical studies showed the expression of bcl-2 in non-tumorous colonic crypts to be restricted to the proliferative cells of the crypt bases (Watson et al., 1996; Bosari et al. 1995). During very early stages of colorectal carcinogenesis, abnormally high levels of Bcl-2 oncoprotein are observed (Mosnier et al., 1996). They are lost during the evolution from adenoma to carcinoma (Watson et al., 1996; Kikuchi et al., 1997) and have no prognostic significance in established tumors (Bosari et al., 1995). Apoptosis is increased in hyperplastic polyps and adenomas but decreased in adenocarcinomas, paralleled by an increased expression of mutant P53 protein (Kikuchi et al., 1997). P53 expression is higher in de-novo-carcinomas than in ex-adenoma carcinomas (Mueller et al., 1996).

Our goal was to develop a technique suitable to detect apoptotic cells on morphologically quotable histological sections and the expression on the mRNA-level of apoptosis-inducing and inhibiting genes.

Norbert Heckmann, Saarland University, Institute of Anatomy and Cell Biology, Homburg, 66421, Germany (*phone* +49-6841-1626101; *fax* +49-6841-1626121; *e-mail* annhec@med-rz.uni-saarland.de)

Materials

Hybridization mixture

- 50% Formamide
- 2×SSC
- 10% dextran sulphate
- 1 µl *E. coli* tRNA
- 1 µl sheared salmon sperm DNA
- A dilution of 1/50 of labeled oligonucleotide probes
- add 50 µl with double distilled water

Polymerase-enzyme-mix

- 0.5 nmol of each dNTP
- 0.25 nmol fluorescein-12-dUTP
- 5 U DNA polymerase I (endonuclease-free)
- add 50 µl with double distilled water

Procedure

Preparation of paraffin-embedded sections

We use sections from tissue routinely fixed in 4% buffered formaldehyde (pH 7.4) and embedded in paraffin. For the pretreatment, all solutions are prepared with double distilled water containing 0.1% diethyl pyrocarbonate (DEPC) to avoid RNase activity. The 5 µm thick sections are dewaxed with xylene and rehydrated in a graded ethanol series down to 50%, and the slides are dried at 50°C. After postfixing for 10 min (4% paraformaldehyde) and washing (PBS-phosphate buffered saline, water) they are incubated in proteinase K solution (20 µg/ml proteinase K in 50 mM tris-HCl, pH 7.6; 5 mM EDTA) for 15-20 min at 55°C. Washing for 5 min (PBS, water) is followed by incubation in 2×SSC (standard saline citrate) at 70°C for 20 min. The sections are dehydrated in an ethanol series of increasing concentration and air-dried.

Preparation of oligonucleotide probes

As probes, DNA oligonucleotides are used. For detecting bcl-2, the 26mer DNA oligonucleotide probe 5′ TCG TAC CCC GTT CTC CCA GCG TGC GC 3′ is used as antisense probe, and the probe 5′ GCG CAC GCT GGG AGA ACG GGG TAC GA 3′ as sense probe (Hell et al., 1995). For detecting p53, the 20mer DNA oligonucleotide probe 5′ TTC TTG CAT TCT GGG ACA GC 3′ was used as antisense probe and the probe 5′ GGT TCA AAG ACC CAA AAC CC 3′ as sense probe (PCR primers, MWG Biotech). The probes are biotinylated at the 3′-end using terminal desoxynucleotidyl transferase. 100 pmol oligonucleotide is mixed with 2.5 nmol biotin-16-dUTP, 1/30 dilution of dATP, 2 µl cobalt chloride solution and 7.5 U terminal transferase; subsequent incubation for 15 min at 37°C. The labeled oligonucleotides are stored at -20°C.

Hybridization

The tissue sections are pre-denatured at 90°C for 15 min and immediately placed on ice. The hybridization mixture is denatured for 5 min at 95°C and immediately placed on ice. 50 µl hybridization mixture is applied to the slides, covered with coverslips and sealed with rubber cement. After denaturation for 30 min at 80°C and overnight hybridization at 37°C, the slides are washed twice in PBS at room temperature.

Detection of apoptosis

Hybridization is followed by in situ nick translation. Fluorescein-12-dUTP is incorporated in situ into DNA nicks by *E. coli* polymerase I.

We applied 50 µl reaction mixture. The slides are covered with coverslips and incubated overnight at 37°C.

Washing steps and detection of mRNA signals

For detection of bcl-2, the specimens are washed twice in 2×SSC at room temperature, twice in 2×SSC at 42°C for 10 min and three times in 25% formamide/2×SSC at 42°C for 15 min. For detection of p53, the specimens are washed twice in 2×SSC at room temperature, twice in 2×SSC at 42°C for 10 min and three times in 33% formamide/2×SSC at 42°C for 15 min.

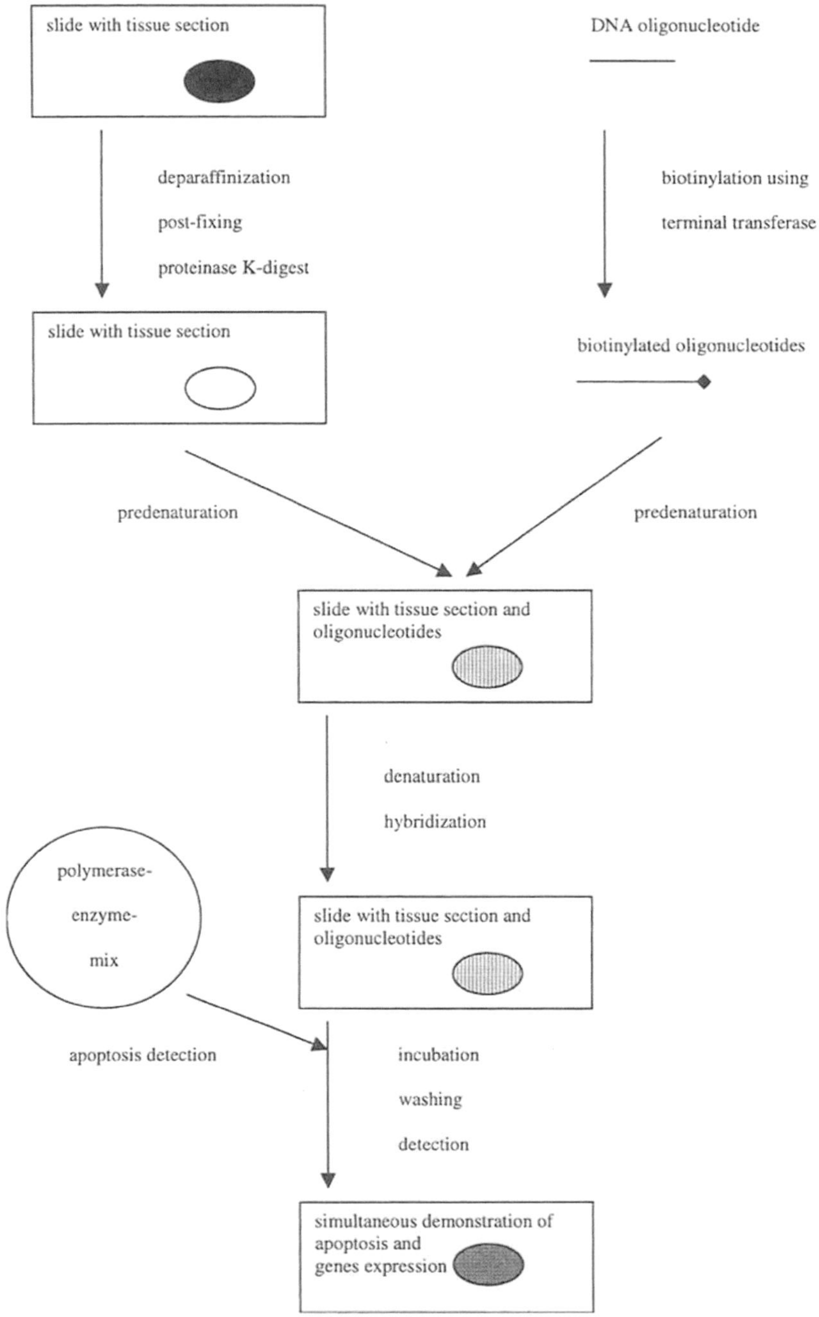

Fig. 1. Technique for simultaneous demonstration of apoptosis and gene expression

After blocking with bovine serum albumin (5% in 4×SSC), specific biotin is detected with Cy3 labeled streptavidin (1:150 in 4×SSC). After washing, the tissue sections are counterstained with antifade solution containing DAPI (1 μg/ml).

HE staining

Unprocessed photographs or digital images are taken from sections with strong positive signals. In a following step, the same slide is stained routinely with hematoxylin and eosin. The same fields of vision are re-examined and photographed with a bright-field microscope for histomorphological analysis.

References

Bosari S, Moneghini L, Graziani D, Lee AK, Murray JJ, Coggi G, Viale G (1995) Bcl-2 oncoprotein in colorectal hyperplastic polyps, adenomas, and adenocarcinomas. Hum-Pathol 26(5): 534-540

Hell K, Lorenzen J, Fischer R, Hausmann ML (1995) Hodgkin cells accumulate mRNA for bcl-2. Laboratory Investigation 73(4):492

Kikuchi Y, Dinjens WN, Bosman FT (1997) Proliferation and apoptosis in proliferative lesions of the colon and rectum. Virchows-Arch. 431(2): 111-117

Mosnier JF, Perret AG, Vindimian M, Dumollard JM, Balique JG, Perpoint B, Boucheron S (1996) An immunohistochemical study of the simultaneous expression of bcl-2 and p53 oncoproteins in epithelial tumors of the colon and rectum. Arch-Pathol-Lab-Med. 120(7): 654-659

Mueller J, Mueller E, Hoepner I, Jutting J, Bethke B, Stolte M, Hofler H (1996) Expression of bcl-2 and p53 in de novo and ex-adenoma colon carcinoma: a comparative immunohistochemical study. J-Pathol. 180(3): 259-265

Watson AJ, Merritt AJ, Jones LS, Askew JN, Anderson E, Becciolini A, Balzi M, Potten CS, Hickman JA (1996) Evidence of reciprocity of bcl-2 and p53 expression in human colorectal adenomas and carcinomas. Br-J-Cancer. 73(8): 889-895

Nucleus Extraction from Cryofixed Tissue

BERND RAUTENSTRAUSS and THOMAS LIEHR

Introduction

The advent of fluorescence in situ hybridization (FISH) rendered it possible to obtain "cytogenetic information" from interphase nuclei (Langer et al., 1981, Pinkel et al, 1986). Karyotypic analysis by direct demonstration of DNA sequences in interphase nuclei (interphase cytogenetics or interphase FISH) can be applied to a wide variety of cellular material, including cytogenetic preparations, fresh material (touch-preparations), cryofixed and paraffin-embedded tissue. In contrary to metaphase FISH, only cosmids, P1-clones, BACs, YACs and satellite probes are suitable for interphase FISH. Detection and quantitation of structural chromosome aberrations, of microdeletions or microduplications and oncogene amplification or deletion of tumor suppressor genes in solid tumors thus became possible.

To date, fresh tumor material (e.g. Liehr et al., 1994) or formalin-fixed and subsequently paraffin-embedded and sectioned tissue (e.g. Neubauer et al., 1994) has mainly been used for interphase cytogenetic studies. Formalin-fixed/paraffin-embedded tissue is readily available, as this kind of tissue fixation is the most common standard technique in clinical practice. However, there are some disadvantages in connection with this material and an increasing number of laboratories now collect archival formalin-fixed/paraffin-embedded and cryofixed tissue samples.

Formalin-fixed/paraffin-embedded tissue is, for example, not suitable for all kinds of immunohistochemical approaches, as specific antigens may be destroyed during the fixation procedure. Moreover, if formalin

✉ Bernd Rautenstrauss, Institut für Humangenetik, Schwabachanlage 10, Erlangen, 91054, Germany (*phone* +49-9131-8532352; *fax* +49-9131-209297; *e-mail* berndwr@humgenet.uni-erlangen.de)
Thomas Liehr, Institut für Humangenetik, Kollegiengasse 10, Jena, 07743, Germany

fixation is performed in unbuffered formalin and/or incubation is too long, the tissue becomes unsuitable for any kind of FISH studies, because DNA is degraded and washed out of the cells (Long et al., 1993). When using sectioned material, some additional problems arise. The most important ones arise during evaluation of the slides: (a) there is the problem that overlapping nuclei, due to the presence of several cell layers, one on top of the other, cannot be evaluated; and (b) the problem that cut nuclei may lead to artificial signal loss in interphase cytogenetic studies (Neubauer et al., 1994, Dhingra et al., 1992).

These problems can be solved using a technique called nuclear extraction, first described in 1983 (Hedley et al., 1983). However, there is only one protocol concerning nuclear extraction from cryofixed tissues in the literature (Liehr et al., 1996, 1997).

Outline

1. Cryofixed tissue is cut into small pieces whilst still frozen, formalin-fixed, washed and treated with proteinase K to digest the cytoplasm and free nuclei. During this treatment, damaged and cut nuclei are eliminated. Interphase nuclei are separated from debris by passing through a nylon mesh. Nuclei are washed in PBS and placed on uncoated, clean slides and fixed. Small tissue pieces (5×2×0.5 mm) are sufficient to produce two to five slides with extracted nuclei suitable for interphase FISH analyses.

2. For a FISH approach, these slides are treated with RNase and pepsin prior to denaturation.

3. Denaturation of the slides is performed for about 12 min at 75°C. Biotinylated and digoxigenated probes are denatured for 5 min and hybridized on the slides for 3 nights at 37°C in a humid chamber. After a postwashing series, detection of the biotinylated probe is performed with an FITC-avidin system which leads to a green signal and of the digoxigenated probe with anti-digoxigenin-rhodamine leading to red signals. After counterstaining of the nuclei with DAPI and addition of antifade solution, slides can be evaluated under a fluorescence microscope.

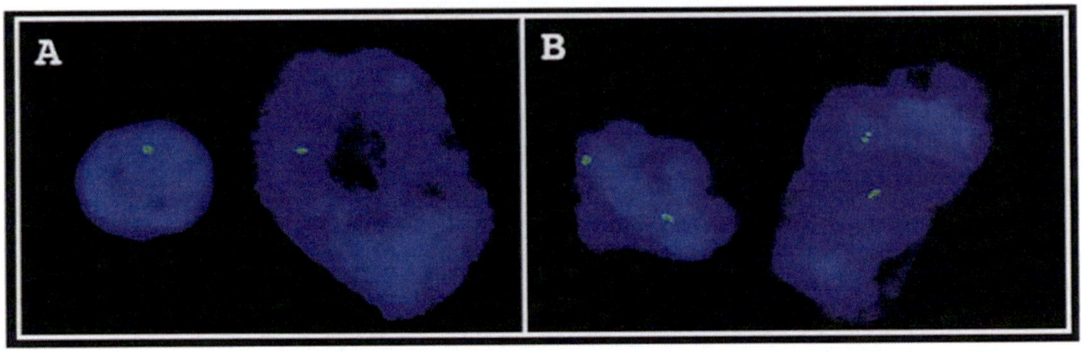

Fig. 1. Results of FISH experiments performed on interphase nuclei extracted from cryofixed meningioma tissue using a chromosome 22 specific cosmid probe are shown. In **A** two nuclei with only one specific spot each are visible, while in **B** two interphase nuclei with two chromosome 22 specific signals are presented. The upper signal in the right nucleus of **B** could be interpreted as two spots. However, according to the evaluation by eye in the microscope, it has to be stated that it is one split signal and the nucleus or the cell was in the S- or G2-phase of cell cycle. Loss of one chromosome 22 is a well-known fact in meningioma and monosomy 22 is indeed the first specific aberration characterized for solid tumors (Zang and Singer, 1967). Images were captured on a Zeiss Axioplan microscope using the ISIS digital FISH imaging system (MetaSystems, Altlussheim, Germany) using an XC77 CCD camera with on-chip integration (Sony)

Procedure

Nuclear extraction

1. Transfer cryofixed tissue from -80 to -20°C freezer for 1 h.
2. Transfer tissue into a glass dish on ice and cut into small pieces (not larger then 1 mm cubed) using a precooled (+4°C) scalpel and forceps (see comment 1).
3. Add 1 ml of formalin-buffer at room temperature (= RT) to the cold tissue pieces and transfer them together with the buffer into a 1.5-ml microtube. Tissue should thaw on addition of formalin-buffer. [Formalin-buffer: 3% (v/v) of acid free formaldehyde (37%; Roth 4979.1), in 1×PBS pH 7.0; make fresh as required].
4. Incubate the tissue in the formalin-buffer for 1-3 h at RT without agitation (see comment 2).
5. Pellet the tissue pieces by centrifugation (3800 *g*, 30 s, RT). If necessary repeat this step.
6. Remove the fluid using a 200 µl Eppendorf pipette, add 1 ml of sterile 0.9% NaCl (w/v) and vortex the microtube.

7. Repeat step 5.
8. Repeat steps 6 and 7.
9. Remove the fluid using a 200 µl Eppendorf pipette, add – depending on the amount of tissue – 0.2-1 ml of PK-solution and vortex the microtube [PK-solution: 5 mg proteinase K (Boehringer 745723), 50 µl 1 M tris-HCl (pH 7.5), 20 µl 0.5 M EDTA (pH 7.0), 2 µl 5 M NaCl, make up to 1 ml with filtered double distilled water; make fresh as required] (see comment 3 as well).
10. Incubate the microtube at 37°C for 30 min. During this time, vortex the microtube every 5 min to promote tissue disaggregation.
11. Transfer the fluid onto a 55 µm nylon mesh (Nytal 55 (SEFAR-AG, Thal, Switzerland; see comment 4). Fluid and nuclei will pass through the mesh by force of gravity and are collected in a 15-ml plastic tube. Nuclei remaining in the mesh are washed out by 4 ml 1×PBS, passed through the mesh and collected in the 15-ml plastic tube.
12. Pellet the extracted nuclei by centrifugation (850 *g*, 8 min); remove the supernatant with the exception of about 300 µl.
13. Resuspend the pellet in 4 ml 1×PBS and repeat step 12.
14. Resuspend the remaining 300 µl of 1×PBS.
15. Distribute the suspension on 2-6 clean dry slides by pipetting the fluid onto the slide surfaces. Place one drop (30-90 µl) of about 1 cm in diameter/slide (see comment 5).
16. Allow drops to dry out on a 40°C warming plate and afterwards at RT overnight.
17. Fix slides in 100 ml formalin buffer in a Coplin jar for 10 min (RT).
18. Replace formalin-buffer in the Coplin jar with 1×PBS. After 5 min of incubation at RT, 1×PBS is replaced by distilled water.
19. Remove the water after 1 min, perform an ethanol series (70%, 90%, 100%, 3 min each) to dehydrate slides and air dry. An evaluation of the success of the nuclear extraction can be performed by phase contrast light microscopy (see comment 6).

Slide pretreatment

As in a conventional FISH approach, a pretreatment of the slides with RNase and pepsin followed by a postfixation with formalin-buffer is required to reduce background (Liehr et al., 1995).

1. Slides with the extracted nuclei are incubated in 2×SSC for 5 min at RT (in a 100-ml Coplin jar on a shaker).
2. Remove slides from the Coplin jar, add 100 µl of RNase solution/slide and cover with a 24×50 mm coverslip [RNase solution: per slide 100 µl 2×SSC plus 1 µl of RNase stock (= 5 µg/µl)].
3. Incubate the slides in a humid chamber for 15 min at 37°C (see comment 7).
4. Put slides back into the Coplin jar with 100 ml 2×SSC (RT) and remove the coverslips with forceps. Leave slides in 2×SSC solution for 3 min with gentle agitation.
5. Discard 2×SSC and replace it with 100 ml 1×PBS (RT) for 5 min (shaker).
6. Replace 1×PBS with 100 ml prewarmed pepsin-buffer (37°C) and incubate the slides for 10 min at 37°C, without agitation (see comment 7). [Pepsin-buffer: Add 1 ml of 1 M HCl to 99 ml of distilled water and incubate at 37°C for about 20 min; then add 50 µl of the pepsin stock solution 10% (w/v) (Serva 31855) and leave the Coplin jar at 37°C; make fresh as required].
7. Replace fluid with 100 ml 1×PBS/$MgCl_2$, incubate at RT for 5 min with gentle agitation. $MgCl_2$ will block the enzymatic activity of pepsin.
8. Postfix nuclei on the slide surfaces by replacing 1×PBS/$MgCl_2$ with 100 ml of formalin-buffer for 10 min (RT, with gentle agitation). (1×PBS/ $MgCl_2$=5% (v/v) 1M $MgCl_2$ in 1×PBS).
9. Formalin-buffer is replaced by 100 ml 1×PBS for 2 min (RT, with gentle agitation).
10. Finally, slides are dehydrated by an ethanol series (70%, 90%, 100%, 3 min each) and air dried (see comment 8).

Fluorescence in situ hybridization (FISH)

1. Add 100 µl denaturation-buffer to the slides and cover with (24×50 mm) coverslips. [Denaturation-buffer: 70% (v/v) deionized formamide, 10% (v/v) filtered double distilled water, 10% (v/v) 20×SSC, 10% (v/v) phosphate buffer; make fresh as required; phosphate buffer: prepare 0.5 M Na_2HPO_4 and 0.5 M NaH_2PO_4, mix these two solutions (1:1) to get pH 7.0, then aliquot and store at -20°C].
2. Incubate slides on a warming plate for 12 min at 75°C (see comment 9).
3. Remove the coverslips immediately with forceps and place slides in a Coplin jar filled with 70% ethanol (4°C) to conserve target DNA as single strands.
4. Dehydrate slides in ethanol (70%, 90%, 100%, 4°C, 3 min each) and air dry.
5. For each slide to be hybridized add 200 ng of labeled YAC-, BAC- or cosmid-DNA or 1.5 µl of biotin or digoxigenin-labeled chromosome specific satellite probes plus 3 µl of 1 µg/µl COT1-DNA (Gibco) to 20 µl of the hybridization-buffer in a 1.5-ml microtube, vortex and spin down. (Hybridization-buffer: Dissolve 2 g dextran sulfate in 10 ml 50% deionized formamide/2×SSC/50 mM phosphate buffer for 3 h at 70°C. Aliquot and store at -20°C).
6. Denature probe-solution at 75°C for 5 min and cool immediately on ice to conserve probe DNA in single strands for satellite probes, or do a prehybridization step at 37°C for 30 min for the other probes (YACs, BACs, cosmids).
7. Add 20 µl of probe-solution onto each denatured slide, put 24×50 mm coverslips on the drops and seal with rubber cement (Fixogum; Marabu).
8. Incubate slides for three nights at 37°C in a humid chamber (see comment 10).
9. Take the slides out of 37°C, remove rubber cement with forceps and the coverslips by letting them swim off in 4×SSC/0.2%Tween (RT, 100-ml Coplin jar) (see comment 11).
10. Postwash the slides 3×5 min in formamide-solution (45°C) followed by 3×5 min in 2×SSC (37°C) in a 100-ml Coplin jar, with gentle agitation.

11. Put the slides in 4×SSC/0.2% Tween (100 ml, RT), for a few seconds.
12. Add 50 µl of solution I to each slide, cover with 24×50 mm coverslips and incubate at 37°C for 30 min in a humid chamber. [Solution I: FITC-avidin (CAMON Vector Laboratories A2011)/4xSSC + 0.2% Tween + 5% BSA (1:300 both Sigma); make fresh as required].
13. Remove the coverslips and wash 3×3 min in 4×SSC/0.2% Tween (RT, with gentle agitation).
14. Add 50 µl of solution II to each slide, cover with 24×50 mm coverslips and incubate at 37°C for 75 min in a humid chamber. [Solution II: Biotinylated antiavidin (CAMON Vector Laboratories BA0300)/anti-digoxigenin-rhodamine (Boehringer Mannheim, 1207750)/ 4×SSC + 0.2% Tween + 5% BSA (1:20:100); make fresh as required].
15. Repeat step 13.
16. Repeat step 12.
17. Repeat step 13.
18. Counterstain the slides with DAPI-solution (100 ml in a Coplin jar, RT) for 8 min. [DAPI-solution: Dissolve 5 µl of DAPI (4,6-diamidino-2-phenylindol.2HCl stock-solution; Serva 18860) in 100 ml 4×SSC/ 0.2% Tween; make fresh as required].
19. Wash slides several times in water for a few seconds and air dry.
20. Add 15 µl of antifade Vectashield (CAMON Vector Laboratories H1000), cover with coverslips and look at the results under a fluorescence microscope.

Comments

1. If only small amounts of cryofixed tissue are available, tissue can be cut on a cryostat. About 15-25 µm cryosections (2-5 mm in diameter) are sufficient for about 5 slides with extracted interphase nuclei.
2. Incubation of the tissue in formalin buffer can be done by agitation as well, with no difference in the quality or quantity of resulting extracted nuclei.
3. If only small amounts of tissue are to be digested (see comment 1) 0.2-0.4 ml of PK-solution are sufficient.

4. Nytal 55 – the nylon mesh has to be cut in 5×5 cm squares. Such a square can be formed into a funnel, pinned by a stapler and used as the required filter.

5. The resulting suspension has a certain turbidity reflecting the number of extracted nuclei. With experience, an assessment of the number of slides on which the suspension should be distributed will be possible. For the first-time user, a distribution of suspension on two slides is suggested.

6. An evaluation of quantity and quality of extracted nuclei was not possible before this step of the protocol, due to crystallization of PBS salts on the slide surface. Now they have been washed away during the steps 17-19.

7. RNase and pepsin pretreatment conditions should be tested in each laboratory on a single slide first. Both RNase and pepsin concentrations can be too stringent, resulting in clean slides without any remaining nuclei.

8. The pretreated slides can be hybridized immediately, or stored at RT for up to 3 weeks. If longer storage is necessary slides are stable at -20°C for several months.

9. Due to the fact that DNA in archival tissues has undergone some fixation steps and has been stored up to several years, a prolonged denaturation time seems useful. Moreover, in other FISH-protocols with denaturation times of 2-5 min only, the maintenance of available metaphase chromosomes is the main aspect, which is of no significance in the actual protocol.

10. Incubation can be stopped – if necessary – after 48 or 96 h as well. While in the first case weaker signals are possible, in the second case some cross-hybridization problems may arise.

11. During the washing steps, it is important to prevent the slide surfaces drying out, otherwise background problems may arise.

References

Dhingra, K, Sahin, A, Supak, J, Kim, SY, Hortobagyi, G, Hittelman, WN (1992) Chromosome *in situ* hybridization on formalin-fixed mammary tissue using non-isotopic, non-fluorescent probes: technical considerations and biological implications. Breast Cancer Res Treat, 23:201-210

Hedley, DW, Friedlander, ML, Taylor, IW, Rugg, CA, Musgrove, EA (1983) Method for analysis of cellular DNA content of paraffin-embedded pathological material using flow cytometry. J Histochem Cytochem, 31:1333-1335

Langer, PR, Waldrop, AA, Ward, DC (1981) Enzymatic synthesis of biotin-labeled polynucleotides: Novel nucleic acid affinity probes. Proc Natl Acad Sci USA, 78:6633-6637

Liehr, T, Stübinger, A, Thoma, K, Tulusan, HA, Gebhart, E (1994) Comparative interphase cytogenetics using FISH on human ovarian carcinomas. Anticancer Res, 14:183-188

Liehr, T, Thoma, K, Kammler, K, Gehring, C, Ekici, A, Bathke, KD, Grehl, H, Rautenstrauss, B (1995) Direct preparation of uncultured EDTA-treated or heparinized blood for interphase FISH analysis. Appl Cytogenet, 21:185-188

Liehr,T, Grehl, H, Rautenstrauss, B (1996) A rapid method for FISH analysis on interphase nuclei extracted from cryofixed tissue. Trends in Genetics, 12: 505-506

Liehr, T, Grehl, H, Rautenstrauss, B (1997) Molecular diagnosis of PMP22-associated neuropathies using fluorescence in situ hybridization (FISH) on archival peripheral nerve tissue preparations. Acta Neuropathol, 94:266-271

Long, AA, Komminoth, P, Lee, E, Wolfe, HJ (1993) Comparison of indirect and direct in-situ polymerase chain reaction in cell preparations and tissue sections. Detection of viral DNA, gene rearrangements and chromosomal translocations. Histochem, 99:151-162

Neubauer, S, Liehr, T, Tulusan, HA, Gebhart, E (1994) Interphase cytogenetics by FISH on archival paraffin material and cultured cells of human ovarian tumors. Int J Oncol, 4:317-321

Pinkel, D, Straume, T, Gray, JW (1986) Cytogenetic analysis using quantitative, high sensitivity, fluorescence hybridization. Proc Natl Acad Sci USA, 83:2934-2938

Zang, KD, Singer H (1967) Chromosomal constitution of meningiomas. Nature, 216:84-85

Preparation of Samples for CGH by DOP-PCR from Cryofixed or Paraffin-Embedded, Microdissected Cells

A comparison of two different PCR amplification methods

VOLKER JUNG and GEORG A. WEBER

Introduction

The development of the "comparative genomic hybridization (CGH)" technique (Kallioniemi et al., 1992) enabled the cytogeneticist to comprehensively analyze chromosomal imbalances in entire genomes. The technique reveals chromosomal regions with relative gains or losses of DNA sequences, by comparing DNA from a test specimen (test DNA) with the DNA from a normal specimen (reference DNA). Test DNA labeled e.g. with biotin is mixed 1:1 with differently labeled reference DNA with digoxigenin and simultaneously hybridized to normal metaphase spreads. For the detection of the hybridized test and reference DNA, different fluorochromes are used, and each is visualized by epifluorescence microscopy with selective filters. Regions within the tumor that are over- or underrepresented in the probe mixture can be identified by an increased or decreased color ratio of the two fluorochromes.

The standard CGH protocol requires 0.5-1.0 μg of genomic DNA, corresponding to approximately 50,000 to 100,000 diploid cells.

Many questions in pathology or in tumor genetics cannot be answered because of the lack of a sufficient amount of fresh material. In many cases, paraffin-embedded specimens are the only available tumor material. Microdissection of different tumor areas of a given tumor from the same tissue section (cryofixed or paraffin-embedded) is an elegant method to investigate focal heterogeneity within the specimen (Aubele et al., 1999, Jung et al., 1999). Microdissected probes, however, are too small

✉ Volker Jung, Saarland University, Institute of Human Genetics, Building 68, Homburg/Saar, 66421, Germany (*phone* +49-6841-1626605; *fax* +49-6841-1626600; *e-mail* volker.jung@uniklinik-saarland.de)
Georg A. Weber, Saarland University, Institute of Human Genetics, Building 68, Homburg/Saar, 66421, Germany (*phone* +49-6841-1626605; *fax* +49-6841-1626600; *e-mail* georg.weber@uniklinik-saarland.de)

to provide enough DNA for standard CGH. With an universal PCR using degenerated oligonucleotides primers (DOP-PCR) and low stringency cycle conditions (Telenius et al., 1992, Müller-Navia et al., 1995, Kuukasjärvi et al., 1997), it is possible to uniformly amplify very small microdissected DNA samples.

Primers used for DOP-PCR (see Fig. 2) contain a variable binding region with a sequence of six randomly combined nucleotides which is flanked by a stringent primer binding region at the 5-end (ten nucleotides) and a main binding region at the 3-end (six nucleotides). The amplification procedure is divided into two steps. During the first step, eight PCR cycles are performed at low stringency conditions (annealing temperature 30°C), followed by a slow temperature increase to the elongation temperature, and primer elongation.

During the second step (30-35 cycles) a more stringent (higher) annealing temperature (Ta=62°C) is used, and all DNA sequences generated during the first eight cycles are exponentially amplified. DOP-PCR amplifications are useful for CGH studies and result in a smear of DNA fragments which is visible on an agarose minigel stained with ethidium bromide.

Two alternative protocols exist for an efficient amplification of the small isolated DNA samples for CGH studies.

The standard protocol combines the two amplification steps in one PCR reaction with the same Taq polymerase (Speicher et al., 1993, Isola et al., 1994, Wiltshire et al., 1995).

The other protocol (Kuukasjärvi et al., 1997) separates the DOP-PCR into two different reactions: The first unspecific step called preamplification is performed with ThermoSequenase followed by adding a Taq polymerase for the second amplification step. In our experience the standard protocol produced sufficient probes for CGH studies with a homogeneous staining pattern of all chromosomes. The probes which were generated with ThermoSequenase seem to be less suitable for CGH studies because the hybridization results in an inhomogeneous staining pattern that is very difficult to interpret.

The quality of the DNA which is isolated from paraffin-embedded tissue depends on two important parameters: First, the use of 4% neutral buffered formaldehyde for the fixation; second, duration of the fixation should be no longer than 1 day.

Outline

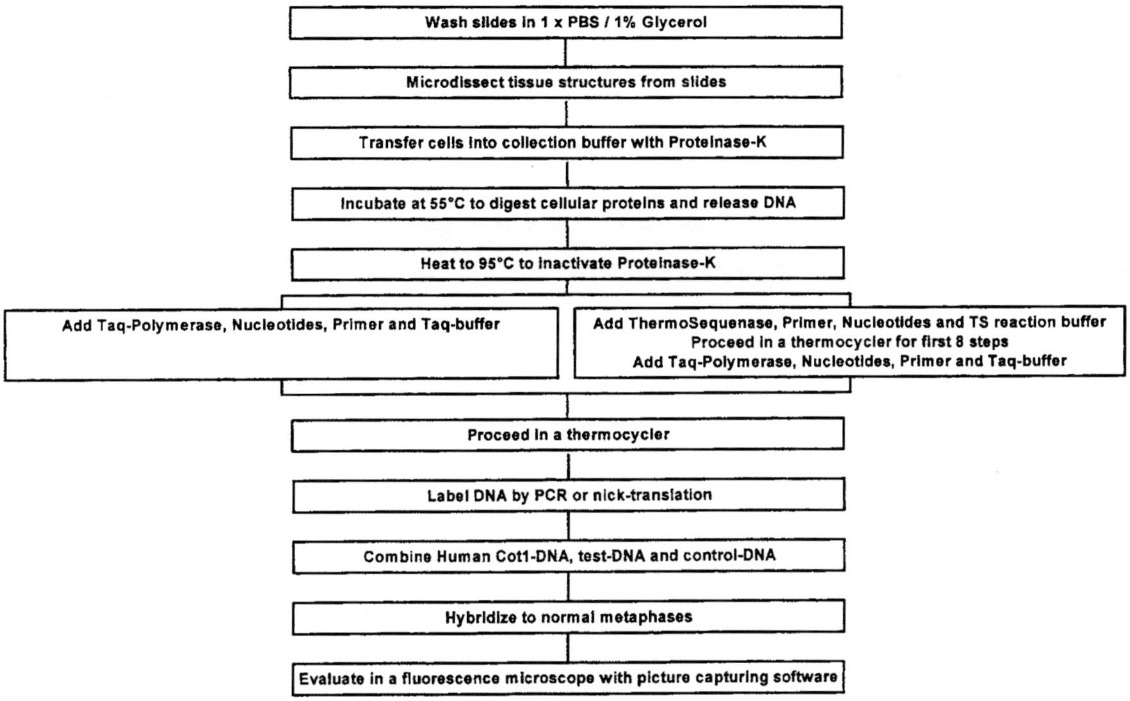

Fig. 1. An outline of the procedure is shown

Materials

- 10-μm-thick cryofixed tissue sections or 10-μm-thick routinely fixed (4% neutral buffered formaldehyde) paraffin-embedded tissue sections
- upright (e.g. Olympus BX 60 with long working distance objectives) or inverted bright-field and epifluorescence microscope equipped with a high-resolution micromanipulator (e.g. Eppendorf)
- image capturing and processing unit (e.g. MetaSystems ISIS) with high resolution b/w CCD camera (e.g. Photometrics)
- pipette puller (e.g. List-Medical)
- low melt borosilicate glass sticks to prepare glass needles (e.g. Hilgenberg Art. No. 7347765)

- collection buffer (see recipe)
- PCR mix (see recipe) containing biotinylated or fluorochrome-labeled nucleotides (e.g. Spectrum Orange- or FITC-dUTP)
- ThermoSequenase
- Mayer's hemalaun stain (Merck 1.09249.0500)
- 0.2 ml thin-walled microcentrifuge tubes
- thermal cycler with heated lid
- mineral oil

Collection buffer

- 50 mM Tris Base
- 1 mM EDTA pH 8.0
- 1% Tween 20
- 2 mg Proteinase K/ml buffer
- store at -20°C in 100 μl aliquots

Hybridization mixture

- 50% Deionized formamide
- 2×SSC
- 10% Dextran sulphate
- 50 mM Na phosphate buffer
- store at -20°C in 500 μl aliquots

Note: Exceptional care must be taken in preparing reagents for microdissection. Each solution should be prepared with ultrapure water (e.g. AMPUWA). All plasticware used should be UV irradiated and the glassware should be autoclaved. It is useful to aliquot all reagents for one-time use and store them at -20°C. When preparing solutions, wear gloves and, if possible, conduct all manipulations under a laminar air flow to minimize the risk of a contamination of extraneous DNA.

Subprotocol 1
Microdissection

Procedure

Preparation of needles for microdissection

1. Place the low melt borosilicate glass sticks into the pipette puller.
2. Pull a sufficient number of glass needles for the experiment planned.
3. Discard each needle after using.

   Note: The pipette puller settings will need to be optimized for any given instrument. The tips are required to be short and sharp for cutting out single cells from tissue sections.

Microdissection of cells from tissue sections

Deparaffinization of paraffin-embedded tissue sections

1. Wash 10-μm-thick routinely fixed (4% neutral buffered formaldehyde) paraffin-embedded tissue sections twice in xylol for 10 min.
2. Rehydrate the sections in a series of 50%-70%-90%-96% ethanol for 2 min each.
3. Wash 1 min in aqua dest.

   Note: Non-neutral formaldehyde or long-term storage of the specimens in formaldehyde during the fixation results in a poor preservation of DNA which is not suitable for an efficient homogeneous amplification with DOP-PCR.

Nuclear staining of cryofixed or paraffin-embedded tissue sections

1. Stain the slides in Mayer's hemalaun (dilution 1:5) for 1 min.
2. Rinse the slides briefly with aqua dest.
3. Incubate the slides for 5 min in tap water at 37°C.
4. Rinse the slides briefly with aqua dest.
5. Transfer the slides into a Coplin jar containing 40 ml phosphate buffered saline (PBS, pH 7.4) and 1% glycerol.

   Note: This step achieves an appropriately sticky consistency of the tissue for the following procedure.

Microdissection of cells from slides

1. Prepare enough microcentrifuge tubes for each experiment.

   Note: Three tubes are needed as controls; one for a reagent blank, one for a sham microdissection and one for a positive control.

2. UV irradiate the glass needles and the 0.2-ml thin-walled microcentrifuge tubes containing 6 µl cell collection buffer with 360 mJ/cm^2 to remove possible extraneous DNA contamination.

3. Cautiously place a glass needle into the micromanipulator with tweezers.

4. Microdissect 20-50 cells from an area of interest.

5. Transfer the cells which adhere to the glass-needle into a microtube containing 6 µl collection buffer.

   Note: All experimenters develop their own "style". The best results of a microdissection come from a tissue which is neither too humid nor too dry. Be careful with the transfer of the microdissected tissue into the collection buffer, the tips may also be broken, but it is important for the following procedures that the needle tip is within the solution.

6. For microdissection of each area of interest, place a new needle into the micromanipulator.

   Note: For a control, touch the needle onto an empty place of the slide (sham microdissection) and transfer the sham dissection to a blank tube. This will serve as a control for any contamination introduced during the microdissection.

7. Overlay with 15-20 µl sterile mineral oil for PCR to avoid evaporation of water (facultative).

8. Incubate cells at 55°C for 4 h to digest cellular proteins and release DNA.

9. Heat to 95°C for 15 min to inactivate proteinase-K.

Subprotocol 2
Alternative protocols for DOP-PCR amplification of released nuclear DNA

Procedure

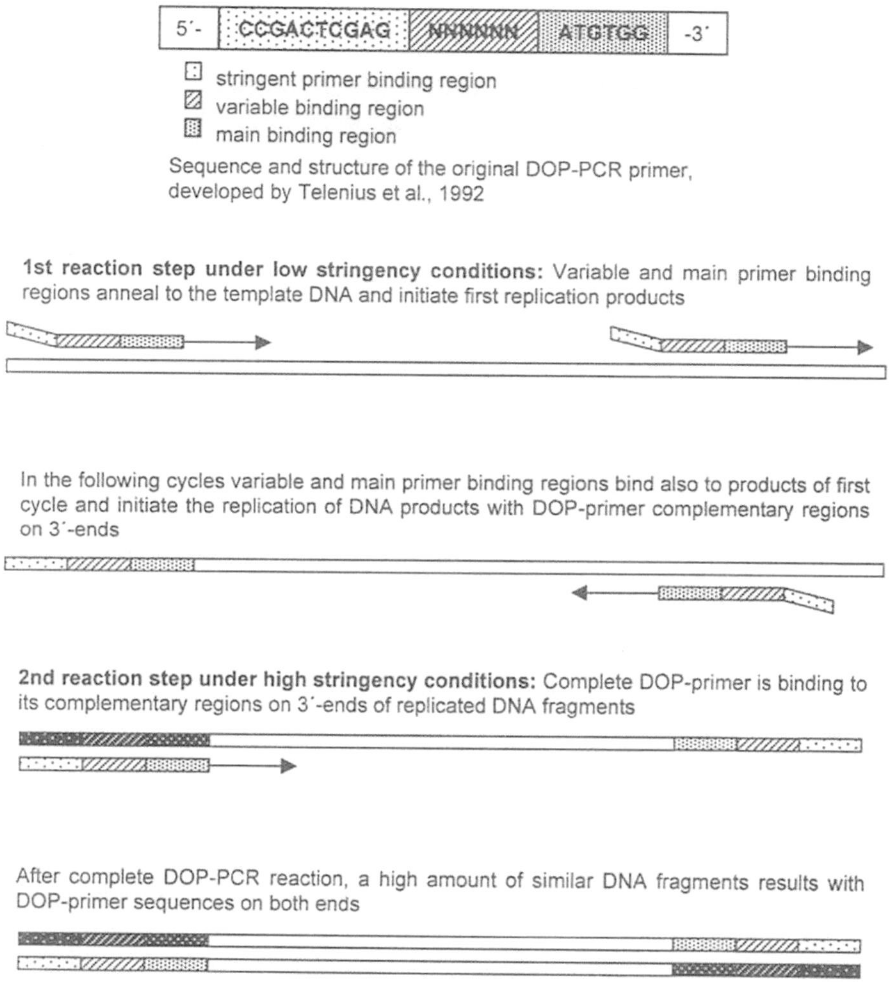

Fig. 2. DOP-PCR reaction mechanism

Standard PCR amplification protocol using Taq polymerase

1. Add to sampled and digested cells:
   - 2.5 µl Taq-buffer
   - 2.5 µl dNTP-mix (2 µM)
   - 2.5 µl DOP-primer (10 µM)
   - 0.5 µl Taq polymerase (2.5 units)
2. Fill with ultrapure water to a volume of 25 µl.
3. Proceed in a thermocycler according to the following PCR program:

94 °C	4 min	
94 °C	1 min	
30 °C	1.5 min	
Up to 72 °C	within 4 min	
72 °C	3 min	(8 cycles)
94 °C	1 min	
56 °C	1 min	
72 °C	3 min	(35 cycles)
72 °C	5 min	
20 °C	endless	

Two-step PCR amplification protocol using ThermoSequenase and Taq polymerase

1. Add to sampled and digested cells:
   - 1 µl ThermoSequenase reaction buffer
   - 1 µl dNTP-mix (2 µM)
   - 1 µl DOP-primer (10 µM)
   - 1 µl ThermoSequenase (4 units)
2. Proceed in a thermocycler according to the following PCR program:

94 °C	4 min	
94 °C	1 min	
30 °C	1 min	
Up to 72 °C	within 4 min	
72 °C	3 min	(8 cycles)
4 °C	endless	

3. Remove tubes from PCR machine and add the following components to each tube:
   - 5 μl Taq-buffer
   - 5 μl dNTP-mix (2 μM)
   - 5 μl DOP-primer (10 μM)
   - 1 μl Taq polymerase (5 units)
4. Fill with ultrapure water to a volume of 50 μl.
5. Proceed in a thermocycler according to the following PCR program:

94 °C	4 min	
94 °C	1 min	
56 °C	1 min	
72 °C	3 min	(30 cycles)
72 °C	5 min	
20 °C	endless	

Quality control of DOP-PCR amplification

1. When PCR is complete, analyze 5 μl of PCR product on a 1% agarose gel containing a low molecular weight DNA marker.
2. Stain the gel with ethidium bromide.
3. The fragment size of the amplified DNA should be in a range between 100 and 1500 base pairs in experimental and positive control lanes.
4. Store amplified DNA at -20°C until use.

Labeling of the PCR-amplified DNA

The DOP-PCR-amplified tumor DNA can be labeled by nick-translation or by PCR labeled directly with fluorochrome conjugated nucleotides (e.g. FITC-16-dUTP, Spectrum-Orange-dUTP) or indirectly with biotin (e.g. Biotin-16-dUTP) or digoxigenin (digoxigenin-11-dUTP).

Note: The labeling with fluorochrome conjugated nucleotides saves time during detection but often results in poor hybridization signals, which cannot be amplified. In our experience the labeling with biotin produces the best results and it is possible to enhance poor hybridization fluorescence by signal amplification.

Nick translation

1. Label an aliquot of 10 µl of PCR-generated pool DNA with a commercially available nick translation kit for in-situ hybridization according to standard procedure.
2. Label 1 µg normal reference DNA from a healthy donor with digoxigenin by nick-translation according to standard procedures.

PCR labeling

3. Transfer 5 µl of the PCR product to a 200 µl PCR reaction tube.
4. Add:
   - 2 µl Taq-buffer
   - 2 µl labeling dNTP-mix (2 µM dATP, dGTP, dCTP, 0.5 µM dTTP)
   - 2 µl labeled nucleotides (2 µM, e.g. bio-16-dUTP)

   Note: If labeled dUTP or dTTP nucleotides are not used, adjust the concentration of the corresponding unlabeled nucleotide to 0.5 µM in labeling dNTP-mix instead of reducing dTTP concentration.

   - 3 µl DOP-primer (10 µM)
   - 0.5 µl Taq polymerase (2.5 units)
5. Fill up to a volume of 25 µl with ultrapure water.
6. Proceed in a thermocycler according to the following PCR program:

94 °C	4 min	
94 °C	1 min	
56 °C	1 min	
72 °C	3 min	(25 cycles)
72 °C	5 min	
20 °C	endless	

Probe combination and hybridization

1. Transfer 10 µl (~ 500 ng) of labeled DNA into an 1.5-ml reaction tube.
2. Add 10 µl (~500 ng) of differently labeled reference DNA and 50 µg human unlabeled Cot1-DNA.
3. Precipitate DNA with a 2.5-fold volume of absolute ethanol and 1/10 volume of 3 M Na-acetate to the probe.
4. Incubate at -20°C for at least 2 h.

5. Centrifuge at 14,000 rpm and 4°C for 30 min.
6. Remove supernatant and wash pellet with 70% ethanol.
7. Spin down again at 14 000 rpm and 4°C for 15 min.
8. Remove supernatant and dry pellet.
9. Dissolve pellet in 10 µl hybridization mixture.
10. Hybridize the probe cocktail to normal human metaphases according to standard hybridization protocols for CGH, using an 18×18 mm coverslip.

Troubleshooting

Problem	Approach
PCR product too weak in control gel	increase number of PCR cycles
Strong PCR product in blank and/or sham control	discard and renew all reagents
No PCR product at all	use a new batch of Taq-Polymerase
High fluorescent background in hybridization	purify the PCR-product before hybridization using e.g. spin column
Weak hybridization signal	amplify signals with secondary antibodies, increase probe concentration
Reduced homogeneity of hybridization/ granulous signals	elongate hybridization time (up to 3 days), modify denaturation temperature and time and the pretreatment of human metaphases (e.g. pepsin digestion)

References

Aubele M, Mattis A, Zitzelsberger H, Walch A, Kremer M, Hutzler P, Höfler H, Werner M (1999) Intratumoral heterogeneity in breast carcinoma revealed by laser-microdissection and comparative genomic hybridization. Cancer Genet Cytogenet 110:94-102

Isola J, DeVries S, Chu L, Ghazvini S, Waldman F (1994) Analysis of changes in DNA sequence copy number by comparative genomic hybridization in archival paraffin-embedded tumor samples. Am J Pathol 145:1301-8

Jung V, Romeike BFM, Henn W, Feiden W, Moringlane JR, Zang KD, Urbschat S (1999) Evidence of focal genetic heterogeneity in glioblastoma multiformed by area specific CGH on microdissected tumor cells. J Neuropath Exp Neurol 58:993-999

Kallioniemi A, Kallioniemi OP, Sudar D, Rutovitz D, Gray JW, Waldman F, Pinkel D (1992) Comparative genomic hybridization for molecular cytogenetic analysis of solid tumors. Science 258:818-21

Kuukasjärvi T, Tanner M, Pennanen S, Karhu R, Visakorpi T, Isola J (1997) Optimizing DOP-PCR for universal amplification of small DNA samples in comparative genomic hybridization. Genes Chromosomes Cancer 18:94-101

Müller-Navia J, Nebel A, Schleiermacher E (1995) Complete and precise characterization of marker chromosomes by application of microdissection in prenatal diagnosis. Hum Genet 96:661-7

Speicher MR, du Manoir S, Schröck E, Holtgreve-Grez H, Schoell B, Lengauer C, Cremer T, Ried T (1993) Molecular cytogenetic analysis of formalin-fixed, paraffin-embedded solid tumors by comparative genomic hybridization after universal DNA-amplification. Hum Mol Genet 2:1907-14

Telenius H, Pelmear AH, Tunnacliffe A, Carter NP, Behmel A, Ferguson-Smith MA, Nordenskjold M, Pfragner R, Ponder BA (1992) Cytogenetic analysis by chromosome painting using DOP-PCR amplified flow-sorted chromosomes. Genes Chromosomes Cancer 4:257-63

Wiltshire RN, Duray P, Bittner ML, Visakorpi T, Meltzer PS, Tuthill RJ, Liotta LA, Trent JM (1995) Direct visualization of the clonal progression of primary cutaneous melanoma: application of tissue microdissection and comparative genomic hybridization. Cancer Res 55:3954-7

Comparative Genomic Hybridization

BURT G. FEUERSTEIN, YUICHI HIROSE, MALGORZATA PELLARIN, and GREGG MAGRANE

Introduction

Comparative genomic hybridization (CGH) allows the entire genome of an organism to be scanned for relative changes in DNA copy number (copy number aberrations) (1-3) in a single experiment. Standard CGH can detect aneuploidies, deletions, and unbalanced translocations at a resolution of 5-10 megabases. This is particularly useful in cases where DNA is unstable and where changes in copy number occur. Thus, tumors are often analyzed by CGH, but other applications could include diagnosis of other chromosome based illnesses.

The method typically uses targets that consist of normal metaphase spreads. The targets are hybridized with unlabeled CoT-1 DNA to block repeat sequences in the targets and with DNA from a test sample (usually tumor) labeled with one tag and DNA from a reference sample (usually normal DNA) labeled with a second tag. The ratio of the amounts of test and reference DNAs hybridizing to each location on the metaphase targets indicates the relative copy number of the test DNA at that location.

Since CGH only measures relative imbalances of genetic material, any chromosome aberration that does not result in a gain or loss of genetic

✉ Burt G. Feuerstein, University of California, San Francisco, Department of Neurological Surgery and Brain Tumor Research Center, Box 1631, San Francisco, CA, 94143-0808, USA (*phone* +415-353-9666; *fax* +415-353-9699; *e-mail* feuer@cc.ucsf.edu), University of California, San Francisco, Department of Laboratory Medicine, and Cancer Genetics Program; UCSF Cancer Center, San Francisco, CA, 94143-0808, USA
Yuichi Hirose, University of California, San Francisco, Department of Neurological Surgery and Brain Tumor Research Center, Box 1631, San Francisco, CA, 94143-0808, USA
Malgorzata Pellarin, University of California, Brain Tumor Research Center, San Francisco, CA, 94143-0808, USA
Gregg Magrane, University of California, Department of Laboratory Medicine, and Cancer Genetics Program, UCSF Cancer Center, San Francisco, CA, 94143-0808, USA

material will not be discerned. Thus, balanced translocations or inversions cannot be detected. In addition, since the technique relies on relative changes in DNA copy number, alterations in ploidy will not be sensed. Thus, a perfectly tetraploid sample cannot be distinguished from a perfectly diploid or triploid sample. Furthermore, since the resolution of the technique is 5-10 megabases, the method cannot identify mutations or small deletions or duplications.

Outline

The basics of the protocol for a CGH experiment are diagrammed in Fig. 1.

Subprotocol 1
Genomic DNA preparation from microdissected paraffin sections

Since CGH evaluates all DNA from a tissue, it is important to isolate DNA from a section of tissue that has disease. If DNA from normal tissue is included in the test sample, the sensitivity of the assay will be correspondingly reduced. After the appropriate tissue is identified, test DNA is isolated and labeled. For experiments on human material, reference DNA is most often prepared from peripheral blood, although DNA from any normal (i.e. nonneoplastic) tissue can be used.

The original descriptions of CGH specify isolation of large molecular weight DNA from frozen tissue. The original extraction protocol included steps to isolate and lyse nuclei, digest protein, and purify DNA with phenol/chloroform/isoamyl alcohol extractions and ethanol precipitations (2,4).

More recently, successful CGH protocols have been developed for degenerate oligonucleotide primer polymerase chain reaction (DOP-PCR) amplified and unamplified DNA isolated from fresh and from fixed tissues preserved in paraffin (5-7). We describe here a recent protocol to microdissect, isolate, amplify, and label DNA from small samples that are either fresh or fixed and preserved in paraffin. We also include a more traditional direct labeling method. It is important to remember that the reference and test DNAs used together in an experiment should be prepared in a similar fashion. Thus, if the test DNA is prepared using DOP-PCR, so should the reference DNA.

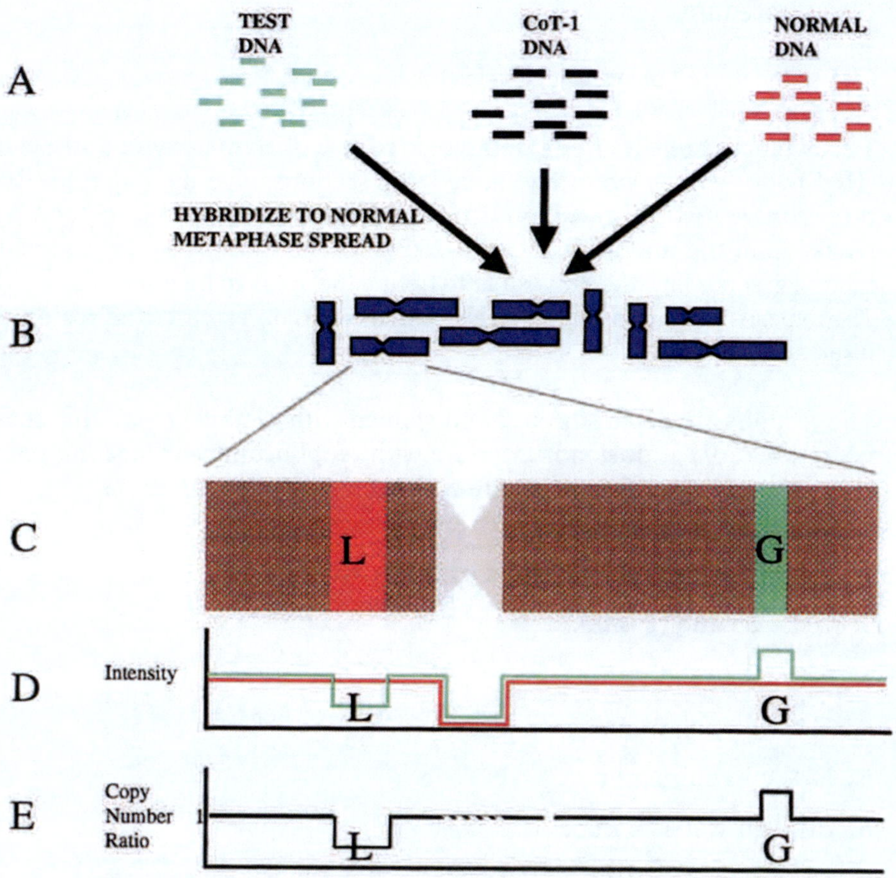

Fig. 1. Comparative Genomic Hybridization. **A** Test DNA, here tagged in green, reference DNA, here tagged in red, and unlabeled CoT-1 DNA are hybridized to **B** normal metaphase chromosomes. **C** The tagged sequences will produce signals at homologous sites on the metaphase chromosome. **D** If the intensity of test sequence (*green line*) measured at a location is less than the intensity of the reference sequence (*red line*) measured there, **E** the copy number ratio is less than one, and there is a loss (*L*) of genetic material at that location. If the intensity of test sequence measured at a location is greater than the intensity of the reference sequence measured there, the copy number ratio is more than one, and there is gain or amplification (*G*)

Procedure

Case selection, sectioning, and staining

Tissue blocks are cut after a pathologic review of Hematoxylin and Eosin (H&E) or other appropriately stained thin sections to verify diagnosis. Cut three consecutive 5-µm sections from a formalin-fixed paraffin-embedded block. Stain the middle slide with H&E and one or both of the rest with methyl green. Mark the H&E slide to define the areas to be microdissected. The experience and attention of the person who determines the region of interest is critical.

Methyl green staining of paraffin sections

The sections are deparaffinized and stained with methyl green. This stain is used because it does not interfere with amplification or labeling reactions. Hematoxylin does interfere with these reactions. After staining, the tissue can be stored for several weeks if necessary.

1. Xylene extraction ×2, 3 min each.
2. 100% ethanol, 2 min.
3. 95% ethanol, 2 min.
4. 70% ethanol, 2 min.
5. 50% ethanol, 2 min.
6. distilled water, 2 min.
7. Dip 3-6 times or leave 15-60 s in 0.1-1% methyl green (Sigma, St. Louis, MO). Start with the lower methyl green concentrations so the material is not too heavily stained.
8. Rinse in distilled water.
9. Incubate in distilled water 1-5 min depending on staining intensity.
10. Air dry upright.

Microdissection and digestion

Reagents

PCR/PK Buffer. 10× PCR buffer (100 mM TrisHCl pH 7.5, 2 mM $MgCl_2$, 500 mM KCl, 1 mg/ml gelatin) supplied with the Taq polymerase kit (Boehringer/Mannheim). PCR/PK buffer is PCR buffer supplemented with 0.5% Tween 20 and 0.5 mg/ml proteinase K (Life Technologies).

Procedures

1. Add PCR/PK Buffer into a 0.5-ml PCR tube. Use 60 μl for a 1-mm^2 sample and vary proportionally if the sample is larger or smaller.
2. Use a fresh blade for each region. Scrape the area corresponding to the marked area on the H&E slide from the methyl green slide. This is done by taking a small drop of PCR/PK buffer with a #15 blade or a #22 needle, placing the buffer onto the sample area, and scraping tissue into the center of the drop. When the sample dries enough to be picked up (don't overdry!), move the scrapings into the PCR tube, overlay with 20 μl mineral oil and parafilm close.
3. Incubate 3 days shaking at 120 rpm, 55 °C. Add fresh proteinase K (1 μg per 2.5 μl original volume) through the oil to the sample twice each day.
4. Remove all parafilm from tubes. Inactivate proteinase K by incubating 10-15 min at 95°C. Remove oil by rolling samples and oil on parafilm and pipette aqueous DNA into a new tube.

DOP-PCR and labeling for microdissected, Tween extracted DNA samples

Reagents

- DOP Primer: 5′-CCGACTCGAGNNNNNNATGTGG-3′ 22mer (Operon)
- DNA Sample: 1 ng of DNA from fixed samples prepared as noted above produces good enough DNA quality for excellent CGH profiles. We have been able to reproducibly obtain faithful CGH profiles using as little as 50 pg of DNA from frozen tissue.
- Buffer A
  - 2.5 μl 600 μM dNTP (300 μM final concentration)
  - 0.5 μl 10μM DOP primer (1 μM final concentration)
  - 1.0 μl 5× Sequenase Reaction Buffer (200 mM Tris-HCl pH 7.5, 100 mM $MgCl_2$
  - 250 mM NaCl) (Amersham, Cleveland, OH)
- Buffer B (Taq reaction. Boehringer Mannheim, Germany)
  - 16.6 μl 600 μM dNTPs (200 μM final concentration)
  - 7.0 μl 10 μM DOP primer (1.4 μM final concentration)
  - 5.0 μl 10× PCR buffer (Boehringer/Mannheim) (1.5 mM $MgCl_2$)
  - 2.0 μl 25 mM $MgCl_2$ (1 mM final concentration)
  - 0.5 μl Taq polymerase (2.5 units)
  - 13.9 μl H_2O

- Buffer C (PCR labeling reaction)
  - 8.3 µl 600µM dNTPs (100µM final concentration)
  - 7.0 µl 10µM DOP primer
  - 6.0 µl $MgCl_2$ (3 mM final concentration)
  - 5.0 µl 10× PCR buffer (1.5 mM $MgCl_2$)
  - 0.5 µl Taq polymerase (1.5 units)
  - 15.7µl H_2O

Low stringency (Sequenase) reaction

1. Dispense 4 µl Buffer A and 1 µl DNA sample to the appropriate tube, and overlay with 3 drops of mineral oil (total reaction volume 5 µl).
2. Denature at 96°C for 10 min.
3. Subject to 5 cycles of 30°C for 5 min, 37°C for 2 min, and 96°C for 2 min. Add 0.65 units of Sequenase at each 30° step. A thermostable sequenase now available (Amersham) may allow one to omit the latter additions of enzyme.

High stringency (Taq) reaction

1. Immediately after the Sequenase reactions have been completed, add 45 µl of Buffer B (total reaction volume 50 µl).
2. Denature at 96°C for 5 min.
3. Subject to 35 cycles of 94°C for 1 min, 56°C for 2 min, and 72°C for 2 min with autoextension of 1 s/cycle.
4. Final extension at 72°C for 5 min.
5. Soak samples at 4°C until retrieval.

PCR labeling reaction

6. Dispense 42.5 µl Buffer C, 2.5 µl digoxygenin-dUTP (50 µM final concentration) (Roche), and 5 µl DOP-PCR product into each tube (total reaction volume 50 µl). Overlay each sample with mineral oil.
7. Denature at 95°C for10 min.
8. Subject to 25 cycles of at 94°C for 70 s, 56°C for 70 s, and 72°C for 3 min with autoextension of 1 s/cycle.
9. Final extension at 72°C for 10 min.
10. Soak samples at 4°C until retrieval.
11. Reference DNA is amplified from 50 ng normal DNA and labeled as described above except that biotin (Boehringer/Mannheim) or fluorescein isothiocyanate (FITC)-dUTP (Du Pont) is substituted for DIG-dUTP.

12. Run 5 µl of sample on a 1-1.5% agarose gel. Samples amplified from fixed tissues are routinely smeared below 1000 bp. Unfixed DNA products usually range from 400-1500 bp.

Direct labeling by nick translation for CGH

Reagents

- 10× A4: 5 µl 10 mM dATP, 5 µl 10 mM dCTP, 5 µl 10 mM dGTP, (Gibco: final, 200 µM), 125 µl 1M Tris, pH 7.2 (final 500µM), 12.5 µl 1M $MgCl_2$ (final 200 µM), 1.7 µl 14.7 M mercaptoethanol (final 100 mM), 0.5 µl 50 mg/ml BSA (final 100 µg/ml), 95.3 µl ddH2O to final volume of 250 µl. Store at -20°C.
- Fluorescent dUTPs: These are available from a number of vendors. For example, Texas Red and FITC are available from DuPont/NEN and Dig-11-dUTP is from Roche. These should be spun down after thawing and before opening.
- DNAPol/DNAase I: This is the 10× enzyme mix from the Gibco Bionick kit (Life Technologies). If this enzyme cuts DNA too vigorously, then vary the amount of enzyme or the incubation time. Standard for fresh DNA is 3 µl of the 10× mix for 60 min. Incorporation decreases significantly at less than 2 µl or 40 min incubation. If the mixture is too active with these minimum amounts try the "slow" mix from the Gibco Nick Translation Kit (Life Technologies) or a mixture of the two ("slow" enzyme can also be ordered separately). Try using 5 µl for 60 min first and adjust conditions as needed. This "slow" enzyme is best for degraded or low molecular weight DNA.
- Fresh DNA: This should be high molecular weight.
- Paraffin DNA: Formalin-fixed DNA resists enzymatic cutting and nucleotide incorporation. We start by using the full 5 µl of fast enzyme for 90 min unless the DNA is easily cut. If DNA volume in Tris/EDTA is greater than 10 µl, more enzyme will be needed to compensate for the extra EDTA in the reaction.

Labeling

1. Prepare reaction mixtures in a total of 50 µl. We use approximately 1 µg fresh DNA and 1 µl of fluorochrome. Add enzymes last, gently vortex, and quickly spin down.
2. Incubate reaction mixtures 60 min at 15-16°C for 60 min.
3. Stop reaction by heating to 70-73°C for 15 min.

4. Run 3-5 μl of probe on a 1% agarose gel. Fresh DNA size should run from 300-2300 bp. Formalin-fixed DNA should run from approximately 200-600 bp.

5. Store labeled probes at -20°C.

Subprotocol 2
Metaphase production

The art of human metaphase preparation is outside the scope of this chapter. We refer the reader to Chapter 3 of this manual for details. However, a few general comments concerning human metaphases for CGH should be made.

After human lymphocyte cultures have been prepared, a cabinet that controls the temperature and the humidity of the environment and allows more reproducible spreads to be produced is used (Thermatron CDS-5 Cytogenetic Drying Chamber, Holland, MI) (8). We use a temperature of 25°C and vary the humidity from 40-55% depending on the quality of metaphase spreads. This can be monitored directly with a phase microscope, starting with the humidity at 55% and decreasing it stepwise until spreads are category II or category III (9, 10). From experience, the optimal humidity is near 47.5%. Since DAPI bands identify chromosomes for analysis, chromosome structure must be preserved in order to utilize them for CGH. The slides are aged for several days before storing in a freezer under nitrogen; the preparations can be used for more than six months.

If the experimenter does not wish to prepare his/her own human metaphases, fairly good spreads prepared specifically for CGH experiments can be purchased (Vysis, Inc. Downers Grove, IL). However, since methods for preparing mouse metaphases are not generally available, we provide a method here that we use for CGH.

Materials

Reagents

- Complete medium: Sterile filter 450 ml Dulbecco's modified Eagle medium (F-12/H-21 (1:1)), 50 ml fetal calf serum, 7.5 ml 1M HEPES, 5 ml 0.2 M glutamine, 5 ml pen/strep (10,000 U/ml penicillin and 10,000 mg/ml strepomycin).
- Hank's Balanced Salt Solution, Calcium and Magnesium Free

- Phosphate Buffered Saline
- STV (0.25% trypsin)
- Methotrexate
- Thymidine
- Colcemid

Procedure

Metaphases from Mouse Embryonic Fibroblasts

Day 1

1. Obtain two to six 12-day-old mouse embryos (C57 Black). If these need to be dissected from the uterus, work in a sterile Petri dish using sterilized dissecting instruments and sterile media. Cultures should all be prepared and maintained using sterile technique and a biological safety hood.
2. Place embryos into a 15 ml centrifuge tube. Add enough complete medium to cover. Using the tip of a 10 ml pipette, smash the embryos against the bottom of the tube repeatedly until a smooth homogenate is obtained. Pass the homogenate through an 18-gauge needle several times.
3. Transfer the homogenate to a sterile Petri dish and run the homogenate through the needle several more times.
4. Place 1 ml of the homogenate into a T-25 cell culture flask. Add 5 ml complete medium to the flask and gently swirl. Cap flask(s) and place horizontally in 37 °C 5% CO_2 incubator. Caps do not need to be loosened.

Day 2

1. Examine cells for healthy attachment to flask using inverted phase contrast microscope.
2. Aspirate complete medium to remove dead and floating debris and wash gently with 8 ml Hank's balanced salt solution, Ca^{++} and Mg^{++} free.
3. Replace with 8 ml complete medium and return flask to incubator.
4. Repeat the Day 2 procedure daily until the cell layer is confluent.

Metaphase production

1. Aspirate complete medium and gently wash cell layer with 8 ml 37°C Phosphate buffered saline, Ca^{++} and Mg^{++} free.
2. Place 3 ml 0.25% trypsin into each flask. Cap the flasks securely and place horizontally in incubator for 5 min.
3. Neutralize the trypsin by adding 6 ml complete medium; repeatedly aspirate and squirt the medium to detach the cells from the flask surface.
4. Split the cells from one T-25 into one T-75 flask and top off with media to 8 ml. Cap the flasks and return to the incubator.
5. Examine cells after about 4 h. They should be 30%-40% confluent. Reexamine cells periodically after18 h.
6. When the culture has reached 70-80% confluence (usually about 24 h), treat with 10^{-7} M methotrexate. For a culture that doubles every 24 h, treat the cells 17 h.
7. Release cells with 10^{-5}M thymidine. Incubate 2.5 h.
8. Add 0.02 µg/ml colcemid. Incubate cells 30 min.
9. Wash cells gently with phosphate buffered saline (37°C). Gently swirl and aspirate.
10. Add 2 ml 0.25% trypsin prewarmed to 37°C and incubate 5 min at 37°C.
11. Add 4 ml complete medium and wash the cells off the side of the flask. Place the cells into a 15-ml tube.

Making metaphase spreads

1. Centrifuge cells at 1000 rpm for 5 min. Remove supernatant, leaving about 100 µl medium.
2. Add 50 µl 0.075 M KCl (prewarmed to 37°C) to the tube and mix cells very gently with a pipette.
3. Slowly add three more 50 µl volumes of 0.075 M KCl to the cells, all the while mixing gently with a pipette.
4. Increase the volume added to 100 µl, adding three volumes while mixing gently. Repeat with 200 µl volumes, then 300 µl, 400 µl and 500 µl volumes. Bring the cell suspension to a total volume of 7 ml, and incubate at 37°C for 15 min.
5. Add 10 drops of fresh Carnoy's solution. Centrifuge cells at 1000 rpm for 5 min. Pour off supernatant.

6. Very slowly add 50 μl of fresh cold Carnoy's fixative (methanol and acetic acid, 3:1) down the side of the tube. Gently resuspend. Add ten more 50 μl volumes, ten 100 μl volumes, and ten 200 μl volumes of Carnoy's, pipetting after adding each volume, finally bringing the total volume to 4 ml. Incubate at 4°C for 20 min.
7. Centrifuge at 1000 rpm for 5 min. Gently pour off supernatant.
8. Repeat the Carnoy's fixing procedure (steps 6-7) twice.
9. After the final centrifugation, resuspend cells in about 200 μl of fixative.
10. Arrange the slides on a flat surface and drop 7 μl of cell suspension a slide. Two drops can be placed on one slide – one on each side. Allow the slides to dry.
11. Examine the slide with a phase contrast microscope to determine if the cell concentration needs adjustment and assess the success of metaphase production.
12. Age the slides at room temperature for 2 weeks. Then store under nitrogen in sealed plastic bag at -20°C.

Subprotocol 3
Hybridization

This phase constitutes the heart of the CGH experiment. Test and reference DNAs labeled with different fluorochromes are prepared as a hybridization mixture by precipitating and resuspending them together with CoT-1 DNA. The metaphases are denatured and incubated together with the hybridization mixture and then washed. This protocol assumes DOP-PCR amplified samples as described above.

Materials

Reagents

- 1× SSC: 0.15 M NaCl, 0.015 M sodium citrate, pH 7.0.
- Master hybridization solution (MM#1): 5 ml formamide, 1 g dextran sulfate, 1 ml 20× SSC, water to final volume of 7 ml. Suspend the dextran sulfate in formamide and quickly add SSC and water mixing vigorously. Heat 1-2 h at 70°C vortexing intermittently until dissolved. pH

to 7.0 if necessary. Bring volume to 10 ml. Store as 1 ml aliquots at -20°C.

- Denaturing solution: 70% formamide/2×SSC. Combine 70 ml formamide and 10 ml 20× SSC. pH to 7.0 if necessary and bring up to 100 ml with water. Store at 4°C. Prepare fresh weekly.
- PN solution: 0.1 M Na_2HPO_4, with 0.1% Non idet P-40. Add 0.1 M $NaH_2PO_4H_2O$/0.1% Nonidet P-40 to 0.1 M $Na_2HPO_47H_2O$/0.1% Nonidet P-40 to bring to pH 8.0. Store up to 1 month at room temperature.
- Blocking buffer: 1% bovine serum albumin in 4×SSC.
- DAPI mounting medium: Stock: Dissolve 1 mg 4',6-diamidino-2-phenylindole (DAPI) in 10 ml H_2O. (Add several drops of methanol to dissolve DAPI before adding water.). Store in foil wrapped tubes at -20 °C in appropriately sized aliquots. Stable for over 1 year.
  Working solution: Add 1-2 µl DAPI stock solution (0.1-0.2 µg/ml final concentration) to 1 ml phenylenediamine antifade mounting medium. Protect from light at -20°C. Replace as solution darkens.
- SSC wash solutions should be changed for each new experiment. The formamide solution can be reused for 3 washes or 20 slides before changing. Wash solutions should also be changed if there is excess unhybridized probe in the previous experiment. Make certain that slides are protected from light when using direct labeled probes.
- CoT-1 DNA quality varies from lot to lot. Often preparations can contain a large amount of very small DNA fragments that are useless for blocking repeat sequences. We have a special batch made, test it at 1 mg/ml, and put a large amount on reserve so that we can use the same lot for multiple experiments.

Procedure

Prepare hybridization mix

1. Mix sequentially 20 µl CoT-1 DNA (1 µg/µl), 40 µl of the labeled test and 20 µl of the labeled reference DNAs. The volumes can vary depending on the efficiency of labeling and the source of DNA. Add 1/10 volume 3 M sodium acetate (pH 5.3) and 2.5 volumes cold (-20°C) 100% ethanol. Vortex vigorously, and centrifuge at 14,000 rpm in a microfuge at 4°C for 30 min.

2. Dry the pellet and resuspend by vortexing vigorously in 7 μl MM1 and 3 μl distilled water. Recentrifuge 15 s in the microfuge.

Prepare metaphase slides immediately before hybridization

1. Inspect the slide to make certain there are enough (more than five) metaphases to evaluate. There should be little or no cytoplasm around the nuclei or metaphases. Outline these regions using a diamond pen.
2. Incubate no more than three slides at a time for 3-8 min in denaturing solution prewarmed to 75°C. Slides that have been aged for longer periods at room temperature or stored for longer periods, require longer denaturation times.
3. Dehydrate slides with successive 2 min incubations in 70%, 80%, 90%, and 100% ethanol, air dry, and preincubate for 1-2 min on a slide warmer set to 37°C.

Hybridization

1. Denature the hybridization mix in an open 80°C water bath for 6 min.
2. Apply 10 μl of the hybridization mix immediately to the marked area of the metaphase slide while on the slide warmer. Remove air bubbles by touching them with the corner of a 22 mm^2 coverslip.
3. Cover each slide with two coverslips. Use a wide-bore pipette or small syringe to seal the edges of the coverslips with rubber cement. Allow 5 min to set.
4. Incubate at 37°C for 1-2 days in a covered humid hybridization chamber.
5. Remove rubber cement from hybridized slides and remove cover slips by placing in 2× SSC (or allow to fall off in first wash).

Washes

1. Wash at 45°C: Three times; each 10-15 min in 50% formamide/2× SSC Two times; each 10 min in 2× SSC

Signal amplification

1. Incubate hybridized metaphase spread in blocking buffer for 5 min at room temperature.
2. Incubate in 1 μg/ml rhodamine-conjugated anti-DIG-antibody (Roche) in blocking buffer for 45 min at room temperature if the control is directly labeled. If the control is labeled with biotin, then simultaneously incubate with both the anti-DIG and avidin-FITC (Boehringer/Mannheim).

3. Wash slides by immersing for 10 min successively in 4× SSC, 4× SSC/0.1% Triton X-100, and again in 4× SSC. Immerse for 5 min each in two changes of PN and air dry.

Mounting

1. Apply ~25 µl of DAPI mounting solution per slide and drop a coverslip that covers the whole slide, taking care to expel any air. Coverslips should be stored in methanol and dried with a kimwipe or cleaned immediately before use.
2. The completed slides can be cleaned but the coverslip should not be disturbed. To clean slides, fold a paper towel in thirds. Place slides in the center of the towel, fold the edges over the slide, and blot excess DAPI by pressing on the covered slides. Constant pressure kept on one end of the coverslip will prevent movement. Then, using a kimwipe to tightly hold one end of the coverslip on the slide, fog the slide by breathing on it and wipe clean with another kimwipe. Repeat with the reverse side of the slide, and store them. Slides are now ready for image acquisition.

Results

Microscopy and imaging

Visual inspection

It is important to directly visualize CGH hybridizations using an appropriately equipped fluorescence microscope. Collecting images from poor quality hybridizations are, at best, a waste of time. At worst, they may generate erroneous results. The experimenter should choose four to six metaphases per experiment with relatively equivalent condensation, excellent spreading (minimal bending, touching, or overlapping chromosomes), negligible fluorescent background, smooth fluorescence distributed along the chromosome, and good DAPI chromosome banding.

Good DAPI banding is critical for chromosome identification. If the chromosomes are overdenatured or overdigested, hybridization may be excellent, but the chromosomes may be unrecognizable. Therefore, the experiment should balance between good hybridization and easy identification of chromosomes.

Excellent blocking of repeat sequences in chromosomes by CoT-1 DNA is crucial (11). The goal of a CGH experiment is to measure a hybridization that reflects the copy number of unique sequences in the test and reference DNAs. If repeat sequences on the metaphase chromosomes are not well blocked, then labeled repeat sequences will contribute fluorescence intensity to the result, and the sensitivity of CGH to changes in copy number of unique sequences will be reduced. If repeat sequences are not well blocked, heterochromatic regions will be brightly stained by one or both fluorochromes. This can often be resolved by changing the lot of CoT-1 DNA or by changing the reference DNA.

If the hybridization signal appears dim, inadequate probe concentration could be the problem. To overcome this, the hybridization can be repeated using more probe, or the DNA can be relabeled. Dim hybridization can also be caused by inadequate illumination. It is critical to make certain that the microscope is properly aligned and equipped with proper excitation and emission filters for the dyes being utilized.

Digital imaging

System equipment (12). Copy number can be grossly evaluated by direct visualization, but image analysis is far more sensitive and specific. Quantitative information to determine DNA copy number can only be obtained using digital imaging. A number of companies provide systems that collect and analyze CGH data. The system operator needs to be well versed in theory and performance: casual use is not recommended. A system generally consists of (1) a high quality fluorescence microscope equipped with high numerical aperture, plan corrected, chromatically corrected objectives; (2) dual or triple-band dichroic mirrors/emission filter blocks, usually in a filter slider or wheel, for accurate color registration between test and reference images. If each position for each color in the test and the reference images are not equivalent, then the numerator and denominator of fluorescence ratios may refer to different chromosome positions. This will produce inaccurate fluorescence ratios. Post processing the image files can produce accurate registration if conventional single-band filter combinations are utilized; (3) a properly selected CCD camera; (4) computer hardware and software to control image acquisition, process data, store data, and display results. The software should provide manual control over the acquisition time for each fluorochrome, the ability to display individual and color composite images, and the capacity to store raw image data in a convenient file format. Systems presently available have specialized software that allow automatic image processing and provide easy manipulation of image data. However, the operator should be familiar with the software and with expected data since he or she must confirm chromosome identification and choose appropriate metaphases and chromosomes for analysis.

Image quality (13). The images collected for CGH analysis should have certain characteristics. These include (1) low signal intensity in the heterochromatic regions of chromosomes 1, 9, and 16. (2) The ratio of the average foreground signal (within the segmentation mask) to the average background signal (outside of the mask) should be >1.5 for each fluorochrome. (3) The coefficient of variation for foreground intensity of the reference DNA image should be <20%. (4) Granular hybridizations should be avoided. Granularity produces rapid changes in intensity in foreground signal. (5) Only well-separated and straight chromosomes should be selected for analysis.

Data analysis The software must perform a number of functions that analyze the data collected by the CCD camera (14). *Segmentation* separates the image into regions that define the chromosomes as foreground. This is often done by thresholding the reference and DAPI images. Various routines are used to remove nuclei, separate chromosomes and refine chromosome edges. *Normalization* equalizes the median intensities of the test and the reference fluorescence intensities by setting each equal to one. This is done by subtracting an estimate of the background around each chromosome from each pixel in the chromosome to determine the corrected foreground intensities. The median intensity for the metaphase is determined from the corrected foreground intensities, and the inverse of that value is used to normalize the fluorescence intensity for each pixel in the metaphase foreground. Since the test and reference fluorescence intensities are each normalized to one, CGH can only detect relative values for alterations in copy number. Figure 2 shows the normalized image for microdissected and DOP-PCR amplified material from a grade II astrocytoma with -1p and -19q. Notice absence of red or green intensity at centromeres and the relatively high red intensity at 1p and 19q (arrows). *Ratio profile.* The normalized fluorescence intensities are integrated along slices orthogonal to the axis of each chromosome. Dividing the integrated test and reference values, point by point along each chromosome, defines the ratio profile. *Average ratio profiles.* The lengths of profiles for each chromosome are normalized to a standard length (15). Then the average fluorescence ratio and its variation among the individual ratio profiles are determined and displayed at each point along the chromosome. Figure 3 is the average ratio profile derived in part from the data shown in Fig. 2. There is loss on 1p and 19q. Note the relatively tight standard deviations.

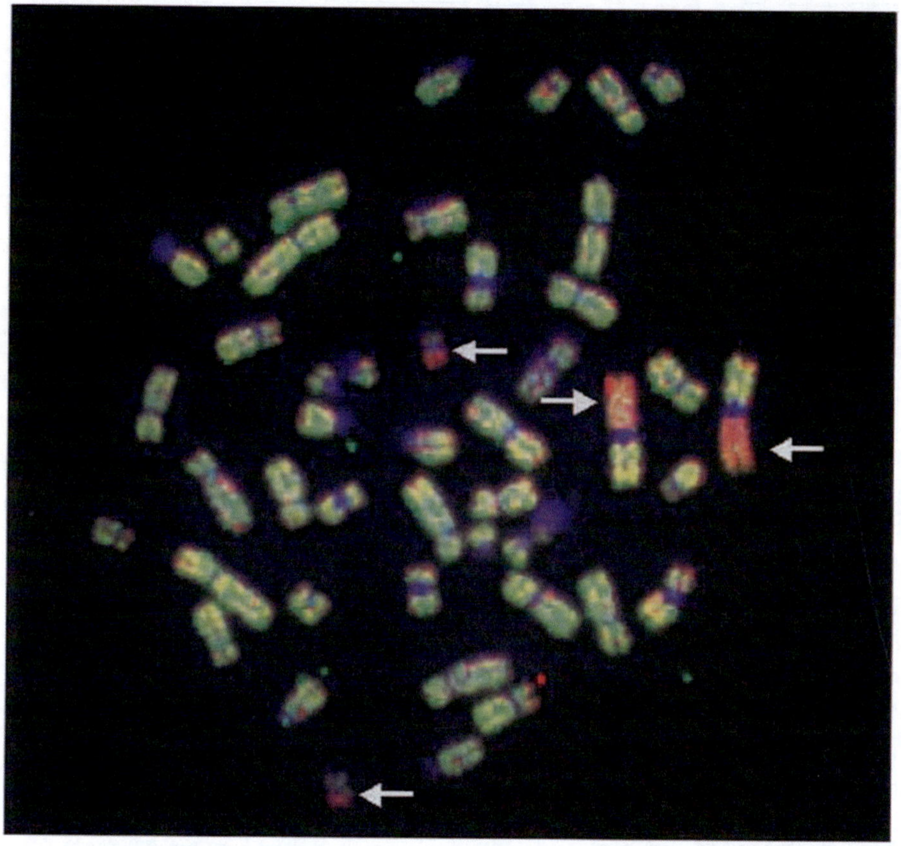

Fig. 2. Normalized fluorescence intensities from DOP-PCR amplified DNA microdissected from a grade II astrocytoma. Red is reference intensity and green is test intensity. The red and green intensities are attenuated at the centromeres and heterochromatic regions. Normalized green intensity is relatively low at 1p and 19q, leading to a redder color at these regions. The tumor had deletions at 1p and 19q

Interpretation

Data interpretation is a crucial part of each CGH experiment. The most important element of interpretation hinges on hybridization quality. As detailed above, fluorescence background, a smooth distribution of fluorescence along the chromosomes, and good DAPI chromosome banding must be evaluated prior to data collection. Centromeric and heterochromatic regions should not be interpreted. Since their unique sequence content is low, only small amounts of probe will hybridize to these regions.

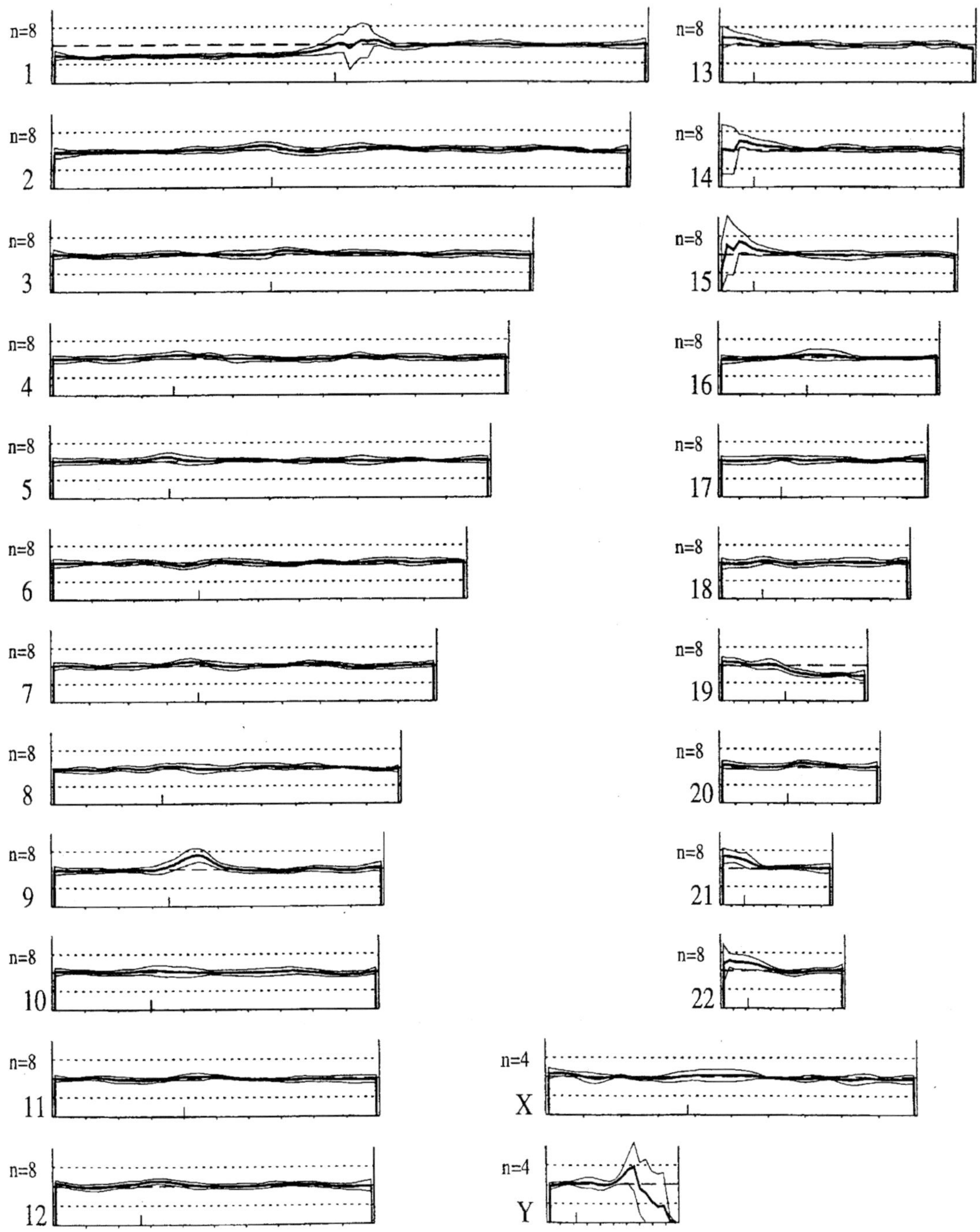
n=8
1
n=8
2
n=8
3
n=8
4
n=8
5
n=8
6
n=8
7
n=8
8
n=8
9
n=8
10
n=8
11
n=8
12
n=8
13
n=8
14
n=8
15
n=8
16
n=8
17
n=8
18
n=8
19
n=8
20
n=8
21
n=8
22
n=4
X
n=4
Y

This causes large fluctuations of the fluorescence intensity ratio in response to relatively small changes in normalized fluorescence intensity.

Controls should be included for each hybridization to monitor the quality of reagents. An important negative control is a normal versus normal hybridization. The result from this hybridization should produce no relative gains or losses. DNA from a known DNA sample with DNAs of interest is an excellent positive control.

After data collection, the quality of averaged profiles should be carefully appraised and unreliable data should be rejected. Three important items to consider are: (1) the coefficient of variation in evaluable regions of the chromosomes should in general be less than 10-15%. Standard deviations larger than this suggest poor hybridization quality. (2) Difficulties in defining chromosome ends may result in false gains or losses at the telomeres. This problem becomes apparent if the fluorescence ratio is abnormal at many telomeres in one experiment.(3) Copy number aberrations at the GC rich regions on chromosomes 1p, 19, and 22 must be interpreted with caution. Copy number aberrations that occur simultaneously at these locations in a single hybridization are particularly suspicious. If the hybridizations employ unamplified DNA, inverse hybridizations can be used to confirm copy number aberrations at these sites. These are repeat hybridizations where the labels for reference and test DNAs have been switched. The resulting ratios should be the inverse of the originals. If the hybridizations employ DNA amplified by DOP-PCR, inverse hybridizations do not work at the present time. In this case, the experiment needs to be repeated and the results confirmed. Other methods of confirmation such as fluorescence in situ hybridization or quantitative PCR should also be considered.

Before copy number aberrations in a sample can be determined, cutoffs that define the difference between normal and abnormal must be established. One method is to first evaluate DNA isolated from a set of normal tissues for copy number aberrations. These fluctuations can define the

Fig. 3. Average ratio profiles from DNA utilized in Fig. 2. Chromosome number from 1-22, X, and Y are to the *left* of each profile. *n* represents the number of chromosomes that were analyzed – 8 for each autosome, and 4 for each sex chromosome. The *X* axis represents distance along the chromosome. The *hatch mark* represents the centromere. The *Y* axis is the normalized fluorescence ratio. The *dashed line* represents a ratio of 1 (seen clearly at 1p and 19q). The *dotted lines* represent ratios of 1.5 and 0.5. Note the relatively small standard deviations over most of each chromosome. Large standard deviations are confined to centromeres and to heterochromatic regions near the centromeres at chromosomes 1, 9, 16, and Y. The relative losses at 1p and 19q are evident

limits of normal. It then remains to quantify how the fluorescence ratio of DNA extracted from genetically abnormal tissue compares with that extracted from the normal tissue. A statistical treatment of this method has been published (16).

Conclusions

Comparative genomic hybridization scans the genome of an organism for relative changes in copy number. DNA from fresh frozen or from fixed and paraffin embedded tissue can be utilized. DNA preparation, labeling, metaphase preparation, hybridization, and analysis are important steps for a successful experiment. Expert and knowledgeable support for these procedures is important to obtain reproducible and accurate results.

References

du Manoir, S., Speicher, M. R., Joos, S., Schrock, E., Popp, S., Dohner, H., Kovacs, G., Robert-Nicoud, M., Lichter, P., and Cremer, T. Detection of complete and partial chromosome gains and losses by comparative genomic in situ hybridization, Hum Genet. 90: 590-610, 1993.

Kallioniemi, A., Kallioniemi, O. P., Sudar, D., Rutovitz, D., Gray, J. W., Waldman, F., and Pinkel, D. Comparative genomic hybridization for molecular cytogenetic analysis of solid tumors, Science. 258: 818-21, 1992.

Kallioniemi, O. P., Kallioniemi, A., Sudar, D., Rutovitz, D., Gray, J. W., Waldman, F., and Pinkel, D. Comparative genomic hybridization: a rapid new method for detecting and mapping DNA amplification in tumors, Semin Cancer Biol. 4: 41-6, 1993.

Mohapatra, G., Kim, D. H., and Feuerstein, B. G. Detection of multiple gains and losses of genetic material in ten glioma cell lines by comparative genomic hybridization, Genes Chromosomes Cancer. 13: 86-93, 1995.

Kuukasjarvi, T., Tanner, M., Pennanen, S., Karhu, R., Visakorpi, T., and Isola, J. Optimizing DOP-PCR for universal amplification of small DNA samples in comparative genomic hybridization, Genes Chromosomes Cancer. 18: 94-101, 1997.

Telenius, H., Carter, N. P., Bebb, C. E., Nordenskjold, M., Ponder, B. A., and Tunnacliffe, A. Degenerate oligonucleotide-primed PCR: general amplification of target DNA by a single degenerate primer, Genomics. 13: 718-25, 1992.

Sanchez-Cespedes, M., Cairns, P., Jen, J., and Sidransky, D. Degenerate oligonucleotide-primed PCR (DOP-PCR): evaluation of its reliability for screening of genetic alterations in neoplasia, Biotechniques. 25: 1036-8, 1998.

Spurbeck, J. L., Zinsmeister, A. R., Meyer, K. J., and Jalal, S. M. Dynamics of chromosome spreading, Am J Med Genet. 61: 387-93, 1996.

Dracopoli, N. C. Cytogenetics. Current protocols in human genetics, pp. 2 v. New York, NY: John Wiley & Sons, 1994.

Donlon, T. A. Practical Approaches to In Situ Hybridization, Karyogram. 12: 3-10, 1986.

Mohapatra, G., Moore, D. H., Kim, D. H., Grewal, L., Hyun, W. C., Waldman, F. M., Pinkel, D., and Feuerstein, B. G. Analyses of brain tumor cell lines confirm a simple model of relationships among fluorescence in situ hybridization, DNA index, and comparative genomic hybridization, Genes Chromosomes Cancer. 20: 311-9, 1997.

du Manoir, S., Kallioniemi, O. P., Lichter, P., Piper, J., Benedetti, P. A., Carothers, A. D., Fantes, J. A., Garcia-Sagredo, J. M., Gerdes, T., Giollant, M., and et al. Hardware and software requirements for quantitative analysis of comparative genomic hybridization, Cytometry. 19: 4-9, 1995.

du Manoir, S., Schrock, E., Bentz, M., Speicher, M. R., Joos, S., Ried, T., Lichter, P., and Cremer, T. Quantitative analysis of comparative genomic hybridization, Cytometry. 19: 27-41, 1995.

Piper, J., Rutovitz, D., Sudar, D., Kallioniemi, A., Kallioniemi, O. P., Waldman, F. M., Gray, J. W., and Pinkel, D. Computer image analysis of comparative genomic hybridization, Cytometry. 19: 10-26, 1995.

Morton, N. E. Parameters of the human genome, Proc Natl Acad Sci U S A. 88: 7474-6, 1991.

Moore, D. H., 2nd, Pallavicini, M., Cher, M. L., and Gray, J. W. A t-statistic for objective interpretation of comparative genomic hybridization (CGH) profiles cytometry 20:183-90, 1997.

Part D

Special FISH Techniques (Fiber-FISH, PNA-FISH, Co-FISH, NU-FISH)

High Resolution FISH of Stretched Chromosome Fibers

BERND RAUTENSTRAUSS and CHRISTINA FUCHS

Introduction

High resolution fluorescence in situ hybridization (FISH), so-called FIBER-FISH, was first described by Heng et al. (1992) for long stretched DNA-fibers and was later subjected to modifications by other investigators (Fidlerova et al. 1994, Heiskanen et al. 1996). The resolution using cosmids on DNA fibers was restricted by the length of the visible, decondensed DNA strand up to 150 to 450 kb. Using earlier protocols, it was not possible to obtain a coherent DNA strand showing a longer section of cosmids in direct orientation. However, freshly prepared lymphocyte suspension could be used (for lymphocyte preparation see Arnoldus et al., 1990, Verma et al., 1989).This method described here is simple, but very efficient. It allows the presentation of cosmids over a region up to 8 Mb in a linear orientation. Thus, the visualization of the transcriptional orientation of genes is possible within a bigger range. Moreover, the modified treatment allows the use of archival suspensions of cultured lymphocytes and is also available for YACs. The main steps are a special treatment of up to two year old lymphocyte suspensions, the disruption of nuclei by a shortened PBS incubation previous to alkaline treatment and the release of chromosome fibers from interphase nuclei by special handling of the slides. With these modifications, we were able to obtain released chromosome fibers, which were suitable for a high resolution two-color FISH. This method has been specially tested for lymphocyte nuclei (Fuchs et al., 1997).

✉ Bernd Rautenstrauss, Institut für Humangenetik, Schwabachanlage 10, Erlangen, 91054, Germany (*phone* +49-9131-8532352; *fax* +49-9131-209297; *e-mail* berndwr@humgenet.uni-erlangen.de)
Christina Fuchs, Institut für Humangenetik, Schwabachanlage 10, Erlangen, 91054, Germany

Fig. 1. Typical high resolution FISH on a stretched chromosome fiber of chromosome 17. The blue color of the DNA results from diamidinophenylindol (DAPI) counterstaining. The images were captured on a Zeiss Axioplan microscope using the ISIS digital FISH imaging system (Metasystems, Altlussheim, Germany using a XC77CCD camera with on-chip integration (Sony). The green signals indicate cosmids specific for the proximal and distal CMT1A-REP elements which are flanking the so-called CMT1A/HNPP region on chromosome 17p11.2-12. The red signal in between is produced by the cosmid indicating the PMP22 gene (Patel et al., 1994, Murakami et al. 1996). The distance between the green signals is 1.4 Mb

Procedure

Preparation of chromosome fibers

1. The lymphocyte suspension is centrifuged (1000 rpm, 10 min, room temperature) and the supernatant is discarded. New fixative (= methanol/glacial acetic acid 3:1, -20°C) is added and if the suspensions are older than one year then the procedure is repeated. The number of cells is determined by a Fuchs-Rosenthal-haemocytometer and the final concentration should be 10^6nuclei per ml (see comment 1).
2. On a clean, moist slide a circle of 2 cm in diameter is drawn with a diamond pencil. In this circle 20 µl of lymphocyte suspension is placed and after air drying for 20 s, slides are incubated in phosphate-buffered saline (PBS) solution at room temperature. Fresh lymphocyte suspensions (up to 6 months) are incubated for 1 min, older ones (6 months and more) only for 30 s (see comment 2).
3. After removing the PBS-solution (nevertheless the slides should not dry out), 100 µl of 0.5 M NaOH/30% ethanol is placed within the circle for 30 s to disrupt the nuclei. Care is taken to avoid movement of the fluid out of the circle at any time. Then 100 µl of fresh methanol is added onto the circle in the same way. After air drying, the slides

are scanned using a light microscope. The disrupted nuclei have a compact short ‘tail’ (=DNA fibers) (see comments 3 and 4).

4. The slides are incubated over night at 60°C and then stored at -20°C (see comment 5).
5. FISH was carried out according to standard protocols (see below).

Fluorescence in situ hybridization (FISH) according to Thomas Liehr

1. Add 100 µl denaturation-buffer to the slides and cover with (24×50 mm) coverslips. [Denaturation-buffer: 70% (v/v) deionized formamide, 10% (v/v) filtered double distilled water, 10% (v/v) 20×SSC, 10% (v/v) phosphate buffer; make fresh as required; phosphate buffer: prepare 0.5 M Na_2HPO_4 and 0.5 M NaH_2PO_4, mix these two solutions (1:1) to get pH 7.0, then aliquot and store at -20°C].
2. Incubate slides on a warming plate for 12 min at 75°C.
3. Remove the coverslips immediately with forceps and place slides in a Coplin jar filled with 70% ethanol (4°C) to conserve target DNA as single strands.
4. Dehydrate slides in ethanol (70%, 90%, 100%, 4°C, 3 min each) and air dry.
5. For each slide to be hybridized, add 200 ng of labeled YAC-, BAC- or cosmid- DNA or 1.5 µl of biotin or digoxigenin labeled chromosome specific satellite probes plus 3 µl of 1 µg/µl COT1-DNA (Gibco) to 20 µl of the hybridization-buffer in a 1.5 ml microtube, vortex and spin down. (Hybridization-buffer: Dissolve 2 g dextran sulfate in 10 ml 50% deionized formamide/2×SSC/50 mM phosphate buffer for 3 h at 70°C. Store aliquots at -20°C.)
6. Denature probe-solution at 75°C for 5 min and cool immediately on ice to conserve probe DNA in single strands for satellite probes or do a prehybridization step at 37°C for 30 min for the other probes (YACs, BACs, cosmids).
7. Add 20 µl of probe-solution onto each denatured slide, put 24×50 mm cover slips on the drops and seal with rubber cement (Fixogum; Marabu).
8. Incubate slides for 3 nights at 37°C in a humid chamber.

9. Take the slides out of 37°C, remove rubber cement with forceps and cover slips by letting them swim off in 4×SSC/0.2%Tween (RT, 100 ml Coplin jar).
10. Postwash the slides 3×5 min in formamide-solution (45°C) followed by 3×5 min in 2×SSC (37°C) in a 100-ml Coplin jar, with gentle agitation.
11. Put the slides in 4×SSC/0.2% Tween (100 ml, RT), for a few seconds.
12. Add 50 µl of solution 1 to each slide, cover with 24×50 mm coverslips and incubate at 37°C for 30 min in a humid chamber. [Solution 1: FITC-avidin (CAMON Vector Laboratories A2011)/4×SSC/0.2% Tween/5% BSA (1:300 both Sigma); make fresh as required].
13. Remove the coverslips and wash 3×3 min in 4×SSC/0.2% Tween (RT, with gentle agitation).
14. Add 50 µl of solution 2 to each slide, cover with 24×50 mm coverslips and incubate at 37°C for 75 min in a humid chamber. [Solution 2: Biotinylated antiavidin (CAMON Vector Laboratories BA0300)/anti-digoxigenin-rhodamine (Boehringer Mannheim, 1207750)/4×SSC/0.2% Tween/5% BSA (1:20:100); make fresh as required].
15. Repeat step 13.
16. Repeat step 12.
17. Repeat step 13.
18. Counterstain the slides with DAPI-solution (100 ml in a Coplin jar, RT) for 8 min. (DAPI-solution: Dissolve 5 µl of DAPI (4,6-diamidino-2-phenylindol.2HCl stock-solution; Serva 18860) in 100 ml 4×SSC/0.2% Tween; make fresh as required).
19. Wash slides several times in deionised water for a few seconds and air dry.
20. Add 15 µl of antifade Vectashield (CAMON Vector Laboratories H1000), cover with cover slips and look at the results under a fluorescence microscope.

Comments

1. A fixative change is necessary for preparing the lymphocyte nuclei for the alkaline disruption. The concentration of nuclei should be at least 10^6/ml. More (up to 10^8) can increase the DNA fiber recovery.
2. If only small amounts of tissue are to be digested (see comment 1) 0.2–0.4 ml of PK-solution are sufficient.
3. There should be no movement of the solutions on the slide. The slides should always be in a horizontal position. Otherwise the DNA fibers are stretched too much. The diamond circle is necessary to avoid movement of solutions on the slide.
4. The slides should always be controlled after the procedure to control the recovery of fibers and the correct size of the fibers. If there are enough cracked nuclei with a 'tail' of DNA fibers, finish the procedure.
5. Slides can be stored up to 1 year.

References

Arnoldus, E.P.J., Wiegant, J., Noordermeer, I.A., Wessels, J.W., Beverstock, G.C., Grosveld, G.C., van der Ploeg, M., Raap, A.K. (1990) Cytogent. Cell Genet. 54, 108-111

Fidlerovà H., Senger, G., Kost, M., Sanseau, P., Sheer, D. (1994) Cytogenet. Cell Genet. 65, 203-206

Fuchs, C., Liehr, T., Rautenstrauss, B. (1997) Trends in Genet. Online, TTO 1114

Heiskanen, M., Peltonen, L., Palotie, A. (1996) Trends in Genet. 12, 379-382

Heng, H.H.Q., Squire, J., Tsui, L.-C. (1992) Proc. Natl. Acad. Sci. 89, 9509-9513

Liehr, T., Thoma, K., Kammler, K., Gehring, C., Ekici, A., Bathke, K.-D., Grehl, H., Rautenstrauss, B. (1995) Appl. Cytogenet. 21, 185-188

Murakami, T., Lupski, J.R. (1996) Genomics 34, 128-133

Verma, R.S., Babu, A. (1989) Manual of Basic Techniques, pp. 5-9, Pergamon Press

Quantitative DNA Fiber Mapping

Heinz-Ulrich G. Weier

Introduction

High resolution physical maps have become indispensable for the positional cloning of disease genes and large scale sequencing projects. Common maps are based on ordered sets of clones from sources such as cosmid, P1/PAC/ BAC, or yeast artificial chromosome (YAC) libraries. The assembly of such maps is facilitated by application of fluorescence in situ hybridization (FISH). Hybridization of non-isotopically labeled probes onto preparations of DNA molecules ('DNA fibers') that were bound with one or both ends to a solid substrate and stretched homogeneously, forms the base of our 'Quantitative DNA Fiber Mapping (QDFM)' technique. Because the DNA fibers are easily accessible to probes and detection reagents, hybridization efficiencies are typically high and allow DNA targets as small as 500-1000 bp to be detected routinely. Quantitative DNA fiber mapping experiments require only standard laboratory equipment and access to a fluorescence microscope. By hybridizing one clone onto another, the extent and orientation of overlaps can be quantitated with near kilobase resolution. To measure the physical distance between non-overlapping DNA fragments, probes are hybridized to DNA fibers representing a larger genomic interval (Fig. 1). This also allows the mapping of expressed sequences (cDNAs) along DNA fibers representing genomic DNA.

Heinz-Ulrich G. Weier, University of California, E.O. Lawrence Berkeley National Laboratory, Department of Subcellular Structure, Life Sciences Division, M.S. 74-157, 1 Cyclotron Road, Berkeley, CA, 94720, USA (*phone* +510-486-5347; *fax* +510-486-5343; *e-mail* ugweier@lbl.gov)

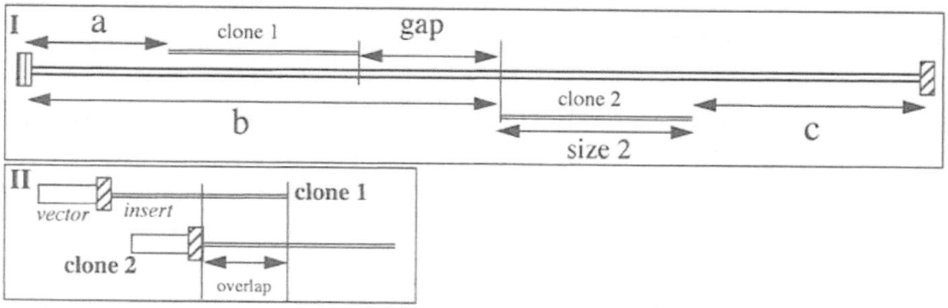

Fig. 1. Physical mapping strategies. I Mapping a small DNA molecule on to a larger molecule; II mapping of clone overlap by pairwise hybridization

Outline

- Isolate DNA
- Pretreat microscope slides or coverslips
- Prepare DNA fibers on glass
- Select probe set
- Denature and hybridize probe set
- Remove unbound probe
- Detect bound probes with antibodies
- Acquire images
- Analyze images
- Save results in spreadsheets for further analysis

Principles and applications

High resolution physical maps have proven indispensable for large-scale, cost-effective gene discovery. Knowledge about the extent of overlap between any two clones and the precise localization of cloned DNA fragments within much larger genomic fragments is needed to assemble such maps. As demonstrated in this chapter, FISH can provide this critical information.

Isolation of DNA from cell nuclei and preparation of some sort of chromatin 'fibers' improves the accessibility of the DNA targets for probes as well as detection reagents, and thus increases the hybridization efficiency. Furthermore, if the DNA molecules can be stretched in some way, they may provide linear templates for visual mapping. FISH applied in the past to various types of crude DNA preparations allowed visualization of probe overlap and it provided some information about the existence and size of gaps between clones. However, none of those techniques provided sufficiently accurate information about the extent of clone overlap or the separation between elements in the map because the chromatin onto which clones were mapped was condensed to varying degrees from site to site.

We demonstrated previously that cloned DNA fragments can readily be mapped by FISH onto DNA molecules prepared by the hydrodynamic action of a receding meniscus and, referring to its quantitative nature, we termed the technique 'Quantitative DNA Fiber Mapping (QDFM)' (Weier et al. 1995). In QDFM, a solution of DNA molecules is placed on a glass or mica surface prepared so that some DNA molecules attach at one or both ends. The DNA solution is then spread over a larger area by placing a coverslip on top, and additional DNA molecules are allowed to bind to the surface. During drying, the molecules are straightened and uniformly stretched by the hydrodynamic action of the receding meniscus. Molecules prepared in this manner are stretched with remarkable homogeneity. A properly stretched molecule should extend about ~2.3 kb/μm, i.e., approximately 30% over the length predicted for a double stranded DNA molecule of the same size (Weier et al. 1995). QDFM can be applied to DNA molecules ranging in size from a few kb to more than 1 Mbp, which allows mapping of small probes with near kilobase resolution onto entire yeast chromosomes and large (mega)YAC clones (Wang et al. 1996, Duell et al. 1997).

Applications of QDFM extend beyond map assembly and can provide valuable information for quality control, clone validation, definition of a minimal tiling path as well as for the sequence assembly process. Furthermore, due to its high hybridization efficiency obtained with DNA fibers, QDFM is also the method of choice for high resolution optical mapping of expressed sequences in genomic intervals defined by the DNA fibers.

Materials

Equipment

- Refrigerated Centrifuge (MP4R, IEC)
- Dry bath (Model 2001, Labline Instruments)
- Fluorescence microscope (Axioskop, Zeiss) equipped with 40× and 63× oil immersion lenses
- Incubator oven (set to 37°C, Precision Scientific)
- Pulsed Field Gel Electrophoresis (PFGE) system (Biorad)
- Shaking Incubators (New Brunswick): 30°C for yeast cell culture, 37°C for culture of *E. coli*
- Thermal cycler for in vitro DNA amplification
- Water bath (Model 188, Precision Scientific)
- Digital Imaging system (optional)

Reagents

- 3-Aminopropyltriethoxy silane (APS, Sigma Chemicals)
- β-Mercaptoethanol (Sigma Chemicals)
- β-Agarase (New England Biolabs (NEB))
- Agarose (Life Technologies)
- Antibodies against digoxigenin, rhodamine-conjugated, made in sheep (Roche Molecular Biochemicals), stock solution is 1 mg/ml in PNM, dilute 1:50 with PNM prior to use. Store at 4°C.
- Antibodies against FITC, made in mouse (DAKO), stock solution is 1 mg/ml in PNM, dilute 1:50 prior to use. Store at 4°C.
- Anti-mouse antibodies, FITC conjugated, made in horse (Vector Labs), stock solution is 1 mg/ml in PNM, dilute 1:50 prior to use. Store at 4°C.
- Anti-avidin antibodies, biotinylated, made in goat (Vector Labs), stock solution is 1 mg/ml in PNM, dilute 1:50 prior to use. Store at 4°C.
- Avidin conjugated to AMCA (Vector Labs), stock solution is 2 mg/ml in PNM, dilute 1:500 prior to use. Store at 4°C.
- Blocking reagent: cat. #1096-176 (Roche Molecular Biochemicals)
- Chloroform/isoamyl alcohol: 24:1 vol:vol. (Life Technologies)

- 4, 6-diamino-2-phenylindole (DAPI) (Calbiochem), 0.05 µg/ml in antifade solution. Store at -20°C.
- dATP, dCTP, dGTP, dTTP: 100 mM each (Roche or Pharmacia). Store at -20°C.
- Digoxigenin-11-dUTP: 1 mM (Roche Molecular Biochemicals). Store at -20°C.
- 10× dNTP mix: dATP, dCTP, dGTP, and dTTP, 10 mM each.
- EDTA (ethylenediamine tetraacetic acid): 0.5 M (pH 8.0) (Life Technologies).
- Ethidium bromide, 10 mg/ml (Life Technologies)
- Fluorescein avidin DCS (avidin-FITC, Vector), dilute stock solution (2 mg/ml in PNM) 1:100 with PNM. Store at 4°C.
- Fluorescein-12-dUTP, 1 mM (Roche Molecular Biochemicals). Store at -20°C.
- Formamide (FA, Life Technologies or Roche). Store at 4°C.
- Glycogen, 20 mg/ml (Roche Molecular Biochemicals). Store at -20°C.
- Geneclean II kit (BIO 101)
- Human COT1 DNA, 1 mg/ml (Life Technologies). Store at -20°C.
- Lambda phage DNA (Roche or Life Technologies). Prepare 2.5 ng/µl in 2×SSC. Store at 4°C.
- Low melting point agarose (Biorad)
- Lysozyme (Sigma Chemicals): prepare stock (50 mg/ml in 10 mM Tris, pH 7.5), and store in aliquots at -20°C. Do not refreeze.
- 10x PCR buffer: 15 mM $MgCl_2$, 100mM TrisHCl, 500 mM KCl, 0.01% gelatin.
- Phenol/Chloroform/Isoamyl alcohol: 25/24/1(vol./vol./vol.)(Life Technologies). Store at 4°C.
- Random-Priming Kit: BioPrime kit (Life Technologies). Store at -20°C.
- Proteinase K (Roche Molecular Biochemicals): 20 mg/ml in 10 mM TrisHCl, pH 7.5. Store at -20°C.

- RNase (Roche Molecular Biochemicals), DNase-free: boil at 100°C for 10 min., aliquot and store at -20°C
- Salmon sperm DNA (3'-5', Boulder, CO), 20 mg/ml. Store at -20°C.
- Sodium dodecyl sulfate (SDS)(Na salt, Sigma Chemicals): 10% in water.
- Thermus aquaticus (Tag) DNA polymerase, 5 U/µl (Perkin Elmer). Store at -20°C.
- Ultrapure water (Mallinckrodt, cat. # H453)
- Yeast artificial chromosome (YAC) library (Research Genetics). Store at -80°C.
- YOYO-1 (Molecular Probes): Stock is 1mM in DMSO. Dilute 1:1000 with water prior to use. Store at -20°C and discard diluted dye after 1 week.
- Zymolase (Sigma Chemicals, 70,000 U/g): prepare 10 mg/ml in 50 mM KH_2PO_4, pH 7.8, 50% glycerol. Store at -20°C.

Buffers and other solutions

- AHC medium (BIO 101): add 36.7 g of AHC powder per liter of purified water, autoclave at 121°C for 15 min.
- AHC agar (BIO 101): add 53.7 g of AHC agar medium per liter of purified water. Autoclave at 121°C for 15 min. Cool to 50°C. Then, mix well and pour plates. Store plates at 4°C.
- Alkaline Lysis (AL) solutions sufficient for 12 preps at the level of 20 ml cell culture:
  - AL Solution I: 50 mM glucose, 10 mM EDTA, 25 mM Tris-HCl pH 8.0, add 4 ml of 0.5 M glucose, 0.8 ml of 0.5 M EDTA and 1 ml of 1 M Tris-HCl to 34.2 ml water. Store at 4°C.
  - AL Solution II: 0.2 N NaOH, 1% SDS. Add 1.4 ml of 10 N NaOH, 7 ml of 10% SDS to 61.6 ml water.
  - AL Solution III: 3 M NaOAc, pH 4.8
- Antifade solution: 1% p-phenylenediamine (Sigma), 15 mM NaCl, 1 mM H_2PO_4, pH 8.0,90% glycerol. Store at -80°C.
- Blocking stock solution: dissolve blocking reagent (Roche Molecular Biochemicals, cat. # 1096 176) in maleic acid buffer (10% w/v) with shaking and heating. Autoclave stock solution and store in aliquots at 4°C.

- Cell fixative: acetic acid/methanol, 1:3 (vol:vol.). Make fresh before use.
- DB 0.5 solution: 0.5 M EDTA (pH 8.0), 1.0% N-lauroyl sarcosine (Sigma), 0.5 mg/ml Proteinase K (Roche Molecular Biochemicals).
- Denaturing solution: 70% FA, 2xSSC, pH 7.0. Prepare fresh at least every 2 weeks. Store at 4°C.
- ES Buffer: 0.5 M EDTA (pH 8.0), 1% sarcosyl.
- Gel loading dye: 1% bromophenol blue in 30% glycerol.
- Hybridization Master Mix: 14.3% w/vol dextran sulfate, 78.6% FA, 2.9×SSC, pH 7.0. For 10 ml MM2.1, mix 1.45 ml of 20×SSC with 0.7 ml ultrapure water, dissolve 1.43 g dextran sulfate (Calbiochem), incubate overnight, then add 7.86 ml formamide. Aliquot in 1.5 ml microcentrifuge tubes and store at -20°C.
- Lysis buffer: 1% Triton X-100, 20 mM TrisHCl, 2 mM EDTA, pH 8.5.
- Maleic acid buffer: 100 mM maleic acid, 150 mM NaCl, adjust to pH 7.5 with concentrated NaOH
- Modified nucleotide mix (10x) for labeling in combination with 1 mM dig-11-dUTP or FITC-12-dUTP: combine 5 µl each of 100 mM dATP, 100 mM dGTP and 100 mM dCTP with 2.5 µl of 1M Tris-HCL, pH 7.5, 0.5 µl 0.5 M EDTA, pH 8.0 (Life Technologies) and 232 µl ultrapure water for a total of 250 µl. Store at -20°C. The final concentration of nucleoside triphosphates is 2 mM each.
- PNM: Dissolve 5 g of non-fat dry milk in 100 ml of PN buffer (PN buffer is 0.1 M sodium phosphate, pH 8.0, 0.1% nonidet-P40), incubate at 50°C overnight and add 1/50 vol. sodium azide, spin at 1000 g for 30 min, aliquot clear supernatant into 1.5-ml tubes and store at 4°C. Spin at 2000 g for 30 s prior to use.
- SCE: 1 M sorbitol, 0.1 M Na citrate, 10 mM EDTA, pH 7.8.
- Slide Blocking Solution (5×SSC containing 2% Blocking Reagent, 0.1% N-lauroyl sarcosine): combine 0.05 g N-lauroyl sarcosine (Na salt, Sigma) and 1 g Blocking Reagent with 12.5 ml of 20×SSC (pH 7.0), add 30 ml water, heat to 60°C while stirring and bring the final volume to 50 ml with ultrapure water, when the Blocking Reagent is dissolved. Aliquot into 1.5-ml tubes, spin at 2000 rpm for 10 min and store at 4°C.
- SSC: 20×SSC is 3 M NaCl, 0.3 M Na_3citrate × $2H_2O$, pH 7.0.

- 10× Taq buffer: 500 mM KCl, 100 mM Tris HCl, pH 8.3, 10 mM $MgCl_2$.
- TBE (Tris/borate/EDTA) buffer, 10× is 890 mM Tris base, 890 mM boric acid, 20 mM EDTA.
- TE (Tris/EDTA) buffer, 1× is 10 mM Tris HCl, 1 mM EDTA, pH 7.4, 7.5 or 8.0.
- TE 50 buffer: 10 mM Tris HCl, 50 mM EDTA, pH 7.8.
- Tris HCl [tris(hydroxymethyl)aminomethane]: 1 M, pH 7.5 or 8.0.

Subprotocol 1
Preparation of aminopropyltriethoxysilane (APS) derivatized slides

The derivatization of glass substrates is among the most critical steps of the procedure. The slides should have the capacity to bind DNA molecules at one or both ends, but allow the molecules to stretch during the subsequent drying.

Procedure

1. Slide preparation

1. Clean glass slides mechanically by repeated rubbing with wet cheesecloth to remove dust and glass particles.
2. Rinse several times with ultrapure water.
3. Immerse slides in boiling ultrapure water for 10 min.
4. Air dry.
5. Immerse slides in 18 M sulfuric acid for at least 30 min to remove organic residues.
6. Immerse in boiling water for 1-2 min.
7. Air dry and store until further use.

2. Silane modification

1. Immerse precleaned dry slides in a solution of 0.1% APS in 95% ethanol for 10 min.
2. Remove slides from the silane solution.
3. Rinse several times with water, and immerse in ultrapure water for 2 min.
4. Dehydrate by immersing in absolute ethanol.
5. Dry upright for 10 min at 65°C on a hot plate.
6. Store slides for 2-6 weeks at 4°C in a sealed box under nitrogen prior to use.

Coverslip silanation is performed similar to the procedure described for slides. Briefly, coverslips are rinsed with distilled water and dehydrated in 100% ethanol. Coverslips are derivatized with a 0.05-0.1% solution of APS in 95% ethanol for 2 min. Coverslips are then rinsed and dried as described above (steps 3-4).

Subprotocol 2
Preparation of high molecular weight (HMW) DNA

The YAC clones used in the examples presented here, were part of the CEPH/Genethon library, which is commercially distributed by Research Genetics. Other clones might be used in their place without changing the protocols. Information for many CEPH of the clones including insert size, STS contents and radiation hybrid or genetic map position is available from the CEPH/Genethon server at URL http://www.genethon.fr/genethon_en.html/ or the Massachusetts Institute of Technology (MIT) server (URL http://www-genome.wi.mit.edu/).

The agarose plug preparation and pulsed field gel electrophoresis using a CHEF electrophoresis system (BioRad) follow standard protocols. Typically, 5 to 15 individual YAC colonies are tested to account for deletions. In most cases, the largest clone carries the least deletion(s).

The P1/PAC/BAC clones typically show far fewer deletions, so that it often suffices to pick 2-3 colonies from a plate, grow the cells overnight, and extract the DNA using an alkaline lysis protocol. The DNA can then be loaded directly onto the PFGE gel using a common gel loading dye.

The DNA is recovered from the low melting point agarose slab gel by excising the appropriate band using a knife or razor blade. High molecular weight DNA is then isolated by β-agarase digestion of the gel slices.

Procedure

1. Pulsed field gel electrophoresis (PFGE)

Preparation of gel plugs containing YACs

1. Spin down cells from 5 ml AHC media at 400 rpm for 6 min. Resuspend cells in 0.5 ml of 0.125 M EDTA, pH 7.8. Spin again and remove supernatant.
2. Resuspend the ~ 70 μl cell pellet in 500 μl of SCE. Mix with an equal volume of 1.5% LMP agarose preheated to 43°C. Quickly pipette up and down, then vortex for 1-2 s to mix. Pipette into plug molds (Biorad) and allow to solidify at room temperature or on ice.
3. Remove plugs from molds, incubate samples in 2 ml SCE containing 100 μl of zymolase and shake at 150 rpm at 30°C for 2.5 h to overnight.
4. Remove SCE and add 2 ml of ES containing 100 μl of proteinase K (20 mg/ml). Shake 5 h to overnight at 50°C.
5. Remove ES and rinse 5 times with 6 ml of TE50 for 30 min each rinse. Store the plugs at 4°C.

PFGE running conditions

1. YACs: voltage gradient, 6 V/cm; switching time, 79 s initially, 94 s final; running time, 38 h; agarose concentration, 1.0% LMP agarose; running temperature, 14°C; running buffer, 0.5×TBE.
2. P1/PAC/BAC clones: voltage gradient, 6 V/cm; switching time, 2 s initially, 12 s final; running time, 18 h; agarose concentration, 1.0% LMP agarose; running temperature, 14°C; running buffer, 0.5×TBE.

For probe production and determination of optimal PFGE conditions: stain the gel with ethidium bromide (EB, 0.5 μg/ml in water), cut out a gel slice containing the target DNA band and transfer slice to a 14 ml polystyrene tube (Cat.# AS-2264, Applied Scientific). Wash slice with ultrapure water for 30 min, and then wash with 1× agarase buffer for 30 min.

For high molecular weight DNA isolation: run duplicate samples on the right and left side of the gel, respectively. After a predetermined run time, cut gel in half, and stain one half with EB. Measure the migrated distance on a UV trans-illuminator, cut out a gel slice at the corresponding position from the unstained half and proceed as described in z.

2. Recovery of high molecular weight DNA from LMP agarose gel slices

1. Melt the gel completely by incubating it for 10 min at 85°C.
2. Transfer the molten agarose to a 43°C water bath.
3. Add 1 µl β-agarase for every 25 µl of molten agarose.
4. Incubate at 43°C for 2 h.
5. Add an equal volume of 200 mM NaCl.
6. Store the sample at 4°C until use.

3. Genomic DNA

Genomic high molecular weight (HMW) DNA is isolated from exponentially growing human cells such as the C32 melanoma cell line (ATCC) or diploid fibroblast cells using standard procedures. Briefly, about 5×10^5 cells are washed in PBS. The cells are then resuspended in 0.5 ml of PBS and mixed with 1.2% low melting point agarose previously melted in PBS and allowed to cool down to 43°C. Aliquots of 100µl are dispensed into plug molds and allowed to set for 30 min at 4°C. Agarose plugs are then placed into DB 0.5 solution and incubated overnight at 50°C. Next, plugs are washed 4-6 times for 30 min each in 50 mM Tris HCl, 1mM EDTA and stored at 4°C. The HMW DNA is released by digestion of the plugs with β-agarase according to the manufacturer's instructions (NEB) as described above.

Subprotocol 3
Immobilization and stretching of DNA molecules

The correct immobilization of DNA molecules is important for the successful stretching as well as the minimization of DNA loss during denaturation and hybridization. We have used different methods for binding and stretching of DNA on APS-pretreated surfaces. The quality of the resulting DNA fibers on glass or mica surfaces (Hu et al., 1996) appears to be determined primarily by the DNA preparation and by properties of the modified surface rather than by the method of DNA stretching. In general, the ideal APS-surface binds the DNA molecules only at their ends, or in the case of circular DNA molecules, at the position of nicks. The remainder of the DNA molecule should be free in suspension. This can be observed in the fluorescence microscope after addition of 1 μM YOYO-1 to the DNA before immobilization.

Procedure

In a typical experiment, 1-2 μl of clonal or genomic DNA are mixed with an equal amount of YOYO-1 (1 μM or 0.1 μM) and 8 μl water. One or two microliter of this diluted DNA is applied to an untreated coverslip, which is then placed DNA side down on the APS-treated slide or coverslip. The DNA concentration can be estimated under the fluorescence microscope using a filter set for FITC, and adjusted as needed. As early as 2 min of incubation at room temperature, the untreated coverslip can be removed slowly from one end, allowing the receding meniscus to stretch the bound DNA molecules ('fibers') in one direction (Hu et al., 1996). Alternatively, the slide or coverslip sandwich can be allowed to dry overnight at room temperature, after which the untreated coverslip is removed. Slides or coverslips carrying DNA fibers are rinsed briefly with water, drained, allowed to dry at room temperature and 'aged' in ambient air at 20°C for 1 week before hybridization.

Subprotocol 4
Probes generated from cloned DNA fragments

A typical QDFM experiment uses several different probes simultaneously. One probe is needed to counterstain the DNA fibers. This probe is usually prepared by labeling DNA from the same batch that was used to prepare the fibers. Probes for sequences to be mapped along the fibers are made such that they can be detected in a different color. Furthermore, it is recommended to include landmark probes that provide reference points by binding specifically to the vector part or the ends of DNA molecules.

Procedure

1. Alkaline lysis protocol and purification of DNA from P1, PAC or BAC clones

This protocol describes the isolation of DNA from ~20 ml overnight cultures using 40 ml Oakridge centrifugation tubes. The protocol can be scaled down to accommodate smaller volumes.

1. Grow culture overnight in ~30 ml LB or TB medium containing the recommended amount of antibiotic.
2. Prepare Oakridge tubes. Write the clone ID on a small piece of tape stuck to the cap. Spin 18.5 ml of culture at 2000 *g* for 10 min at 4°C and discard the supernatant.
3. Resuspend the pellet in 2340 µl of AL Solution I, then add 100 µl of lysozyme stock to each tube. Incubate tubes for 5 min at room temperature. Then, place the tubes on ice.
4. Add 5.2 ml of AL Solution II. The mixture should now become clear. Mix gently by inverting the tubes several times. Incubate for 5 min on ice.
5. Add 3.8 ml of AL Solution III and mix gently by inverting the tubes several times. Incubate for 10 min on ice.
6. Spin for 15 min at high speed (11,500 rpm/14,000g).
7. Transfer 10.4 ml of supernatant into a new Oakridge tube, add 5.8 ml of isopropanol and mix gently by inverting the tubes several times. Use the old cap [with the ID sticker] on the new tube.
8. Spin for 5 min at ~10,000 *g* and discard the supernatant. Watch the pellet!
9. Wash the pellet in cold 70% ethanol. Let the pellets dry briefly, i.e., for ~20-40 min at 20°C to 37°C.
10. Resuspend the pellet in 0.8 ml of TE buffer and split the volume into two 1.5 ml microcentrifuge tubes.
11. Add 400 µl phenol/chloroform/isoamyl alcohol to each tube. All centrifugations during the following phenol/chloroform extraction are done at 12,000 *g*.
12. Vortex for 15 s and spin down for 3 min.

13. Remove most of the bottom layer and spin again for 3 min.
14. Transfer the top layer to new microcentrifuge tubes and add 400 µl chloroform/isoamyl alcohol (24:1, vol.vol.)
15. Vortex well for 15 s, spin down for 3 min and remove most of the bottom layer followed by a second centrifugation for 3 min.
16. Transfer top layer to a new microcentrifuge tube, add 2.5 volumes, i.e., 1 ml 100% ethanol and let the DNA precipitate for 30 min at -20°C.
17. Spin down for 15 min, discard the supernatant and wash the pellet in ice cold 70% ethanol, spin again briefly, remove supernatant and air dry the pellet.
18. Resuspend the pellet in 20-40 µl TE, pH 7.4 containing 10 µg/ml RNase.
19. Incubate 30 min at 37°C [in water bath]; then, store at -20°C until used.

2. Preparation of DNA from Yeast Artificial Chromosome (YAC) clones

Retrieve the desired yeast clone containing the YAC from the library and grow it on AHC agar for 2-3 days at 30°C. Pick colonies from the plate and culture the clones in up to 35 ml AHC media at 30°C for 2-3 days.

DNA extraction, phenol purification and alcohol precipitation

1. Centrifuge cells (in ~35 ml AHC media) at 2000g at 4°C for 5 min.
2. Decant the supernatant and resuspend cells in 3 ml total of 0.9 M sorbitol, 0.1M EDTA, pH 7.5, containing 4 µl β-mercaptoethanol, followed by addition of 100 µl of zymolase (2.5 mg/ml), and then incubate at 37°C for 60 min.
3. Pellet the cells at 2000 *g* and 4°C for 5 min and decant supernatant.
4. Resuspend pellet in 5 ml of 50 mM Tris, pH 7.4, 20 mM EDTA. Add 0.5 ml of 10% SDS and mix gently. Incubate at 65°C for 30 min.
5. Add 1.5 ml of 5 M potassium acetate and place on ice for 60 min.
6. Spin at 12,000 *g* for 15 min at 4°C, and transfer the supernatant to a new tube.
7. Mix the supernatant gently with 2 volumes of 100% ethanol by inverting the tube a few times. Spin at 5000 rpm (2000g) for 15 min at room temperature.

8. Prepare 1.5 ml microcentrifuge tubes (4 tubes per clone with ID on top).
9. Decant supernatant and air dry the pellet. Resuspend pellet in 3 ml of 1×TE, pH 7.5.
10. Transfer the DNA solution to four 1.5 ml microcentrifuge tubes.
11. Add an equal volume of phenol/chloroform/isoamyl alcohol (25:24:1, pH 8.0), vortex well and spin at high speed (10,000 *g*) for 3 min.
12. Transfer the top layer to new 1.5 ml microcentrifuge tubes and add an equal volume of chloroform/isoamyl alcohol (24:1). Vortex well and centrifuge at high speed (10,000 *g*) for 3 min.
13. Transfer the top layer to new 1.5 ml microcentrifuge tubes. Add 40 μl of RNase (1 mg/ml, DNAse free) to each of the four tubes and incubate at 37°C for 30 min.
14. Add 1 volume of isopropanol and gently mix by inversion. Centrifuge at high speed (10,000 *g*) for 20 min.
15. Decant supernatant and wash pellet with 1 volume of cold 70% ethanol, and centrifuge at high speed (10,000 *g*) for 3 min.
16. Decant the 70% ethanol and air dry the pellet.
17. Resuspend pellet in 30 μl 1×TE, and measure DNA concentration after the pellet is completely dissolved.

Subprotocol 5
Generation of probes by in vitro DNA amplification

In vitro DNA amplification using the polymerase chain reaction (PCR) is a very efficient method to synthesize probe DNA. It can be applied to amplify a particular DNA sequence, such as a part of the cloning vector, or with mixed-base primers to perform arbitrary amplification of virtually any sequence of interest. As illustrated in the following paragraphs, the former amplification can be applied to prepare DNA landmark probes, while the latter allows the preparation of probes to counterstain the fibers.

Procedure

1. Cloning vector-specific probes

The generation of P1/PAC-, BAC- and YAC-vector probe DNA takes advantage of the access to published vector sequences. PCR primers are typically designed to amplify fragments of 1100-1400 bp of vector sequence. Several such oligonucleotide pairs have been designed in several laboratories including ours and are used in either single pairs or combinations. The PCR usually follows standard conditions, i.e., a Tris-HCl buffer containing 1.5 mM $MgCl_2$ and 1 unit Taq DNA polymerase per 50 µl reaction is used, annealing temperatures range from 50°C to 60°C.

On the other hand, the YAC cloning vectors pJs97 and pJs98, cloned in plasmid vectors (BRL), can be used to prepare probes useful to determine the orientation of the YAC insert (Duell et al., 1997). For this purpose, plasmid DNA is extracted using the above alkaline lysis protocol or a commercial kit and labeled by random priming as described below.

2. Mixed base oligonucleotide primed PCR

The DNA probes for counterstaining the YAC DNA fibers are generated by mixed base oligonucleotide primed PCR (sometimes referred to as degenerate oligonucleotide primed PCR or 'DOP-PCR') (Cassel et al., 1997). An aliquot of the HMW DNA obtained by PFGE for fiber preparation is PCR amplified for a total of 42 cycles with oligonucleotide primers that anneal about every 200-800 nucleotides. In our preferred scheme, we use two different DNA amplification programs. Initially we perform a few manual PCR cycles using T7 DNA polymerase to extend the oligonucleotide primers at a relatively low temperature. Next, DNA copies prepared in these first cycles are amplified using the thermostable Taq DNA polymerase and a rapid thermal cycling scheme.

In the first amplification stage, T7 DNA polymerase ('Sequenase II', Amersham Pharmacia Biotech) is used in 5-7 cycles to extend the mixed base primer JUN1 (5-CCAAGCTTGCATGCGAATTCNNNNCAGG-3, N=ACGT) that is annealed at low temperature. Briefly, 2-3 µl of HMW DNA solution are removed from the bottom of each tube and PCR amplified using the following conditions: denaturation at 92°C for 3 min, primer annealing at 20°C for 2 min and extension at 37°C for 6 min. Sequenase must be added after each denaturation. (See Chapter 29 by Fung et al. for details.)

In the second amplification stage, 20 µl of the reaction product are resuspended in a 200 µl Taq amplification reaction buffer and amplified with primer JUN15 (5-CCCAAGCTTGCATGCGAATTC-3) with the following PCR conditions: denaturation at 94°C for 1 min, primer annealing at 50°C for 1 min, and extension at 72 °C for 2 min, repeated for 35 cycles. After precipitation of the PCR products in 1.2 vol of isopropanol, the products are resuspended in 30 µl of TE buffer. Subsequently, 1.5 µl of this solution is labeled in a 25 µl random priming reaction incorporating FITC-12-dUTP or biotinylated dNTPs.

Subprotocol 6
Probe labeling via random priming and hybridization

Labeling of DNA by random priming is a reliable method and, in our laboratory, is applied routinely to label DNA fragments from 100 bp to several hundred kb. The procedure involves an initial thermal denaturation of the DNA to allow the random oligonucleotides ('primers') to anneal. Thus, restriction or hydrolysis of large molecules is not necessary. Several companies now offer kits for random priming reactions. Slight differences exist with regard to enzyme activity, amount of random primers and cost per reaction.

Procedure

Measurement of DNA concentration

The concentration of PCR products can be estimated from the agarose gels run to confirm target amplification. If a sufficient amount of clonal or genomic DNA is available, one or two microliters can be used to accurately determine the concentration using Hoechst 33258 fluorometry using a TK100 fluorometer (Pharmacia).

Random priming

1. Add 250 ng of DNA to water to a final volume of 7 µl in a 0.5-ml microcentrifuge tube.
2. Boil DNA at 100°C for 5 min, then quickly chill on ice.

3. For labeling with either dig-dUTP or FITC-dUTP, add:
   - 2.5 µl 10× Modified Nucleotide Mixture
   - 3.25 µl 1 mM dTTP
   - 1.75 µl dig-11-dUTP or FITC-12-dUTP (1 mM, ROCHE MOLECULAR BIOCHEMICALS, #1093 088)
   - 10 µl 2.5× Random primers (BioPrime kit, Life Technologies part# YO1393)

   Note: For labeling the DNA with biotin, add 2.5 µl 10× dNTP mix provided with the BioPrime kit (containing biotin-14-dCTP), 5 µl water, and 10 µl 2.5× random primers).

4. Mix well, add 0.5 µl DNA polymerase I Klenow fragment (40 units/µl, Life Technologies, part #YO1396) and incubate in a water bath at 37°C for 60-120 min.
5. Add 2.5 µl of 10× stop buffer (Life Technologies, part #YO1107, part of the BioPrime kit).
6. Store probe at -20°C until used.

Subprotocol 7
Fluorescence in situ hybridization (FISH)

All hybridizations are carried out overnight at 37°C in a moist chamber. Fiber hybridizations include a comparatively low concentration of a biotin- or FITC-labeled DNA probe prepared from the high molecular weight DNA that is used to prepare the fibers. This counterstain highlights the otherwise invisible DNA fibers and allows competitive displacement by the probes to be mapped along the DNA fiber (Weier et al. 1995, Duell et al. 1997). Additionally, one or several cloning vector-specific probes are included to allow a determination of the orientation of the insert.

Procedure

The hybridization procedure is very similar to protocols used with metaphase spreads:

1. Hybridization mix: combine 1 µl of each probe, 1 µl of human COT1 DNA (optional), 1 µl of salmon or herring sperm DNA, and 7 µl of hybridization master mix.

2. Apply the hybridization mixture to the slide and coverslip.
3. Denature the slide at 88-92°C for 90 sec on a hot plate.
4. Transfer the slide to a moist chamber and incubate overnight at 37°C.

Washing and detection steps are not much different from protocols used for FISH in interphase and metaphase cells that have been described in detail (Weier et al. 1995, Duell et al. 1997, 1998; Wang et al. 1997):

1. After hybridization, wash the slide three times in 2×SSC at 20°C for 10 min each.
2. Incubated the slide with 100 μl PNM buffer or blocking stock solution under a plastic coverslip at 20°C for 5 min.
3. The slide is then incubated at room temperature for 30 min with 100 μl PNM buffer containing AMCA-avidin (Pharmacia), anti-digoxigenin-rhodamine (Roche Molecular Biochemicals) and a mouse antibody against FITC (DAKO).

   Note: If only two labels are used, i.e., biotin and digoxigenin, bound probes are detected with avidin-FITC DCS (Vector) and anti-digoxigenin-rhodamine, respectively.
4. The slide is washed two to three times in 2×SSC for 15 min each at 20°C with constant motion on a shaking platform.
5. If necessary, signals are amplified using a biotinylated antibody against avidin raised in goat (Vector) followed by another layer of AMCA-avidin, a Texas Red-labeled antibody against sheep raised in rabbit (Sigma) and a horse-anti-mouse antibody conjugated to FITC (Vector) (Wang et al. 1996).
6. The slide is mounted in 8 μl of DAPI (0.05 μg/ml in antifade solution) and covered by a 22×22 mm coverslip.

Subprotocol 8
Digital image acquisition and analysis

Although not a prerequisite for QDFM, digital image acquisition and computer-assisted analysis greatly facilitate the analysis of hybridization images. Since QDFM is based on simple measurements of distances between probe hybridization domains, the analysis can alternatively be performed on images recorded on film and either printed or projected on a screen.

Procedure

Images are acquired using a standard fluorescence microscope (Zeiss Axioskop) equipped with 63×, 1.25 N.A. and 40×, 1.2 N.A. objectives, and a filter set for excitation and simultaneous observation of DAPI, Texas Red/rhodamine, FITC and CY5 fluorescence, respectively (Chroma-Technology). Current filters are capable of excitation in single bands centered around 360, 405, 490, 555, and 637 nm, and visualization in multiple bands in the vicinities of 460 nm (blue), 520 nm (green), 600 nm (red) and 680 nm (infrared). Images are collected using a CCD camera (Xilix, Hamamatsu or Photometrics) connected to a computer workstation (Weier et al., 1995).

For determination of map positions, interactive software is available for either Apple Macintosh, IBM/PC or SUN computers that allows the user to trace DNA fibers by drawing a segmented line and then calculates the length of the line in pixels (Wang et al. 1996; Duell et al. 1997). The pixel spacing is known from the microscope objective used in the experiment (use a 63× objective for molecules up to 100 kb, a 40× objective for larger molecules) and is converted into µm (or kb using the factor of 2.3 kb/µm). After measuring all relevant distances along the DNA fibers in triplicate, the results in the form of lists are imported into Microsoft Excel spreadsheets and used to calculate average values for each fiber and mean values and standard deviations for individual experiments.

Results

Solid substrates for QDFM are prepared in batches of 20-50 by derivatization of standard microscope slides, coverslips or sheets of mica with APS, which results in primary amino groups on the surface (Weier et al. 1995; Hu et al. 1996). For DNA fiber stretching, a solution of target DNA molecules onto which probes are to be mapped is placed on an untreated coverslip and spread by placing the coverslip upside-down on the APS-derivatized glass or mica surface. Binding of DNA to the substrate and the stretching effect can be monitored by staining the DNA with YOYO-1 prior to deposition. This also allows the rejection of batches of slides that bind DNA too tightly. Following DNA binding and stretching, the coverslips are removed, the slides are rinsed briefly with double-distilled water, air dried and stored at 4°C.

The DNAs from plasmid, cosmid, P1/PAC and BAC clones are isolated using an alkaline lysis protocol and inserts are sized by PFGE. Digestion of DNA with a rare cutting restriction enzyme produces linear high molecular weight DNA molecules, but the alkaline lysis procedure typically provides sufficient amounts of nicked circular or randomly broken DNA suitable for QDFM (Wang et al. 1996). In general, the DNA is loaded onto a 1.0% low melting point agarose gel and electrophoresed for about 15 h. The band containing the desired linear or circular DNA is then excised from the gel, and the gel slice is digested with agarase. Similarly, YAC DNA from various clones is purified by PFGE. The integrity of DNA molecules can be assessed by microscopic inspection of aliquots of DNA stained with 0.5 μM YOYO-1, before high molecular weight DNAs are used for DNA fiber or FISH probe preparation or stored at 4°C in 100 mM NaCl.

The density of DNA molecules after DNA fiber stretching can be adjusted by altering the concentration of the DNA molecules prior to binding. Figure 2 shows the typical density of hybridized lambda DNA molecules. In experiments depositing circular P1 and BAC DNA molecules, the fraction of intact DNA molecules sometimes reaches ~80%. While binding of DNA molecules in their circular form helps to maintain their integrity, it interferes with DNA fiber stretching, and the molecules are found to be stretched to varying degrees (Fig. 3B). Mapping onto circular molecules can thus be used for a rough estimation of overlap, and mapping on linear fibers for high precision measurements. This can be done in a single experiment, because some circular DNA molecules are sheared during deposition, thus providing randomly broken linear DNA molecules (Fig. 3B).

Quantitative DNA Fiber Mapping can facilitate the construction of high resolution physical maps comprised of any combination of cosmid, P1, PAC or BAC clones in two ways: if a low resolution map is available, for example, in the form of a YAC contig, individual clones can be mapped directly onto DNA fibers prepared from the larger clones (Weier et al. 1995, Cheng and Weier 1997). Alternatively, a high resolution map can be constructed by measuring the extent and orientation of overlap between individual clones (Fig. 1). In most experiments, the applied scheme will be determined by the sources of the clones and might combine both schemes. Figure 3 shows some typical examples of QDFM applications.

Figure 3A shows the mapping of a P1 clone (red) onto a colinear YAC clone (green). Precise localization of the region of overlap is facilitated by probes that mark specifically the ends of the YAC molecules (red).

Preparation and purification of the circular DNA molecules is simple and fast. An example of mapping an exon specific plasmid clone onto circular BAC molecules is depicted in Fig. 3C. The use of circular DNA molecules results in a dense deposition of circular DNA molecules in the presence of linear fragments of different sizes (Fig. 3B). The largest circles are stretched to about ~2.3 kb/μm, but smaller, more condensed molecules can also be analyzed using the extent of the vector-specific green ~7 kb domain on BACs as standard for normalization. The linear fragments

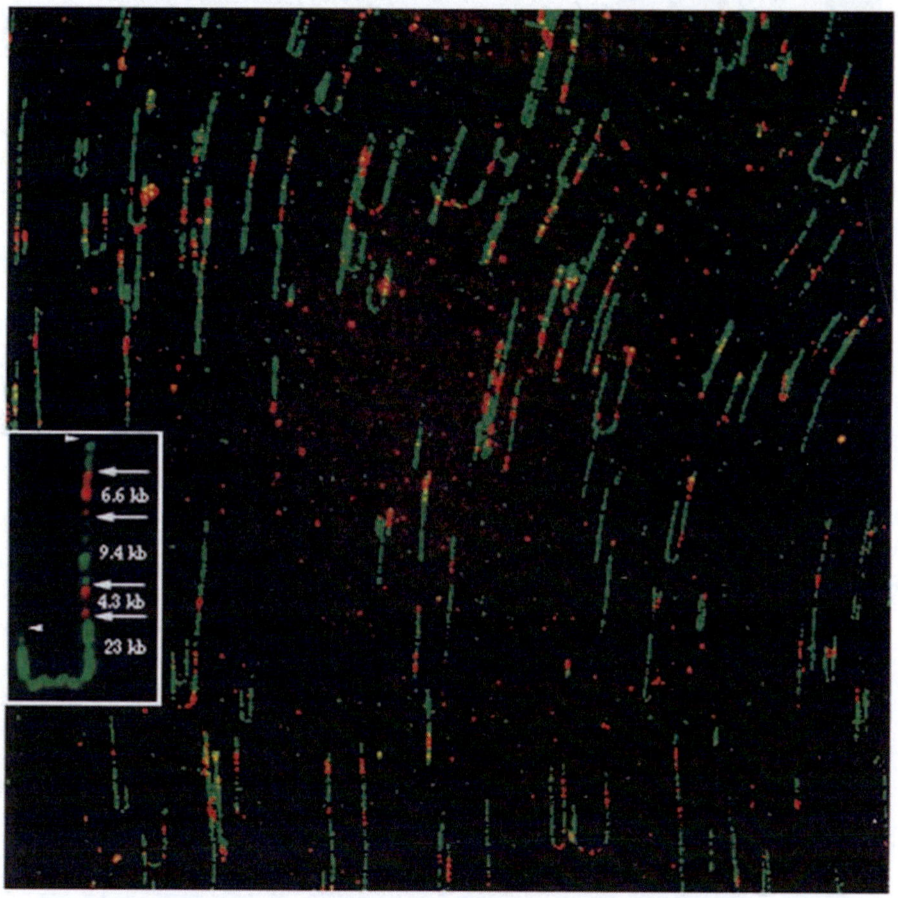

Fig. 2. Quantitative DNA Fiber Mapping (QDFM) using phage DNA molecules. Lambda DNA molecules immobilized on APS-derivatized glass slides were hybridized with a mixture of biotin- and digoxigenin-labeled lambda DNA restriction fragments. The molecules show specific labeling after incubation with avidin-FITC and rhodamine-labeled antibodies against digoxigenin. The *insert* shows a typical molecule

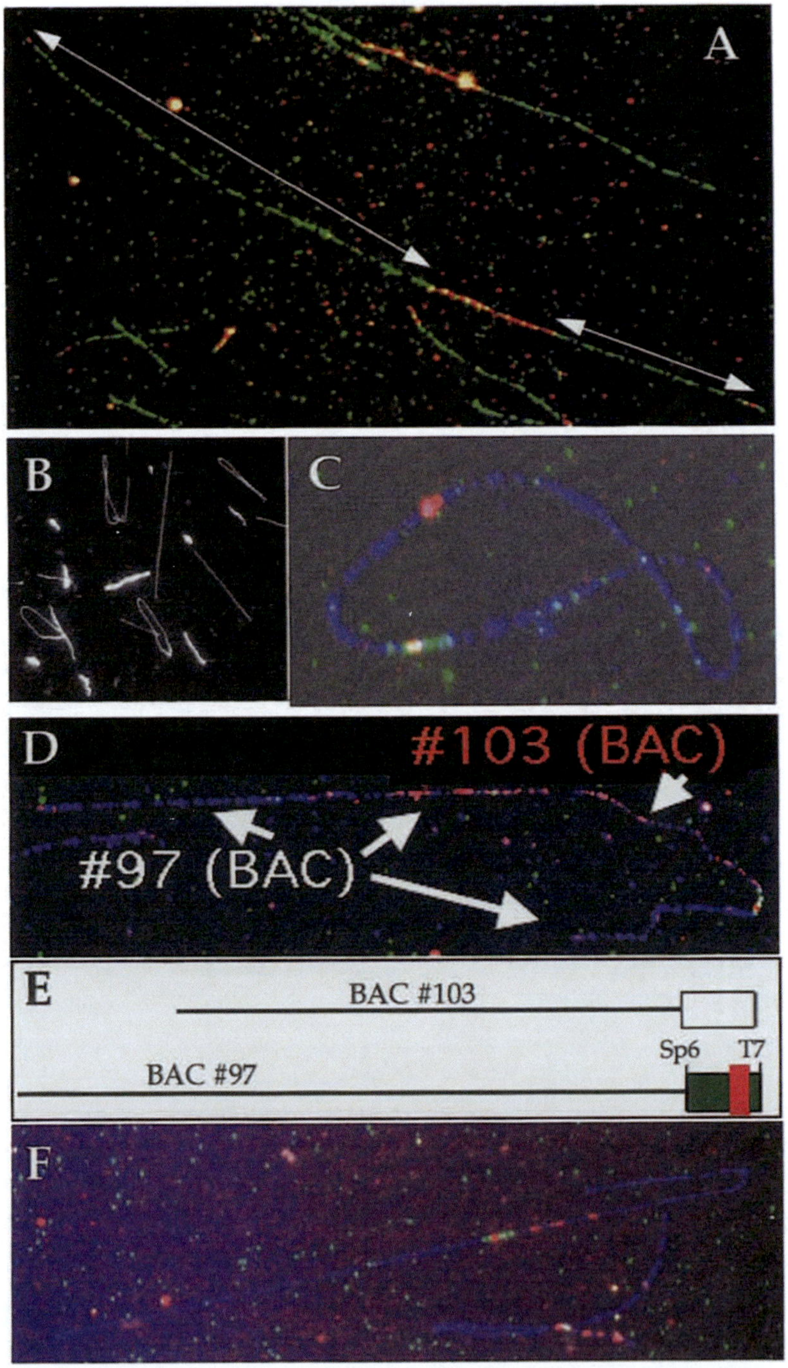
A
B
C
D
#103 (BAC)
#97 (BAC)
E
BAC #103
Sp6
T7
BAC #97
F

found on the same slides are stretched more homogeneously, thus providing DNA fibers without need for normalization.

Expressed sequences can be mapped easily by QDFM, if each individual target extends for a few hundred base pairs or more. A common approach hybridizes small genomic DNA fragments of 1-2 kb that contain known exons onto larger genomic DNA molecules. If the cDNA sequence and some information about intron-exon boundaries are available, such small DNA fragments can rapidly be generated from genomic DNA using PCR. Figure 3C demonstrates this by mapping exon 2 of the human Band 4.1 gene onto a homologous BAC molecule. This allowed the localization of the ~2 kb exon with near kilobase precision.

Figures 3D,E illustrate the application of QDFM for measurement of BAC clone overlap. In such experiments, the DNA fiber is counterstained by a probe detected in blue. A BAC vector-specific probe (green) and a PCR-generated probe of ~1300 bp (red) that binds close to the T7 end of the vector are included to highlight the vector part of the DNA fibers (Figs. 3C,D). The two BAC clones shown in Fig. 3D overlap by approximately 80 kb and the overlapping region is close to the SP6 promoter in the BAC vector (Fig. 3D,E). These results are summarized schematically in Fig. 3E.

A rapid approach to studying the genomic organization of genes relies on direct mapping of expressed sequences. The probe DNA is isolated from cDNA clones, labeled and hybridized onto genomic DNA fibers. In the presence of blocking DNA, the cDNA probes will bind specifically to their complementary DNA targets, i.e., exons and 5- or 3-untranslated regions (UTRs) along the DNA fiber. This leaves non-coding regions (introns, 5- and 3-flanking DNA) unstained (Fig. 3F). Using FISH conditions similar to those

Fig. 3. Quantitative DNA Fiber Mapping (QDFM) using large insert human genomic DNA clones. **A** Mapping P1 clones along YAC molecules. The *arrows* indicate the distance from the respective ends. **B** Circular DNA molecules excised from a PFGE gel purified and stained with YOYO-1 revealed closed circular DNA molecules in the presence of linear molecules of different length. **C** Physical mapping of small cDNA clones in larger genomic intervals. DNA fibers (blue) prepared from a BAC clone were hybridized with a ~2 kb insert of a plasmid containing exon 2 of the human Band 4.1 gene (red). **D, E** Determination of overlap between linked BAC clones. A probe prepared from BAC # 103 (red) was hybridized onto DNA fibers prepared from BAC #97 (blue). **E** shows a schematic representation. **D** Mapping the genomic organization of expressed sequences. Here, BAC DNA fibers (blue) were hybridized with a ~5 kb cDNA probe (red). Three hybridization domains representing larger exons and the 3 UTR were detected on the DNA fibers. The green probe in **C**, **D** and **F** delineates the BAC vector

applied to hybridization and probe detection on metaphase chromosomes, this system works well with cDNA probes of several kb.

Troubleshooting

- **Slide pretreatment**
  Slides from different manufacturers or even of the same brand may produce very different qualities of fibers. Use a large batch of slides from one manufacturer; avoid slides that are painted on one end, since the paint might come off during pretreatment. Slides that have a sand-blasted area at one end are preferable.

- **Homogenous stretching of DNA molecules**
  Different procedures have been described to stretch DNA molecules. In our hands, stretching involving a hydrodynamical force (meniscus) at 20°C or 4°C has proven most reproducible. There is, however, no need to wait until the preparation has dried to completion. Once the DNA molecules have bound to the substrate, the coverslip can be lifted to exert the hydrodynamic stretching force (Hu et al. 1996).

- **Immunocytochemical signal amplification**
  Never let the slides or part of them dry out during the immunocytochemical signal amplification. It is important to just drain the liquids from the slides, and then rapidly apply the next solution such as a blocking solution or the antibodies. If the slides are allowed to dry out, the level of background staining will increase to unacceptable levels.

- **Image acquisition**
  Most fluorochromes fade very quickly. Thus, minimize the exposure of slides to the excitation light. Don't waste time studying the beauty of the molecules.

- **Image analysis**
  Always measure additional segments of the molecule such as the vector segment, since these might provide additional information about the extent and homogeneity of DNA stretching.

- **Errors**
  Relative standard deviations are typically in the order of 5%. Higher standard deviations provide a simple means of controlling the procedure. They signal it is necessary to check the data analysis results for

operator errors and undesirable images such as broken molecules or insufficiently stretched fibers.

Acknowledgements. This work was supported by a grant from the Director, Office of Science, Office of Biological and Environmental Research, U. S. Department of Energy, under Contract DE-AC03-76SF00098 and by the 'Training Program in Genome Research' sponsored by the University of California Systemwide Biotechnology Research and Education Program (#S96-03). Additional Support was provided by grants from United States Army Medical Research and Material Command, United States, Department of the Army (DAMD17-99-1-9250, DAMD17-00-1-0085) and the National Institute of Health (1R01 CA80792).

References

Cassel MJ, Munné S, Fung J, Weier H-UG (1997) Carrier-specific breakpoint-spanning DNA probes for pre-implantation genetic diagnosis [PGD] in interphase cells. Hum Reprod 12:2019-2027

Cheng J-F, Weier H-UG (1997) Approaches to High Resolution Physical Mapping of the Human Genome. Biotechnology International, C. F. Fox and T.H. Connor, (Eds.), Universal Medical Press, San Francisco, pp. 149-157.

Cohen D, Chumakov I, Weissenbach J (1993) A first-generation physical map of the human genome. Nature 366:698-701

Duell T, Wang M, Wu J, Kim U-J, Weier H-UG (1997) High Resolution physical map of the immunoglobulin lambda variant gene cluster assembled by quantitative DNA fiber mapping. Genomics 45:479-486

Duell T, Nielsen LB, Jones A, Wang M, Young SG, Weier H-UG (1998) Construction of two near-kilobase resolution restriction maps of the 5' regulatory region of the human apolipoprotein B gene by Quantitative DNA Fiber Mapping (QDFM). Cytogenet Cell Genet 79:64-70

Hu J, Wang M, Weier H-UG, Frantz P, Kolbe W, Olgletree DF, Salmeron M (1996) Imaging of single extended DNA molecules on flat (aminopropyl)triethoxysilane-mica by atomic force microscopy. Langmuir 12:1697-1700

Wang M, Duell T, Gray JW, Weier H-UG (1996) High sensitivity, high resolution physical mapping by fluorescence in situ hybridization [FISH] on to individual straightened DNA molecules. Bioimaging 4:1-11

Weier H-UG, Wang M, Mullikin JC, Zhu Y, Cheng J-F, Greulich KM, Bensimon A, Gray JW (1995) Quantitative DNA fiber mapping. Hum Mol Genet 4:1903-1910

Weier H-UG, Munné S, Fung J (1999) Patient-specific probes for preimplantation genetic diagnosis (PGD) of structural and numerical aberrations in interphase cells. J Assist Reprod Genet 16:182-189

Suppliers

5 Prime -3 Prime, 5603 Arapahoe, Boulder, CO 80303, USA

Amersham Pharmacia Biotech, Amersham Pharmacia Biotech, Inc., 800 Centennial Avenue, P.O. Box 1327, Piscataway, NJ 08855-1327, USA; phone 1-732-457-8000, fax: 1-732-457-0557; http://www.amersham.com.au/

Applied Spectral Imaging Inc., 2120 Las Palmas Drive, Suite D, Carlsbad, CA 92009, USA; phone (800) 611-3466, fax (760)929-2842

BIO-RAD Laboratories, 2000 Alfred Nobel Drive, Hercules, CA 94547-9980, USA; phone (510) 741-1000, fax (510) 741-5800, http://www.bio-rad.-com/index1.html

Calbiochem, 10394 Pacific Center Court, San Diego, CA 92121 USA; phone (800) 854-3417, fax (800)776-0999

Fisher Scientific, 9999 Veterans Memorial Drive, Houston, TX 77038-2499, USA; phone (800) 766-7000, fax (800) 926-1166

Life Technologies, P.O. Box 9418, Gaithersburg, MD 20898, USA; phone (301) 840-8000, fax (301) 670-8539

International Equipment Company (IEC), 300 Second Avenue, Needham Heights, MA 02194, USA, phone (617) 449-8060, fax (617) 444-6743

Labline Instruments, 15th and Bloomingdale Ave., Melrose Park, IL 60160-1491, USA; phone (800) 522-5463, fax (708) 450-0943, http://www.labline.-com, email: value@labline.com

New Brunswick Co., Inc., 44 Talmadge Rd., Edison, NJ 08818-4005, USA; phone (908) 287 1200, fax (908) 287 4222, email bioinfo@nbsc.com, http://www.nbsc.com

New England Biolabs, Inc., 32 Tozer Road, Beverly, MA 01915-5599, USA; phone (800) 632-5227, fax (978) 921-1350, http://www.neb.com

Nikon, Inc., 1181 Chess Drive, Suite H, Foster City, CA 94404-1109, USA; phone (650) 513-0542, fax (650)513-0540

Perkin Elmer, 850 Lincoln Center Drive, Foster City, CA 94404, USA; phone (800) 345-5224, fax (415)572-2743

Pharmacia Biotech Inc., 800 Centennial Ave., P.O. Box 1327 Piscataway, NJ 08855-1327, USA; phone (800) 526 3593, fax (800) 329 3593, http://www.biotech.pharmacia.com

Research Genetics, Inc., 2130 Memorial Pkwy, SW, Huntsville, AL 35801, USA; phone (205) 533 4363, fax (205) 536 9016, http://www.resgen.com/

Roche Molecular Biochemicals, 9115 Hague Rd., P.O. Box 50414, Indianapolis, IN 46250-0414, USA; phone (800) 428-5433; fax (800) 428-5433

Sigma, P.O. Box 14508, St. Louis, MO 63178, USA; phone (800) 325-3010, fax (800)325-5052

Thermatron Industries, Inc., 291 Kollen Park Drive, Holland, MI49423, USA; phone (616) 393-4580, fax (616)392-5643

Vector Laboratories, Inc., 30 Ingold Road, Burlingame, CA 94010, USA; phone (415) 697-3600, fax (415)697-0339

VWR Scientific, 1310 Goshen Pkwy., West Chester, PA 19380, USA; phone (800)932-5000, fax (610) 436-1761; email solutions@vwrsp.com; http://www.vwrsp.com/

Vysis, Inc., 3100 Woodcreek Drive, Downers Grove, IL, USA: phone (800) 553-7042, fax (630) 271-7138

Abbreviations

AFM	atomic force microscopy
BAC	bacterial artificial chromosome
CEPH	Centre des Études du Polymorphisms Humain, Paris, France
LMP	low melting point
PCR	polymerase chain reaction
PFGE	pulsed field gel electrophoresis
QDFM	quantitative DNA fiber mapping
UTR	untranslated region
YAC	yeast artificial chromosome

PNA-FISH

William M. Strauss

Introduction

Peptide nucleic acids (PNAs) are nucleic acid mimics that contain a pseudo-peptide backbone composed of charge neutral and achiral N-(2-aminoethyl) glycine units to which the nucleotides are attached via a methylene carbonyl linker (Nielsen et al. 1991; Egholm et al. 1992; 1993). PNAs hybridize with high affinity to complementary DNA sequences, forming PNA-DNA complexes via Watson-Crick or Hoogsteen binding (Leijon et al. 1994). The PNA/DNA duplex has a higher melting temperature, Tm, than the DNA/DNA duplex. For instance, a typical 15-mer PNA/DNA melts at 69°C whereas the corresponding DNA/DNA duplex melts at 54°C (Egholm et al. 1993). In addition to the high thermal stability of complexes, PNA-DNA binding is highly sensitive to mismatches (Nielsen et al. 1993; Egholm et al. 1993; Jensen et al. 1997). The Tm changes caused by single base mismatch averaged 15 and 11°C, respectively, in PNA/DNA and DNA/DNA duplexes for a 15-mer oligo. These novel characteristics of short PNA oligomers obviously lead to development of PNA fluorescence in situ hybridization (PNA-FISH).

Lansdrop et al. (1996) was the first to use directly fluorescein-labeled $(C3TA2)_3$ PNA 18-mer for fluorescence in situ hybridization on telomeres. Since then, PNA-FISH has been applied in detecting trinucleotide repeats (Taneja 1998) and centromeric repeats (Chen et al. 1999, 2000). PNA-FISH can discriminate between two CENP-B sequences that differ by a single base pair in mouse and human centromeres. Interestingly, PNA probes have been used in a flow cytometry/FISH method for determining the replication history and potential replication capacity of several different types of cells from the same tissue (Hultdin et al. 1998; Rufer et al. 1998).

William M. Strauss, Harvard Medical School, Beth Israel Deaconess Medical Center, Harvard Institute of Human Genetics, 4 Blackfan Circle, Boston, MA, 02115, USA (*phone* +617-432-8475; *fax* +617432-8476; *e-mail* wstrauss@hihg.med.harvard.edu)

Elegant features of PNA/DNA duplex formation have made PNA-FISH more powerful than conventional DNA-FISH. First, the higher affinity of PNA-DNA duplex and greater accessibility of short and charge-neutral PNA oligomers (Egholm, et al. 1992, 1993) lead to very strong FISH signals. Second, these signals are highly saturable, relatively stable, and reproducible. Third, PNA-FISH also shows a very low level of background. Superior hybridization and discrimination ability have brought a potential application in quantitative FISH (Q-FISH, Lansdorp et al. 1996; Slijepcevic et al. 1997; Zijlmans et al. 1997; Martens, 1998; Hande et al. 1998).

Outline

I. Chromosome preparations

II. Post-fixation in methanol:acetic acid (3:1) for 1-2 h

III. Dehydration in 70%, 80%, 90%, and 100% ethanol
   - Air dry
   - 1xPBS at 37°C for 5 min

IV. Denature the chromosomes in 4% formaldehyde at 37°C for 1.5-2 min

V. Dehydrate slides in the ethanol series and air dry

VI. Dissolve PNA in a pre-warmed hybridization mix at 37°C for at least 15 min

VII. Apply 15-25 μl of hybridization mix to each slide

VIII. Post-denature chromosomal preps at 80°C for 3 min in a hot plate

IX. Hybridization at 37°C for 2 h

X. Post-hybridization washes at RT (room temperature) (70% formamide 2×15 min and then TNT 3×5 min)

XI. Slides counterstained with DAPI or PI for subsequent microscope examination

Materials

- PNA (PerSeptive Biosystems, Framingham, MA)
- 0.1% aqueous trifluoroacetic acid (TFA, Sigma, St. Louis, MO)
- 10X phosphate-buffered saline (PBS) pH 7.2 (Life Technologies, Gaithersburg, MD)
- 70%, 80%, 90%, and 100% ethanol series
- 37% formaldehyde (Fisher Scientific)
- 100% formamide (Fisher Scientific) for post-hybridization washing
- 100% deionized formamide (Oncor, Gaithersburg, MD)
- Tween-20 (Sigma, St. Louis, MO)
- 100 mM Tris-pH 7.2; 5 M NaCl
- 10% blocking reagent (Boehringer-Mannheim, Indianapolis, IN)
- µg/ml of 4,6-diamidino-2-phenylindole-dehydrochloride (DAPI) (Oncor, Gaitherburg, MD)
- 0.6 µg/ml of propidium iodide (PI) (Oncor, Gaitherburg, MD)

Buffers

- 1× hybridization mix: Add 35 ml of deionized formamide, 5 ml of 10% blocking reagent, and 5 ml of 100 mM Tris pH 7.2. Adjust pH to 7.2 by adding 1:5 diluted HCl. Aliquot into 0.5 ml each and store at minus 20°C for up to 6 months.
- 70% formamide /1×PBS: For 5 slides, mix 105 ml of formamide, 15 ml of 10×PBS, and 20 ml of dd. H_2O. Bring pH to 7.2 by adding 1:10 diluted HCl and volume to 150 ml. Make freshly.
- TNT washing buffer: 50 mM Tris pH 7.5/ 150 mM NaCl/0.05%(v/v) Tween-20. Make freshly.

Equipment

- Moist chamber
- Glass slides
- Glass coverslips (22×22, 24×30, 24×40 mm)
- 4 glass staining dishes
- Blunted forceps

- One set of slide racks
- 37°C water bath
- 37°C incubator
- PCR with slide holder or heat block
- -20°C freezer

Procedure

Proper design and handling of PNAs

1. Because of the higher Tm of PNA/DNA duplex, shorter 12- to 18-mer PNAs are widely used in FISH studies. High purine (G > A)-rich sequences are to be avoided owing to difficulty in synthesis. Significant primer dimers should also be eliminated. A variety of fluorescent dyes can be labeled at either the 5 or 3 site of PNA molecules. They include rhodamine, fluorescein, and Texas Red etc. In comparison with fluorescein-labeled PNAs, rhodamine-labeled PNA is dissolved more easily. A short incubation of the PNA oligomer at 37-50°C will increase its solubility.

2. It is recommended that PNA should be lyophilized for shipping and storage. Because PNA oligomers have an affinity for glass surfaces and polystyrene, polypropylene or polyethylene materials are recommended for the handing and storage of PNA. Upon arrival from PerSeptive Biosystems, 0.2-2 ml of 0.1% TFA is added to dissolve PNA in order to reach an initial PNA concentration at 20 μM. After the PNA is completely dissolved, it is necessary to re-quantify PNA using a UV spectrophotometer at 260 nm. PNA can be heated at 50°C for up to 30 min to improve its solubility. Finally, PNA is aliquoted in 5 or 50 μl (equivalent to 0.1 or 1 nmol) volumes, speed-vacuum dried, and stored at -20°C.

Prepare metaphase or interphase cells

3. Metaphase chromosomes or interphase nuclei are made by standard cytogenetic procedures (Lawrence et al. 1990) and stored at -20°C if not being used.

Post-fix chromosomal preparations

4. We strongly recommend a post-fixation for all slides. After removal from the freezer or being freshly made, slides are simply fixed again in methanol:acetic acid (2-3:1) for 1-2 h at room temperature, dehydrated in a cold 70, 80, 90, 100% ethanol series, and air dried up to overnight. If necessary, slides can be treated with pepsin. We found

that this treatment is particularly valuable in interphase FISH, resulting in cleaner and stronger signals in telomeres or centromeres. Alternatively, slides can be treated 1 mg/ml of pepsin pH 2.0 for 10 min at 37°C. This step can, however, be omitted for PNA-FISH on telomeres, centromeres, or trinucleotide repeats (Chen et al. 1999).

Denature the chromosomes

5. After washing in 1×PBS at 37°C for 5 min, slides are fixed in 4% formaldehyde in 1×PBS, pH 7.2 at 37°C for 1.5 min for human chromosomes and 2 min for mouse chromosomes. After fixation, the slides are immediately dehydrated with a cold ethanol series and air dried for at least 10 min.

PNA-DNA hybridization

6. A total of 0.1 nmol of PNA is needed for 10 to 15 slides. First, 0.1 nmol of PNA is dissolved in 200 µl of pre-warmed 1×hybridization mix containing 70% deionized formamide, 500 (50-1000) nM of fluorochrome-labeled PNA probe, 1% (w/v) blocking reagent in 10 mM Tris pH 7.2 (no Na+). PNA hybridization mix is incubated at 37°C or 50°C in the dark for at least 15 min. 15-20µl of PNA (500 nM) is transferred to each slide. A coverslip of 20×20 mm or 20×30 mm is applied and sealed with rubber cement. The DNA on the slide is denatured again at 80°C for 3 min and then hybridized in a dark, wet chamber at 37°C (20-40°C) for 2 h (0.5-24).

Post-hybridization washes

7. From this step, excessive light should be avoided to prevent fluorescence breaching. After hybridization, slide preparations are washed three times with 70% formamide/10 mM Tris pH 7.2 for 10 min each and with 50 mM Tris/150 mM NaCl/0.05% Tween-20 pH 7.5 for 5 min each. The slides are then dehydrated in an ethanol series and air dried in the dark. Chromosomes are counterstained with either 0.1 µg/ml of 4,6-diamidino-2-phenylindole-dehydrochloride (DAPI) in antifade for rhodamine- and fluorescein-labeled probes or 0.6 µg/ml of propidium iodide (PI) for fluorescein-labeled probes. It is recommended to analyze slides afterwards. Fluorescent signals can, however, be well retained at -20°C for a few weeks.

PNA-FISH short protocol

Make your favorite chromosome preparations.

Pre-hybridization preparations

1. Post-fixation in methanol:acetic acid (3:1) for 1–2 h.
2. Dehydration in 70%, 80%, 90%, and 100% ethanol.
3. Air dry (store desiccated until needed) Then:
4. 1×PBS at 37°C for 5 min.
5. Post-fixation of the chromosomes in 4% formaldehyde at 37°C for 1.5–2 min.
6. Dehydrate slides in the ethanol series and air dry.

Hybridization

Total of 0.1 nmol of PNA is needed for 10-15 slides.

1. 0.1 nmol of PNA is dissolved in 200 μl of pre-warmed 1×hybridization mix (70% de-ionized formamide, 500 (50-1000) nM of fluorochrome-labeled PNA probe, 1% (w/v) blocking reagent in 10 mM Tris pH 7.2 (no Na+)).
2. PNA hybridization mix is incubated at 37°C or 50°C in the dark for at least 15 min.
3. 15-20 μl of PNA (500 nM) is transferred to each slide. A coverslip of 20×40 mm or 20×30 mm is applied and sealed with rubber cement.
4. The DNA on the slide is denatured again at 80°C for 3 min (PCR program FISH1) and then hybridized in a dark, wet chamber at 37°C (can be adjusted from 20-40°C) for 2 h (can be adjusted from 0.5-24 h).

Post-hybridization washes

From this step, excessive light should be avoided to prevent fluorescence bleaching. After hybridization:

1. Slide preparations are washed 3×10 min with 70% formamide/10 mM Tris pH 7.2.
2. Then slides are washed 3×5 min with TNT buffer (50 mM Tris/150 mM NaCl/0.05% Tween-20 pH 7.5).
3. The slides are then dehydrated in an ethanol series (70%, 80%, 90%, 100% -30°C and then 100% at room temp.) and air dried in the dark.
4. Chromosomes are counterstained with either 0.1 μg/ml of 4,6-diamidino-2-phenylindole-dehydrochloride (DAPI) in anti-fade for rhoda-

mine- and fluorescein-labeled probes or 0.6 μg/ml of propidium iodide (PI) for fluorescein-labeled probe.

It is recommended to analyze the slides immediately afterwards. Fluorescent signals can, however, be well retained -30°C for a few weeks.

Troubleshooting

Loss of PNA

a. Do not use polystyrene materials for handling or storing PNA.

b. Keep PNA in dd. H_2O, not in neutral pH buffer that may cause PNA precipitation at low temperatures.

c. Dissolve PNA in pre-warmed hybridization buffer and place PNA in 37-50°C for at least 15 min.

No signals or weaker signals

a. Use telomeric or centromeric repeat PNA as a positive control. If control signals are seen clearly.

b. Check the concentration of PNA by scanning at 260 (230-650) nm. Examine the second peak for labeled dye.

c. Make sure hybridization and washing procedures have been executed properly. Lower hybridization temperature from 37°C to 30°C or even room temperature or decrease formamide concentration from 70% to 40% for hybridization and washing, particularly for shorter PNA 12-mer or 15-mer or AT-rich sequences.

High background

a. Too much PNA probe. Check PNA concentration.

b. Increase stringency of hybridization and washing such as a higher concentration of formamide, increased washing temperature, and longer time.

c. Add a blocking step in the procedure, simply by treating slides with 1×blocking solution at 37°C for 5 min, air dry, and go to hybridization step.

d. During the process for making chromosomal slides, make sure any medium (mainly serum) is completely washed away using 1×PBS.

References

Chen C, Hong Y-K et al. (1999). Single Base discrimination of cen P-B repeats on mouse and human chromosomes with PNA-FISH. Mammalian Genome 10:13-18

Chen C, Wu B et al. (2000). Chromosome identification and sequence-specific structural analysis with short DNA oligomers. Mammalian Genome 11:384-391

Egholm, M , O. Buchardt et al. (1993). PNA hybridizes to complementary oligonucleotides obeying the Watson-Crick Hydrogen bonding rules. Nature 365: 566-568

Egholm, M., O. Buchardt, et al. (1992). Peptide nucleic acids (PNA): Oligonucleotide analogues with an achiral peptide backbone. J. Am. Chem. Soc. 114: 1895-1897

Hande, M. P., P. M. Lansdorp, et al. (1998). Induction of telomerase activity by in vivo X-irradiation of mouse splenocytes and its possible role in chromosome healing. Mutant Res 404(1-2): 205-14

Hultdin, M., E. Gunlund, et al. (1998). Telomere analysis by fluorescence in situ hybridization and flow cytometry. Nucleic Acids Res 26(16): 3651-6

Lansdorp, P. M., N. P. Verwoerd, et al. (1996). Heterogeneity in telomere length of human chromosomes. Human Mol. Gen. 5(5): 685-691

Lawrence JB, Singer RH, McNeil JA (1990) Interphase and metaphase resolution of different distances within the human dystrophin gene. Science 249, 928-932

Leijon, M., A. Graslund, et al. (1994). Structural Characterization of PNA-DNA Duplexes by NMR. Evidence for DNA in a B-like Conformation. Biochemistry 33: 9820-9825

Martens, U. M., J.Mark, et al. (1998). Short telomeres on human chromosome 17p. Nature Genetics: 76-80

Nielsen, P. E., M. Egholm, et al. (1991). Sequence-selective recognition of DNA by strand displacement with a thymine-substituted polyamide. Science 254: 1497-1500

Nielsen, P. E., M. Egholm, et al. (1993). Peptide nucleic acids (PNAs): Potential Antisense and Anti-gene Agents. Anti-Cancer Drug Design 8: 53-63

Rufer, N., W. Dragowska, et al. (1998). Telomere length dynamics in human lymphocyte subpopulations measured by flow cytometry. Nature Biotechnology 16: 743-747

Taneja, K. L. (1998). Localization of trinucleotide repeat sequences in myotonic dystrophy cells using a single fluorochrome-labeled PNA probe. BioTechniques 24(3): 472-476

Abbreviations

DAPI	of 4,6-diamidino-2-phenylindole-dehydrochloride
FISH	fluorescence in situ hybridization
PBS	phosphate-buffered saline
PI	propidium iodide
PNA	peptide nucleic acid
Q-FISH	quantitative FISH
TFA	trifluoroacetic acid

Strand-Specific Fluorescence in Situ Hybridization: CO-FISH and COD-FISH

JULIANNE MEYNE, SUSAN M. BAILEY, and EDWIN H. GOODWIN

Introduction

The ability to prepare single-stranded target DNA allows innovative uses of FISH technology for studies of chromosome organization. Standard FISH methods require functionally single-stranded DNAs at the beginning of the hybridization step to allow nucleotide base pairing between the probe and the complementary chromosomal target sequence. This usually involves denaturation of double-stranded probe and target DNAs to induce a temporary separation of the DNA strands. The strand-specific FISH method involves selective removal of one of the strands from the DNA helixes within metaphase chromosomes and/or interphase nuclei. The result, for the purposes of hybridization, is single-stranded target DNA. Single-stranded probes can then be hybridized to the single-stranded target DNA without the denaturation step, resulting in strand-specific hybridization. In the context of this chapter, "single-stranded probe" refers to nucleotide sequences without complementary strands present, not denatured double-stranded probes.

Using the CO-FISH method, single-stranded probes of repetitive DNA sequences arranged in head-to-tail tandem arrays along the chromosome or sequences present as single copies in the genome will hybridize to only one chromatid of a metaphase chromosome. Probes of DNA sequences, such as Alu, that are present in both directions on each DNA strand will hybridize to both chromatids. Thus, the method can be used to determine

✉ Julianne Meyne, Los Alamos National Laboratory, Bioscience Division, MS-M888, Los Alamos, NM, 87545, USA (*phone* +505-667-2697; *fax* +505-667-2891; *e-mail* co-fish@telomere.lanl.gov)
Susan M. Bailey, Los Alamos National Laboratory, Bioscience Division, MS-M888, Los Alamos, NM, 87545, USA
Edwin H. Goodwin, Los Alamos National Laboratory, Bioscience Division, MS-M888, Los Alamos, NM, 87545, USA

the orientation of repetitive sequences along the chromosome and has been designated CO-FISH for Chromosome Orientation-FISH (Goodwin and Meyne, 1993). When combined with a telomere sequence probe, the procedure can be used to determine both the orientation and the short

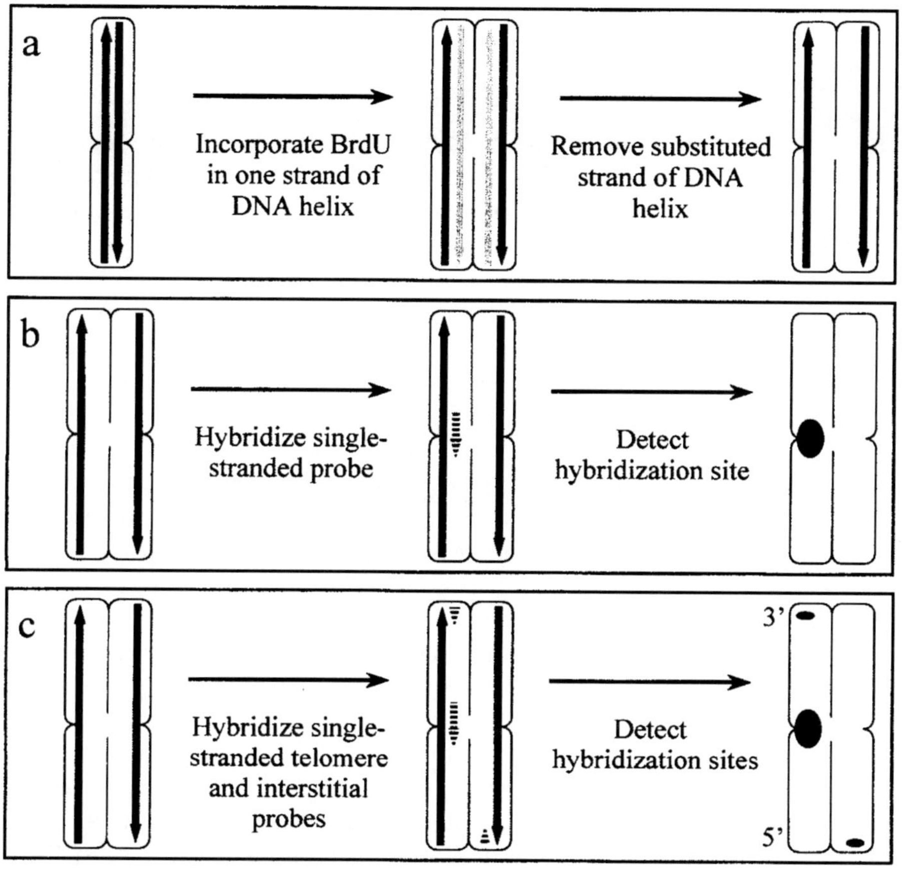

Fig. 1. Schematic illustration of our strategy for CO-FISH and COD-FISH. **a** Preparation of single-stranded target DNA: The original strands of the DNA helix of a G1 chromosome are shown as *black arrows* indicating 5' to 3' direction of each strand. BrdU incorporation into newly replicated DNA strands is illustrated by *gray arrows*. Selective removal of the substituted strands results in single-stranded target DNA in each chromatid. **b** CO-FISH: Hybridization of a single-stranded DNA probe is represented by the *striped arrow*. The detected hybridization site is shown as a *black oval*. **c** COD-FISH: Hybridization of the antisense strand of an interstitial DNA sequence probe is represented by the *striped arrow* and hybridization of the C-rich single-stranded telomere sequence probe is represented by the *striped arrowheads*. The detected hybridization sites are shown as *black ovals*

arm-to-long arm direction of DNA sequences along chromosomes. This variation of the method is called COD-FISH (Chromosome Orientation and Direction-FISH; Meyne and Goodwin, 1995). Multiple color and sequential hybridizations are easily accomplished on slides prepared using the CO-FISH method. Because no denaturation step is required the chromosomes maintain their morphology, even after several hybridizations.

Outline

Figure 1 summarizes the strategy of our approach for obtaining single-stranded DNA for CO-FISH. The procedure is effective for both metaphase chromosomes and interphase DNA. Cultured cells are grown for one round of replication in medium containing the nucleotide analogue bromodeoxyuridine (BrdU). Because of the semiconservative nature of DNA synthesis, BrdU will be substituted for thymidine at many sites within the newly replicated DNA strands. Thus, each newly replicated chromosome contains sister chromatids with singly substituted complementary DNA strands. That is, BrdU partially replaces thymidine in one DNA strand of each chromatid and the substitution takes place in opposite strands of the two sister chromatids (Fig. 1a). At this point, the cells are prepared for microscopy. Any of the standard metaphase chromosome preparation methods or whole cell fixation methods compatible with FISH can be used for CO-FISH. Cells can also be grown directly on coverglasses or microscope slides and fixed in place.

At this point, the newly replicated DNA strands are removed. The prepared slides are stained with the fluorescent DNA-binding dye Hoechst 33258 and then exposed to long-wave ultraviolet light. The UV exposure induces numerous strand breaks that preferentially occur at BrdU substitution sites. The nicked DNA strands produced by this treatment serve as a substrate for digestion by exonuclease III. This enzyme excises nucleotides by starting at the sites of nicks and proceeding along the DNA strand. This process effectively removes the BrdU substituted strand from each chromatid, while leaving the unsubstituted strand essentially intact. It is important to note that not only is each chromatid now single-stranded, but the single DNA strands remaining in the sister chromatids are complementary.

Strand-specific hybridization of CO-FISH slides requires that the probe must also be single-stranded. The complement of DNA sequences present as single copies in the genome or of repetitive sequences arranged as a head-to-tail tandem array will only be present on one of the DNA strands.

Thus, single-stranded probes of these sequences will hybridize to only one chromatid, as shown in Fig. 1b. Repetitive sequences that are present in both directions within the human genome will hybridize to both chromatids.

The COD-FISH method uses knowledge of the structural organization of most eukaryotic telomeres to determine the direction of DNA sequences hybridized to slides prepared using the CO-FISH method. The eukaryotic telomere sequence is a tandem repeat, $(TTAGGG)_n$, with the C-rich sequence at the 5' end and the G-rich sequence at the 3' end of each chromosomal DNA strand. Co-hybridization of a single-stranded telomere sequence probe with a single-stranded probe of an interstitial DNA sequence to CO-FISH prepared chromosomes allows determination of the strand direction of the interstitial sequence. For example, in Fig. 1c the C-rich strand of the telomere sequence probe hybridizes to the G-rich 3' end of each chromatid. The antisense strand of an interstitial DNA sequence probe hybridizes to the chromatid with the telomere signal on the short arm of the chromosome. This information is used to determine the short arm-to-long arm chromosome direction of the sense strand of the interstitial sequence (see Meyne and Goodwin, 1995).

Because multiple, sequential hybridizations can be done without significant degradation of chromosome morphology, the CO-FISH method is also useful for preparing single-stranded target DNA for hybridization with denatured double-stranded probes. However, a double-stranded probe will always hybridize to both chromatids, as each strand of the probe will bond with its complementary site.

Materials

- Cells in culture and appropriate growth reagents and equipment
- 10^{-3} M bromodeoxyuridine stock solution in water. Aliquots of this solution can be kept frozen for several months, but must be protected from direct light at all times.
- Phosphate buffered saline
- Colcemid
- 75 mM potassium chloride hypotonic solution
- Methanol
- Acetic Acid

- Microscope slides and coverslips
- Hoechst 33258
- Ultraviolet light source, long-wave (~365 nM, see comment 1)
- Exonuclease III with buffer. Most companies supply the appropriate buffer with the enzyme, although any buffer recommended for use with exonuclease III can be used.
- Labeled, single-stranded DNA probe (see comment 2)
- Formamide
- 20×SSC, pH 7, stock solution (1×SSC = 0.15M sodium chloride and 0.015 M sodium citrate)
- 2×SSC solution prepared from 20×SSC stock solution
- PN Buffer made by titrating 0.1M solutions of dibasic and monobasic sodium phosphate to give a solution at pH 8, then add Triton X-100 to final concentration of 1%
- Fluorescent counterstain, such as DAPI
- Antifade solution
- Fluorescence microscope with camera and/or image analysis system

Procedure

Bromodeoxyuridine incorporation

Any type of cell culture with a reasonable rate of growth can be used for CO-FISH. It is desirable to know the approximate doubling time of the cells to determine the optimal time required for most of the cells to complete one round of DNA replication.

1. Add bromodeoxyuridine stock solution to cell culture medium to achieve a final concentration of 10^{-5} M. BrdU concentration may need to be adjusted for some cell cultures. Other nucleotide analogues may also be used but results may vary.
2. Incubate the cell culture for a period of time equal to about one doubling time of the cells being used.

3. Add Colcemid to collect cells at metaphase. We use 0.1-0.2 μg/ml final concentration for the last 2-4 h of culture for human fibroblast cultures.
4. Harvest the cell culture using standard methods. For example, for metaphase chromosome preparation of human fibroblast cultures, we use a 15 min hypotonic treatment in 75 mM KCl solution followed by fixation in 3:1 methanol-acetic acid. The cell suspension is dropped onto clean glass microscope slides and allowed to air dry. Slides are best if aged at least 24 h before use.

Preparation of single-stranded DNA

1. Stain the slides with a solution of 0.5 μg/ml Hoechst 33258 in 2×SSC for 15 min at room temperature.
2. Rinse briefly in distilled water and air dry.
3. Place 2–3 drops of 2×SSC on each slide and cover with a 22×50 mm #1 grade glass coverslip.
4. Expose to long-wave UV light for 30 min.
5. Remove the coverslip, rinse briefly in distilled water and air dry.
6. Place 50 μl of exonuclease III solution (3 U/μl in buffer supplied by manufacturer) onto the slide and cover with a 22×50 mm glass coverslip. We have used exonuclease III from a variety of sources with equal success.
7. Incubate at room temperature for 5 min. Some cell types or slides more than a few weeks old may require longer times of exonuclease incubation.
8. Remove the coverslip, rinse briefly in distilled water and air dry.

Hybridization and detection of probe

No denaturation step is necessary because the DNA is single-stranded. The hybridization mix, incubation and detection procedure should be the same as for standard FISH with the same probe.

1. Prepare probe hybridization mix, place it on the slide and cover with a coverslip.

2. Incubate in moist chamber for the time appropriate for the probe used.
3. Remove coverslip and wash slide in 5 changes of 2×SSC at 42°C. This washing procedure works well for synthetic oligomer probes. Other probes may require an initial high stringency wash in a formamide solution, at the concentration used for hybridization, to reduce background fluorescence.
4. Transfer slide to a solution of PN buffer for about 5 min before counterstaining with DAPI and mounting in antifade solution.

Multiple hybridizations

Additional hybridizations can be done immediately after completing the post-hybridization wash steps. Alternately, the slide can be mounted with antifade and viewed under fluorescence before additional hybridizations are done. In the latter case, the coverslip is removed and the slide is washed in 2–3 changes of 2×SSC. The slide is drained of excess SSC (but not allowed to dry) and the next hybridization mix is placed on the slide for incubation and processing. For hybridization of single-stranded probes for both strands of a sequence, each labeled for detection with different fluorochromes, it is often best to perform sequential hybridizations, rather than placing both probes in the same hybridization mix. This is especially true for short repetitive sequence probes. We have frequently done 3–5 sequential hybridizations without degradation of chromosome morphology or significant increases in background.

Analysis of fluorescence

It is important to keep in mind that the target DNA is single-stranded. The fluorescence signal may be less intense than for standard FISH using the same probe. For most repetitive DNA sequence probes, such as alpha-satellite and classical satellite human DNA, the intensity of the fluorescence is easily observed directly through the microscope. For other probes an image analysis system will be advantageous.

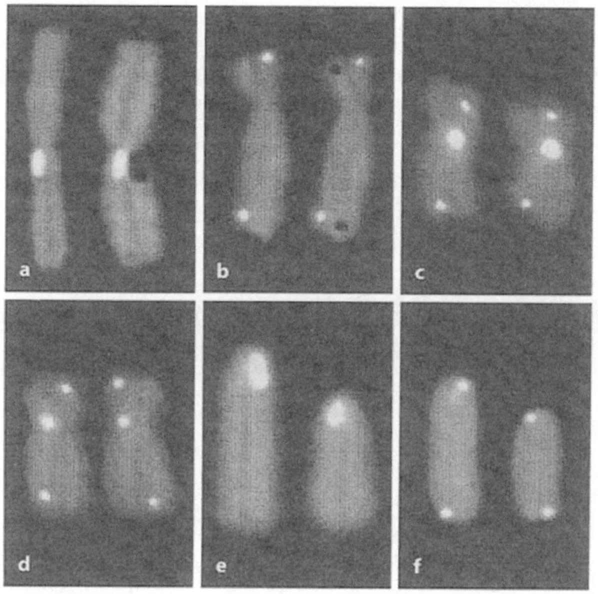

Fig. 2. Examples of CO-FISH and COD-FISH on human and mouse chromosomes. The hybridization sites appearing as *white areas* indicate probes labeled with FITC, while probes labeled with Texas Red have been pseudocolored *black*. **a** Human chromosome-1 hybridized with repetitive sequence probe specific for the long arm heterochromatic region. Chromosome on left is from a slide hybridized with only one strand of the sequence. Chromosome on the right was sequentially hybridized with probes from both strands of the sequence, one strand was labeled with FITC and the other with Texas red. **b** Human chromosome-5 hybridized with the telomere repeat probe. On the *left* is a chromosome after hybridization with the C-rich strand. On the *right* is the same chromosome after a second, sequential labeling with the G-rich strand. **c** Homologues of human chromosome-X hybridized with both the C-rich strand of the telomere sequence and a single-stranded probe of an alpha-satellite DNA sequence specific for the centromeric region of chromosome-X. **d** Homologues of human chromosome-8 hybridized with both the C-rich strand of the telomere sequence and a single-stranded probe of an alpha-satellite DNA sequence specific for the centromeric region of chromosome-8. The change in the positions of the telomere hybridization sites relative to the centromeric site indicate the presence of a pericentromeric inversion. **e** Mouse chromosomes hybridized with a single-stranded probe of the mouse major satellite. **f** Mouse chromosomes hybridized with a single-stranded probe of the telomere repeat

Results

CO-FISH and COD-FISH have been used in a variety of studies of chromosome structure and organization. Figure 2 shows images of chromosomes after strand-specific hybridizations with various repetitive DNA probes. Additional images and details of using CO-FISH and COD-FISH for cytogenetic analysis of inversions, isochromosomes and lateral asymmetry are available in the references listed at the end of this chapter. Recently, the CO-FISH method has been used in combination with whole-chromosome paint probes for measuring chromosome aberration frequencies in first-versus-second division cells for biological dosimetry (Jordan et al., 1999).

Comments

1. Long-wave UV (about 365 nM) is preferable for inducing nicks in the Hoechst 33258 stained DNA, as short-wave UV can nick unsubstituted DNA. The exposure time will vary with the type and wavelength of the UV lamp used. If a sunlamp is used, place a thin sheet of plastic between the UV source and the slides to protect the DNA from dimer formation that will interfere with removal of the nicked DNA strand. Use caution to avoid optic exposure when using UV light. An enclosed light source is best. We use a Stratalinker 1800 UV Irradiator (Stratagene, La Jolla, CA). Alternately, wear eyeglasses or a face shield specifically designed for use with UV light.
2. There are a variety of methods for preparation of single-stranded DNA probes, including oligonucleotide synthesis and PCR. We have worked primarily with repetitive DNA sequences and find that synthetic oligonucleotide probes provide the brightest signals with the lowest background levels. This is especially true when probes are labeled directly with fluorochrome-tagged molecules.

References

Bailey SM, Meyne J, Cornforth MN, McConnell TS, Goodwin EH (1996) A new method for detecting pericentric inversions using COD-FISH. Cytogenetics and Cell Genetics 75:248-253

Bailey SM, Goodwin EH, Meyne J, Cornforth MN (1996) CO-FISH reveals inversions associated with isochromosome formation. Mutagenesis 11:139-144

Goodwin E, Meyne J (1993) Strand-specific FISH reveals orientation of chromosome-18 alphoid DNA. Cytogenetics and Cell Genetics 63:126-127

Goodwin EH, Meyne J, Bailey SM (1993) Strand-specific in-situ hybridization reveals long-range molecular order in repetitive DNA. Cytogenetics and Cell Genetics 63:253-253

Goodwin EH, Meyne J, Bailey SM, Quigley D (1996) On the origin of lateral asymmetry. Chromosoma 104:345-347

Jordan R, Edington J, Evans HH, Schwartz JFL (1999) Detection of chromosome aberrations by FISH as a function of cell division cycle (harlequin-FISH). Biotechniques 26:532-534

Meyne J, Goodwin EH, Bailey SM, Tennyson RB, Moyzis RK (1993) Strand direction of human repetitive DNA-sequences determined by strand-specific fluorescent in-situ hybridization. American Journal of Human Genetics 53:580-580

Meyne J, Goodwin EH (1993) Chromosome orientation and direction of repetitive sequences determined by strand-specific fluorescent in-situ hybridization. Cytogenetics and Cell Genetics 63:253-253

Meyne J, Goodwin EH, Moyzis RK (1994) Chromosome localization and orientation of the simple sequence repeat of human satellite-I DNA. Chromosoma 103:99-103

Meyne J, Goodwin EH (1995) Direction of DNA-sequences within chromatids determined using strand-specific FISH. Chromosome Research 3:375-378

Pandita TK, Derubeis D (1995) Spontaneous amplification of interstitial telomeric bands in Chinese-hamster ovary cells. Cytogenetics and Cell Genetics 68:95-101

The Combined Use of Nuclease in Situ Digestion and FISH (NU-FISH)

Jaime Gosálvez, Joaquina de la Torre,
Carmen López-Fernández, and José Luis Fernández

Introduction

Metaphase chromosomes or even interphase nuclei are highly organised structures made up of proteins and DNA and have a high degree of complex packaging. However, this does not preclude the "in situ" cleavage of DNA on conventionally fixed cytological preparations with active nucleases, i.e. fixed chromatin is susceptible to "in situ" digestion. DNA extraction can be monitored by staining the cytological preparations with DNA-specific dyes. This approach may furnish information on the presence/absence of enzyme receptive sites along the chromosomes. The removal of specific DNA fragments results in the production of a chromosome banding pattern, which can be used to analyse the differences between and resemblance of distinct chromosomal regions and/or of genomes of related species.

After almost 15 years of experience using endo- and exo-nucleases to induce the selective removal of DNA from fixed and unfixed chromatin, we know that DNA extractability depends on a series of factors (López-Fernández et al. 1991; de la Torre and Sumner 1994). All of them must be carefully evaluated before comparing the results obtained from slightly different experimental conditions. However, the interpretation of the results obtained after nuclease "in situ" digestion and post-staining with DNA-specific dyes is far from straightforward. This situation substantially

✉ Jaime Gosálvez, Universidad Autónoma de Madrid, Departamento de Biología, Unidad de Genética, Madrid, 28049, Spain (*phone* +34-91-397-8210; *fax* +34-91-397-8344; *e-mail* jaime.gosalvez@uam.es)
Joaquina de la Torre, Universidad Autónoma de Madrid, Departamento de Biología, Unidad de Genética, Madrid, 28049, Spain

Carmen López-Fernández, Universidad Autónoma de Madrid, Departamento de Biología, Unidad de Genética, Madrid, 28049, Spain
José Luis Fernández, Centro Oncológico de Galicia, Unidad de Genética, Avenida

changes when the DNA remaining on the slide after "in situ" digestion is "in situ" hybridised with specific DNA sequences. This approach reveals more accurately the action of the nucleases on targeted DNA sequences and interpretations concerning the effect of nucleases on the chromatin can be made in parallel with the chromosome assignment of the probe.

Two main types of nucleases have been extensively used to produce a selective "in situ" DNA extraction/modification: Restriction Endonucleases (RE) and Modifying Enzymes (ME).

Outline

So far, three different approaches have been used for Nuclease (NU) digestion in combination with FISH (NU-FISH). The general steps for these subprotocols are outlined in Fig. 1.

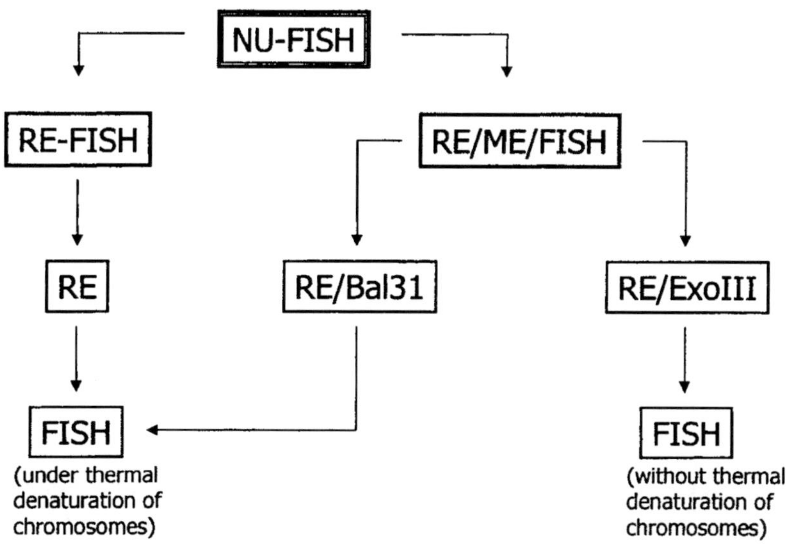

Fig. 1. Outline of the combined use of nuclease in situ digestion and FISH (NU-FISH)

Procedure

Restriction endonucleases

REs recognise specific DNA targets. REs such as AluI, MboI, Sau3A, MspI, HpaII, HaeIII, DdeI have been extensively used for "in situ" DNA digestion (Gosálvez et al. 1997a); all of them have been able to induce longitudinal differentiation of the chromosomes of different species.

In situ digestion with REs

1. Chromosome preparations are made following standard procedures of spreading or squashing. The slides are left at room temperature for at least 2 h in order to ensure that they are completely dried before digestion.
2. Digestion of chromosomes with REs is always performed on freshly dried preparations (no more than 12-24 h after having been prepared). The standard procedure was described by Mezzanotte et al. (1985). However, slight modifications of the standard procedure may be required for each type of chromosome.
   10-50 units of the RE diluted to 20-50 μl in the buffer recommended by the supplier are usually used per slide. The enzyme solution is placed on the slide and covered with a coverslip to spread the enzyme evenly.
3. The slides are incubated in a humid box for a selected time and at the temperature specified by the manufacturer. Usually, overnight incubation at 37°C is a good standard for most REs. In order to prevent evaporation during long-term incubations, the coverslip can be sealed with rubber solution.
4. Control experiments are performed under identical conditions by treating the preparations with the incubation buffer without the enzyme.
5. The reaction is stopped by washing the slides at room temperature for 5 min in 2×SSC (pH 7).
6. Those preparations to be used for FISH are dehydrated in an ethanol series (70%, 85%, 100%) and air dried.

Notes

Chromosomes obtained from different sources (insects, fish, amphibia, anura, birds, mammals and plants) have been used by different authors. Preparations are obtained following standard procedures of fixation and current techniques of spreading and squashing. Unfixed chromatin is also susceptible to RE-"in situ" digestion (Burkholder 1989, Gosálvez et al. 1993). Standard prepared chromosomes or chromatin fibre can be used.

Any RE has the potential to produce selective DNA cleavage on fixed chromatin and, in turn, to provoke DNA loss. However, since DNA removal is more effective when short DNA fragments are produced, the use of four, five or six DNA base pair cutters is preferable. In fact, after cleaving with REs with more than 6 bp targets, long DNA fragments are produced that remain tangled in the chromatin mass.

Modifying enzymes

Modifying enzymes (ME) alter the DNA or RNA according to a specific polynucleotide structure. So far, only Exonuclease III and Bal 31 have been used in combination with REs, for "in situ" hybridisation purposes.

Exonuclease III

Exonuclease III (Exo III) produces a stepwise removal of nucleotides from the 3'-ends of double-stranded DNA, thus giving rise to single-stranded DNA that can be used as a target for DNA probing.

1. Chromosomal DNA is nicked with REs according to the subprotocol described for RE-digestion but using shorter incubation times (15-30 min).
2. After RE digestion, the slides are washed in 2×SSC (pH 7), dehydrated in an ethanol series (70%, 85%, 100%) and air dried.
3. The slides are incubated with 150U of ExoIII in 50 μl of buffer for 10 min at 37°C.
4. After ExoIII digestion, the slides are washed in 2×SSC (pH 7), dehydrated in an ethanol series (70%, 85%, 100%) and air dried.

Notes

The action of ExoIII produces enzymatic denaturation of DNA; thus, single-stranded DNA (ssDNA) areas will already be available for "in situ" hybridisation of selected probes, i.e., no further heat denaturation of the chromosomes will be required.

Bal 31 Nuclease Bal 31 degrades both the 3' and the 5' termini of duplex DNA, causing a progressive shortening of double-stranded DNA (dsDNA).

1. Chromosomal DNA is digested overnight according to the standard subprotocol described for RE digestion.
2. After overnight RE digestion, the slides are washed in 2×SSC (pH 7), dehydrated in an ethanol series (70%, 85%, 100%) and air dried.
3. The slides are incubated with 5U of Bal 31 in 50 µl of buffer for 10 min at 37°C.
4. After Bal 31 digestion, the slides are washed in 2×SSC (pH 7), dehydrated in an ethanol series (70%, 85%, 100%) and air dried.

Notes The action of Bal 31 decreases the amount of double-stranded DNA (dsDNA), and so, less target DNA will be available for FISH. Thermal denaturation of the slides will be required before "in situ" hybridisation of the selected probes.

FISH after NU in situ digestion

After extracting and/or modifying the DNA on fixed chromosomes by the action of NUs, the DNA remaining on the slide can be "in situ" hybridised with any DNA probe.

As pointed out above, preparations treated with ExoIII are not thermally denatured, since this enzyme already produces ssDNA for the hybridisation. All other NU-digested preparations require gentle denaturation in 50% formamide/2×SSC (pH 7) for 105 s at 67°C. Usually, standard FISH protocols use higher temperatures and formamide concentrations for chromosome denaturation. However, since DNA has previously been "in situ" cleaved, aggressive denaturation would result in massive DNA loss. Besides this slight modification, the protocols for FISH after NU-digestion are those currently used for each material or DNA probe. See the articles included in this volume for specific FISH procedures.

Results

The present section aims to show a number of classical topics with regard to genome organisation in eukaryotes where the combined use of "in situ" digestion with Nucleases and FISH have provided valuable information.

Checking differences in satellite DNAs

Differences in chromatin organisation have been inferred by using NU-FISH in human lymphocytes. Extended or condensed chromatin in interphase nuclei and chromosomes were "in situ"-digested by the restriction endonuclease Alu I; the remaining DNA was hybridised with alphoid probes specific for chromosome 1 and chromosome X (DIZ5 and DXZ1 loci) and with a classical satellite DNA probe specific for chromosome 9 (D9Z1 locus). With this experiment we analysed the differential DNA removal produced by Alu I at specific repetitive DNA sequences with known restriction site frequency and distribution. Fluorescent hybridisation signals were quantified by high-resolution digital image analysis. The results showed that the specific repetitive DNA at D9Z1 locus (classical) remained resistant to digestion (compare Fig. 2a, a'). On the other hand, although restriction sites at the D1Z5 locus were at least as frequent as at the DXZ1 locus, they showed different hybridisation intensities after digestion; while DNA at the DXZ1 locus was extensively removed (compare Fig. 2b, b'), that at the D1Z5 locus was only partially removed (compare Fig. 2c, c'). These results indicate that chromatin organisation within constitutive heterochromatin partially prevents DNA extraction and, specifically, that alphoid DNA sequences seem to be differentially organised in different chromosomes. The same results were obtained from condensed and extended chromatin, suggesting that higher-order chromatin organisation does not influence "in situ" DNA cleavage and removal by Alu I (compare Fig. 2a-c with d-g) (Fernández et al. 1997).

Checking differences between tissue-specific cells

When fixed chromosomes obtained from human amniotic fluid cells and peripheral blood lymphocytes were "in situ"-digested with ExoIII followed by "in situ" random priming extension, an R-banding pattern was obtained in both cases. However, this pattern was more evident in lymphocytes than in amniocytes. In addition, constitutive heterochromatin of chromosomes 1, 16, Yq, and mostly the pericentromeric region of chromosome 9 were far more intensely labelled in amniocytes than in lymphocytes. By using a specific classical satellite DNA probe, FISH showed that this differential labelling was due to the greater sensitivity to ExoIII digestion of the constitutive heterochromatin of chromosome 9 in amniocytes. This experiment shows cryptic and qualitative differences in constitutive heterochromatic regions between different cells (Fernández et al. 1995).

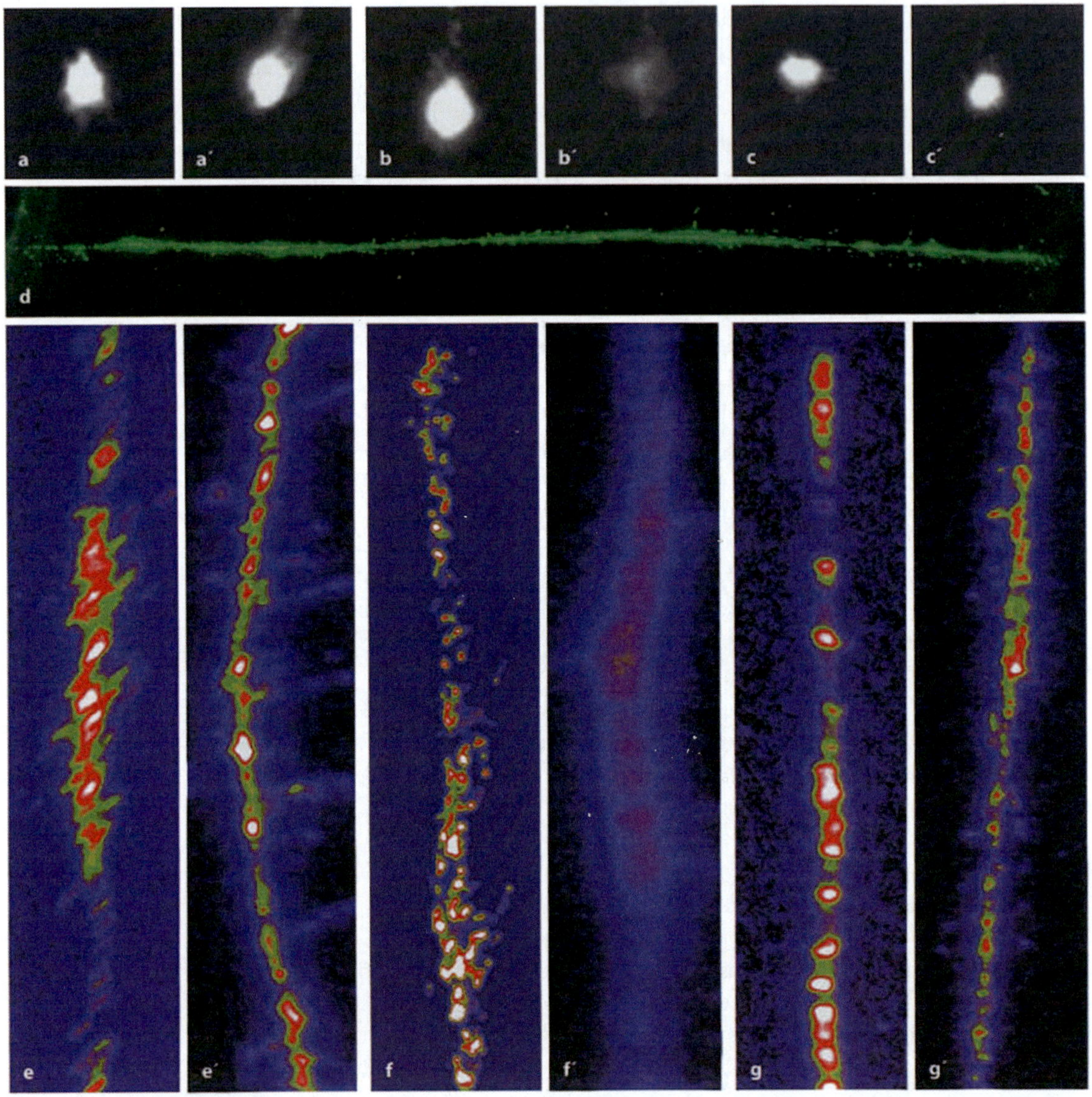

Fig. 2. NU-FISH in standard and extended chromatin fibre in human chromosomes using highly repetitive DNA probes. **a-c** Selected 9, X and 1 FISH signal from normally condensed nuclei after NU-FISH using Alu I and D9Z1, DXZ1 and D1Z5 as DNA probes. **d** Complete 10× extended chromatin fibre and selected regions in colour-intensity map (**e-g**) to show the effect of NU-FISH. Same conditions of digestion and hybridisation as in **a-c**. Note that only the X chromosome shows a reduction in fluorescence intensity. Prime figures correspond to Alu I-digested chromosomes. (All images except **d** with permission of S. Karger AG)

Checking differences in telomere DNA sequences

Telomere-like Chinese hamster interstitial pericentromeric DNA sequences and human telomeres are differentially affected by "in situ" digestion with ExoIII or nuclease Bal 31 (Fig. 3a-f). Quantification of the FISH signal produced by a telomeric probe (poly-TTAGGG) by digital image analysis, showed that ExoIII was 2.6 times more active in interstitial (Chinese hamster) than in terminal telomeric (human) chromosome regions. On the other hand, nuclease Bal 31 was 2.3 times more effective in terminal than in interstitial telomeric sequences. Both results support the hypothesis that the same DNA sequences can be differentially organised, depending on the chromosome region where they are located and probably on their functional role (Fernández et al. 1998).

Checking DNA methylation

Tandem REs that recognise the same restriction site but differentially cleave DNA according to the pattern of DNA methylation can be employed for the "in situ" digestion of fixed chromatin. For example, isoschizomers such as Msp I or Hpa II, which recognise CCGG sites, act differentially on DNA depending on whether the internal cytosine is methylated or not. While Msp I cleave CCGG sites independently of methylation, Hpa II does not cleave this site if the internal cytosine is methylated. RE-FISH using this pair of isoschizomers and post-hybridisation with rDNA probes, permits the visualisation of DNA methylation at CCGG sites.

For example, the grasshopper *Eyprepocnemis plorans* has a family of extra B-chromosomes containing long repeats of silent rDNA. When these chromosomes are digested with MspI and HpaII and the amount of fluorescence recovered after FISH using rDNA probes is measured, higher values of fluorescence are detected in B-chromosomes digested with HpaII than in MspI-digested Bs (Fig. 3g-i). This result indicates less severe DNA removal with HpaII because of internal cytosine methylation. In fact, when the same experiment is performed on Southern-blotted DNAs, differences in DNA methylation are observed at these sites (Gosálvez et al. 1997b).

Acknowledgements. This work was supported by the Consejo de Seguridad Nuclear, grants from FIS and SAF (Spain) and European Community.

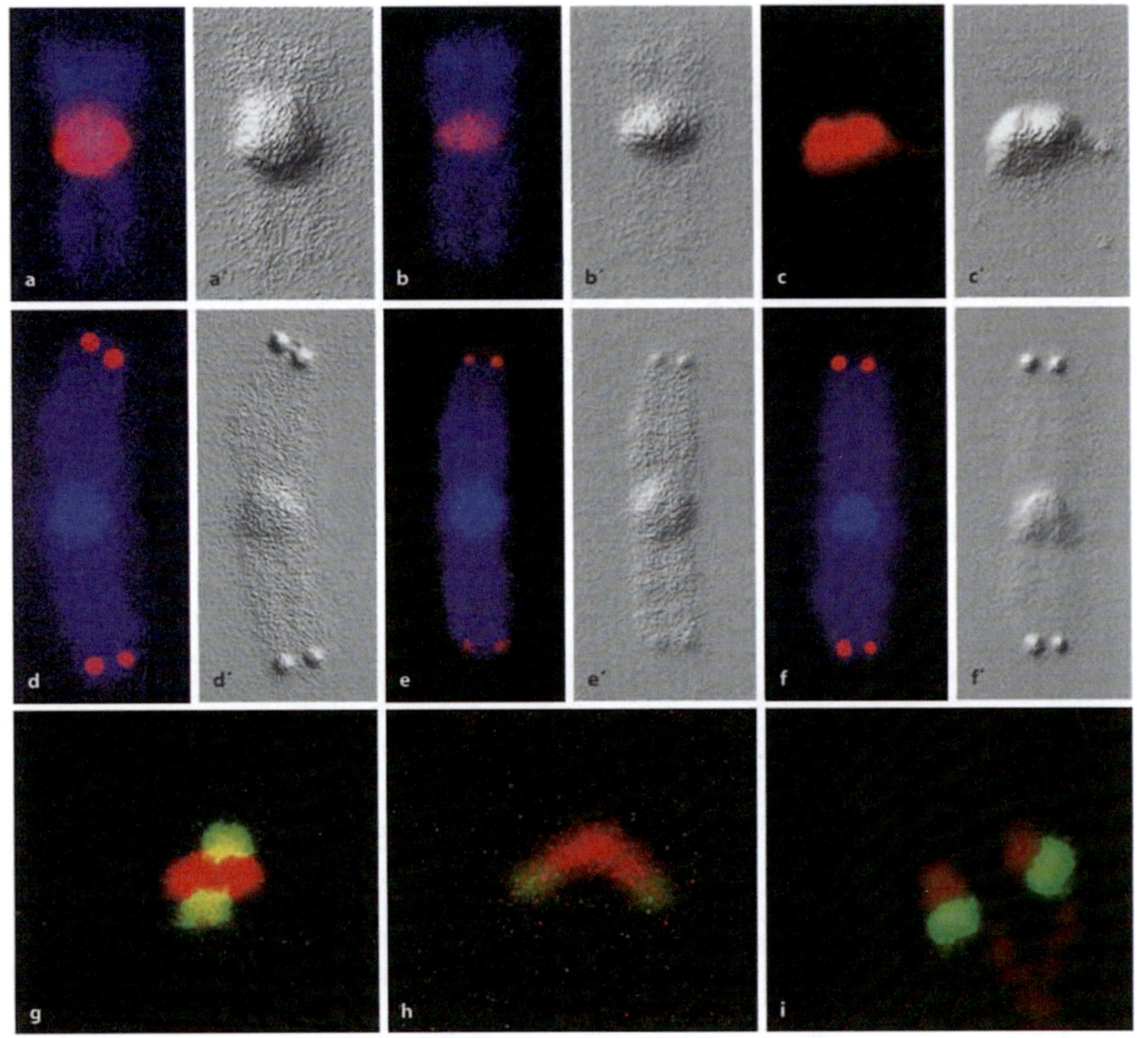

Fig. 3. a-f/a'- f' Selected chromosomes accompanied by a pseudo-3D intensity profile (prime figures) from Chinese hamster Don cells (**a-c/a'- c'**) and human lymphocytes (**d-f/d' – f'**) hybridised with a telomeric probe after "in situ" digestion with nucleases. **a/a'** Selected chromosome from a Chinese hamster cell digested with AluI alone (control). **b/b'** Selected chromosome from a Chinese hamster cell digested with AluI+ExoIII. **c/c'** Selected chromosome from a Chinese hamster cell digested with AluI+Bal31. ExoIII shows high enzymatic activity; produces a large amount of ssDNA in interstitial telomere-like sequence areas of Chinese hamster chromosomes, yielding a strong hybridisation signal. Bal31 shows a low enzymatic activity; removes little of the dsDNA, so after heat denaturation, the amount of target susceptible to being hybridised is only very slightly reduced and the FISH signal is very strong and closer to that of its control. **d/d'** Selected chromosome from a human lymphocyte digested with AluI (control). **e/e'** Selected chromosome from a human lymphocyte digested with AluI+ExoIII. **f/f'** Selected chromosome from a human lymphocyte digested with AluI+Bal31. ExoIII has low enzymatic activity; produces a reduced amount of ssDNA in terminal telomeric sequences of

References

Burkholder GD (1989) Morphological and biochemical effects of endonucleases on isolated mammalian chromosomes in vitro. Chromosoma 97:347-355

Fernández JL, Campos A, López-Fernández C, Gosálvez J, Goyanes V (1995) Difference in constitutive heterochromatin behaviour between human amniocytes and lymphocytes detected by a sequential "in situ" exonuclease III digestion-random primer extension procedure. J Med Genet 32:32-35

Fernández JL, Goyanes V, Ramiro-Díez J, Gosálvez J (1998) Evidence of a differential organization of chromatid containing terminal or interstitial (TTAGGG)n repeats by in situ digestion with nucleases. Cytogenet Cell Genet 82:195-198

Fernández JL, Valverde D, Goyanes V, Buño I, Gosalvez J (1997) AluI in situ digestion of human alphoid and classical satellite DNA regions. High-resolution digital image analysis of FISH signal from condensed and extended chromatin. Cytogenet Cell Genet 76:94-100

Gosálvez J, López-Fernández C, Fernández JL, Goyanes V (1993) In situ oligonucleotide priming to detect induced DNA breaks under electron microscopy. TIG 9:156-157

Gosálvez J, López-Fernández C, Goyanes V, Mezzanotte R (1997a) Chromosome differentiation using nucleases. An overview. In: Henriques-Gil N, Parker JS, Puertas MJ (eds) Chromosomes Today, vol. 12. Chapman and Hall, London, pp 23-49

Gosálvez J, Mason PL, López-Fernández C (1997b) Differentiation of individuals, populations and species of Orthoptera: the past, present and future of chromosome markers. In: Gangwere SK, Muralirangan MC, Meera Muralirangan (eds) The Bionomics of Grasshoppers, Katydids and their Kin. CAB International. University Press, Cambridge, pp 355-383

López-Fernández C, Gosálvez J, Ferrucci L, Mezzanotte R (1991) Restriction endonucleases in the study of eukaryotic chromosomes. Genetica 83:257-274

Mezzanotte R, Manconni PE, Rotzzo C, Ennas MG, Ferrucci L (1985) The banding pattern produced by restriction endonucleases. Bas Appl Histochem 29:115-120

Torre J de la, Sumner AT (1994) Analysis of chromosomes with Restriction Endonucleases and DNase hypersensitivity. In: Gosden JR (ed) Chromosome Analysis Protocols. Methods in Molecular Biology, vol. 29. Humana Press, New Jersey, pp 123-139.

human lymphocyte chromosomes, yielding a low hybridisation signal. Bal31 shows a high enzymatic activity; removes a considerable amount of dsDNA in the target area, so after heat denaturation, a faint FISH signal is evident compared with that of its control. **g-i** Methylation detection after NU-FISH on the B-chromosomes of *Eyprepocnemis plorans*. **g** Control chromosomes (without previous RE-digestion) after FISH using an rDNA probe. **h-i** B-chromosomes digested with MspI (**h**) and with HpaII (**i**) and probed with rDNA. Note the reduction of the fluorescence intensity only on the B-chromosomes digested with MspI. (All images except **g-i** with permission of S. Karger AG)

DNA Breakage Detection-FISH (DBD-FISH)

José Luis Fernández, Vicente Goyanes, and Jaime Gosálvez

Introduction

DBD-FISH (DNA Breakage Detection- FISH) allows DNA breaks in single cells to be detected and quantified, either in the whole genome or within specific DNA sequence areas. This simple technique couples microgel embedding procedures (Fairbain et al. 1995), alkali unwinding assays (Rydberg 1975) and FISH.

Cells trapped within an agarose matrix over a slide are immersed in an alkaline unwinding solution that produces single-stranded DNA (ssDNA) motifs beginning from the ends of the molecule or internal strand breaks. After neutralisation and protein removal in lysing solutions, the microgel is dehydrated and cells incubated with DNA probes. The amount of hybridised probe in a target sequence is related to the amount of melted ssDNA generated by the unwinding step, which is in turn proportional to the degree of local DNA breakage.

Outline

The general steps for DBD-FISH are outlined in Fig. 1. This is a flexible technique, allowing several variants depending on the experimental pur-

✉ José Luis Fernández, Centro Oncológico de Galicia, Laboratorio de Genética Molecular y Radiobiología, Avda. de Montserrat s/n, La Coruña, 15009, Spain (*phone* +34-981-287499; *fax* +34-981-287122; *e-mail* genetica@cog.es, and Unidad de la Mujer H. USP Santa Teresa, La Coruna, Spain)
Vicente Goyanes, Hospital Materno Infantil Teresa Herrera, Departamento de Genética, La Coruña, 15009, Spain, and Unidad de la Mujer H. USP Santa Teresa, La Coruna, Spain
Jaime Gosálvez, Universidad Autónoma de Madrid, Unidad de Genética, Facultad de Biología, Madrid, 28049, Spain

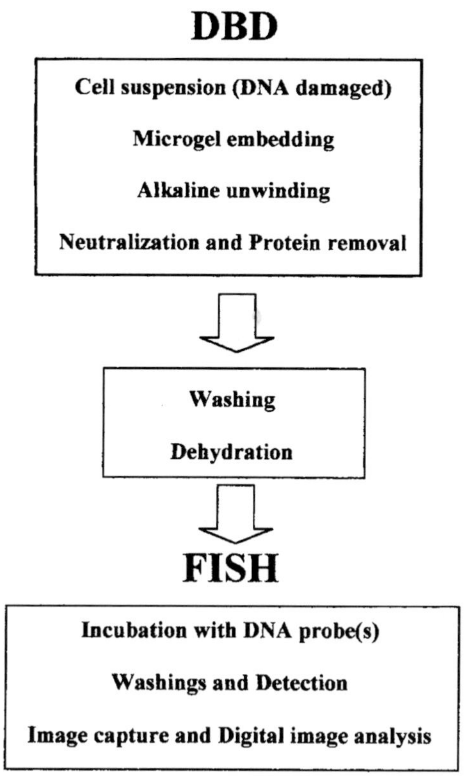

Fig. 1. Outline of the DBD-FISH protocol

pose, e.g. in composition, conditions and order of incubation in the unwinding and/or lysing solutions.

Materials

- Standard and low melting point agarose
- Epifluorescence microscope with appropriate filters and objectives (e.g. Leica DMRB)
- High-sensitivity CCD digital camera (e.g. Ultrapix-1600, Astrocam, Cambridge, UK)
- Image analysis software (e.g. Visilog 5.1, Noesis, France)

Solutions

- Alkaline unwinding solution (pH 12.2), freshly prepared

- 0.03 M NaOH
- 1 M NaCl
- Neutralising and lysing solution 1 (pH 7.5), stored at 4°C, used at room temperature
  - 0.4 M Tris HCl
  - 0.8 M DTT (Irritant)
  - 1% SDS (Harmful if swallowed. Irritant)
  - 0.05 M EDTA
- Neutralising and lysing solution 2 (pH 7.5), stored at 4°C, used at room temperature
  - 0.4 M Tris HCl
  - 2 M NaCl
  - 1% SDS
- Neutralising solution (pH 7.5), stored at 4°C, used at room temperature: 0.4 M Tris HCl
- Isotonic saline solution, stored at 4°C: 0.9% NaCl
- TBE buffer (pH 7.5), stored at room temperature
  - 0.09 M Tris borate
  - 0.002 M EDTA

Procedure

DBD-FISH for detection and quantification of X-ray induced initial DNA breakage (Fernández et al. 1998).

Microgel on slide

1. 1% low melting point agarose in water is kept at 37°C in a water bath.
2. Isolated cells in suspension (e.g. buffy coat or cells in culture medium) are gently mixed with the agarose to achieve the latter at a final concentration of 0.7%, at 37°C.
3. 50 μl of the suspension are pipetted onto a glass slide coated with a thin dry film of 0.65% standard agarose, and immediately covered with a glass coverslip (24×60 mm). Coating has been performed by immersing vertically for 2s the glass slide in the liquid standard agarose solution. After solidifying horizontally on a cold plate, it was dried at 80 °C. Other systems such as "sandwich" gels on frosted slides (Fairbain et al. 1995) can also be applied.

4. The slide is left at 4°C for 5 min to allow gel solidification.

DNA damage induction

5. Half of the slide is irradiated with a Philips RT-100 machine, while kept on a cold plate to avoid DNA repair. The other half (protected or out of the irradiation field) serves as a control with undamaged cells.

   Note: Cells may be damaged by physical or chemical agents either in the gel or while in a monolayer or suspension. We prefer to produce DNA breakage in cells embedded in microgels, avoiding possible mechanical DNA breaks by manipulation, or repairing when mixing with agarose at 37°C.

6. Immediately, the coverslip is removed by pulling carefully from an overhanging end. Immersion in abundant cold (4°C) isotonic saline solution, without buffer, for 5-10 min, aids coverslip removal.

Transformation of DNA breaks into ssDNA

7. The slide is horizontally immersed in abundant alkaline unwinding solution, freshly prepared, at 22°C for 2-5 min. Time, temperature and NaCl or NaOH concentration strongly influence the amount of ssDNA generated (Rydberg 1975). Alkali-labile sites are also transformed into ssDNA by this solution.

Neutralising and protein removal

8. The slide is transferred horizontally to abundant neutralising and lysing solution 1 for 20 min and then to abundant neutralising and lysing solution 2 for 10 min, both at room temperature. 0.4 M Tris, pH 7.5, neutralises the alkali and stops DNA denaturation, and DTT breaks protein disulphide bonds thereby allowing their extensive removal by detergent and highly concentrated salt solutions. Though unwinding solution partially lyses cells, specific lysing steps improve the accessibility of DNA probes to ssDNA.

   Note: Once the alkali is neutralised, DNA breaks produced afterwards, e.g. by manipulation, are irrelevant since they are not transformed into ssDNA, and so will not be detected by the probes.

Washing and dehydration

9. The slide is left horizontally in abundant TBE buffer for 15 min, at room temperature.

10. After immersion in 70%, 90% and 100% ethanol baths, 2 min each, at room temperature, the slide is air-dried by incubation at 37°C for 15-30 min. The agarose film turns into a very thin film that is well-attached to the slide. If necessary, the slides can be stored dry, in the darkness, at room temperature, for a week (and probably more), before hybridisation.

Probe hybridisation

11. All kinds of probes may be hybridised (whole genome, chromosome painting or subpainting, YACs, satellite, euchromatic, cosmids, etc.) as usual, overnight, in a moist chamber.

    Note: To avoid gel damage the use of rubber solution is not recommended for sealing glass coverslips.

Washing and detection

12. The slides are incubated in cold isotonic saline solution for coverslip removal. Washes are performed in accordance with the necessary stringency of the probe, with very gentle agitation.
13. Blocking, fluorochrome conjugated antibody and antibody washes are performed as set out in current protocols. No amplification of the signal is necessary since protein removal improves probe and reagent accessibility. Directly labelled probes may be used.
14. Cells are DAPI counterstained (1 μg/ml in Vectashield, Vector), and signals visualised with an epifluorescence microscope using 100× or 40× objectives and appropriate fluorescence filters.

Digital image analysis

15. Images are captured using a high-sensitivity CCD camera. In our case we use an Ultrapix 1600, which distinguishes over 16,000 grey levels, and allows subtraction of the dark current image and corrections for non-uniform sample illumination. Groups of 50-150 16 bit digital images are taken in similar conditions, stored in the file format of the camera (.apf) and then exported to .img files.
16. Image analysis is performed using a routine designed with the Visilog 5.1 software. This allows for thresholding the area of interest, background subtraction and measurement of the informative parameters. These data are exported as .txt files to Excel software for statistical analysis.

Results

As the dose of ionising radiation increases, DNA breakage level and alkali-labile sites increase, and more ssDNA is generated by the alkaline unwinding solution. Thus, more probe hybridises and fluorescence increases. The probe selects the chromatin area to be analysed (Figs. 2, 3a-c). An inverse result would be expected if repair were allowed for. Three simple parameters may be used for quantification of the FISH signal:

- The surface area: number of pixels. As DNA breakage increases, DNA loops of nucleoids progressively relax from supercoiling and the surface of the signal increases.
- The mean fluorescence intensity: average pixel grey count.
- Whole fluorescence intensity: surface × mean fluorescence intensity. This is the sum of the grey intensities of all pixels of the signal. Thus, it corresponds to the whole amount of hybridised probe.

  Note: The same whole fluorescence intensity value may be obtained by a low surface with a high grey mean level, and by a great surface with a low grey mean level.

Comments

Initially, we recommend incubation times of 2.5 min in the unwinding solution. In our conditions, shorter times do not allow the diffusion of the alkali into the deeper cells within the gel, whereas longer times may produce much melted ssDNA that is extended from breaks very distant from the probed locus.

The unwinding step before lysis may "catch" induced breaks immediately, without increasing the background breakage level by manipulation or during incubation. Nevertheless, lysis can be carried out before the unwinding, e.g., to obtain additional data on chromatin structure, as in sperm cells (Fig. 3d,e).

Areas of hybridisation (e.g., of an alphoid DXZ1 locus probe) appear as clearly separate, non-contiguous, spots after a high dose of X-rays (Fig. 3h). Neither double-strand break frequency nor DNA fragmentation after dehydration and gel drying can explain this. Possibly, the neutralising step produces partial and irregular renaturation of the ssDNA motif generated from a DNA break by the alkali. The renaturation should take place within the ssDNA motif and with the adjacent complementary strand in short areas of homology, so the melted DNA curls up and becomes tangled into a ball, trapping regions of ssDNA that hybridise with the probe, thereby giving the appearance of a spot (Fig. 3h).

Acknowledgements. This work was supported by the Consejo de Seguridad Nuclear, grants from FIS and SAF (Spain) and European Community (FIGH-CT-1999-00009, TELORAD).

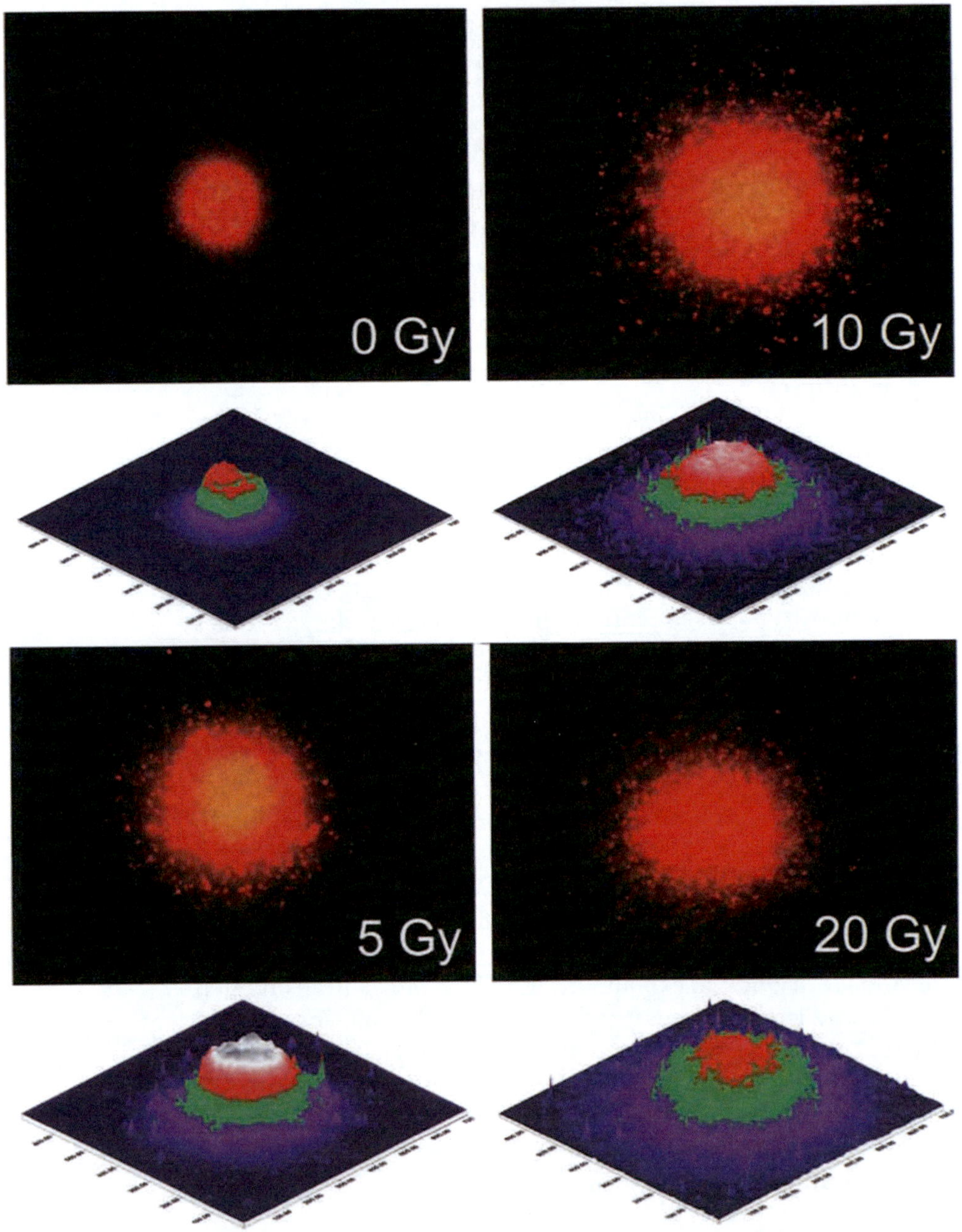

Fig. 2. Dose-response in DNA breakage in the whole genome of human blood leukocytes exposed to X-rays, after hybridisation of a whole genome probe. Each signal is accompanied by a pseudo-3D representation of the surface, and the fluorescence intensity using a colour code

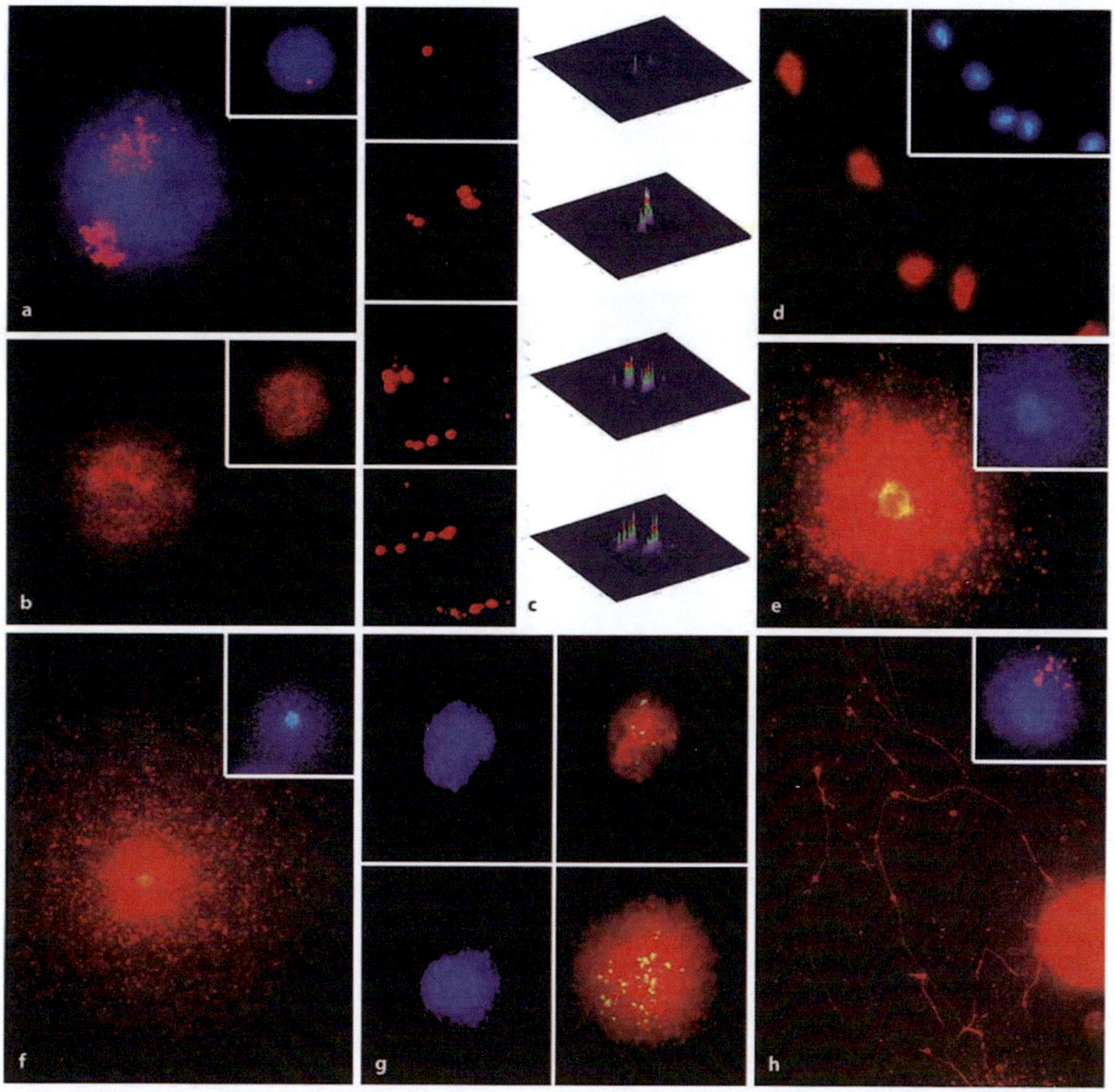

Fig. 3. Human blood leukocytes exposed to 20 Gy of X-rays and probed with a whole chromosome painting probe for chromosome 8 (**a**), and with a cosmid probe for c-erbB-2. (**b**). An unirradiated nucleus appears in the *upper right-hand corner* of each image. **c** Dose-response in the alphoid DXZ1 loci area of a female, with a pseudo-3D representation of the signal (0-5-10-20 Gy of X-rays, from upper to lower). **d, e** Unirradiated human sperm cells initially treated with the unwinding solution and then with the lysing solutions (**d**), or initially lysed and then exposed to the unwinding step (**e**). Hybridisation with a whole genome probe shows faintly labelled sperm cells in the first case, except in chromocentres, whereas they appear very strongly labelled when the unwinding step is performed after protein removal. The previous protein extraction would promote the accessibility of the alkali to a peculiar DNA, with abundant alkali-labile sites. These same cells appear DAPI counterstained, in the *upper right-hand corner* of each image. **f** Apoptotic human neutrophil

References

Fairbain DW, Olive PL, O'Neill KL (1995) The comet assay: a comprehensive review. Mutation Res 339:37-59

Fernández JL, Goyanes VJ, Ramiro-Díaz J, Gosálvez J (1998) Application of FISH for in situ detection and quantification of DNA breakage. Cytogenet Cell Genet 82:251-256

Rydberg B (1975) The rate of strand separation in alkali of irradiated mammalian cells. Radiat Res 61:274-287

---

leukocyte after 30 s of unwinding and hybridisation with a whole genome probe. DAPI counterstaining is shown in the *upper right-hand corner*. **g** Dual hybridisation with a whole genome probe (red) and a telomeric probe (yellow-green), in unirradiated (*top*) and 20 Gy irradiated (*below*) Chinese hamster Don cells. DNA breakage is simultaneously detected in the whole genome and within the interstitial telomeric sequence areas, in the same cell. DAPI counterstaining is presented on the left side. **h** Upper right-hand corner shows the signal obtained in the alphoid DXZ1 locus after irradiation with 20 Gy of X-rays. The signal is composed of dispersed spots, without continuity among them. The extended chromatin fibre from a nucleus hybridised with a whole genome probe reveals that the size of these spots is bigger than those making up the continuous fibre-FISH signal

Detection of Radiosensitivity by Means of Three-Color FISH

SUSANN NEUBAUER

Introduction

Fluorescence in situ hybridization (FISH) using whole chromosome painting (wcp) probes has been shown to be a reliable technique for studying chromosomal changes. Early investigations in radiation cytogenetics have been performed using classical cytogenetic analysis of Giemsa-stained (e.g. Schmid and Bauchinger, 1969) or GTG-banded chromosomes (e.g. Barrios et al. 1989). The application of FISH with wcp probes has extended the spectrum of observed aberrations. Now the reliable detection of translocations, insertions and complex rearrangements is possible in addition to the well-known acentric fragments, dicentrics and ring chromosomes. Even though this approach can be done by painting one single chromosome only (e.g. Natarajan et al. 1992) the simultaneous labeling of chromosomes 1, 2 and 4 in three different colors allows a more differentiated analysis of complex chromosome aberrations. Using this three-color FISH technique where the three selected chromosomes represent about 22% of the DNA-content of the human genome a sensitive method for studying radiation-induced chromosomal aberrations is available (Gebhart et al. 1996).

Up to now, radiation induced chromosomal aberrations were analyzed using different specific wcp probes (Natarajan et al. 1992), by single-color-FISH studying each human chromosome separately (Cigarran et al. 1998), and even using multi-color FISH (Greulich et al. 2000). The principal idea of the three-color FISH method presented here was to develop a simple and rapid assay for the detection of increased radiosensitivity in cancer patients (Neubauer et al. 1996). Performing the three-color FISH method, it is possible to detect and quantify increased radiosensitivity on the one

Susann Neubauer, Gynäkologen-Medizinische Genetik, Bankgasse 3, Nürnberg, 90402, Germany (*phone* 0049-911-2061024; *fax* 0049-911-2061031;
e-mail genetik@praenatal.com)

hand by quantifying the mis- and unrepaired break-events visible as chromosomal aberrations, and on the other hand by the frequency of complex aberrations (Neubauer et al. 1997).

Outline

1. Heparinized peripheral blood of each patient is divided into (at least) three portions. One sample remains unirradiated and serves as a control for spontaneous frequency of aberrations, the other two (or more) samples are irradiated in vitro with 0.7 and 2.0 Gy of X-rays.
2. After exposure, classical lymphocyte cultures are set up and grown at 37°C for 48 h.
3. 4 h before standard metaphase spreads are prepared, colcemid is added to induce a mitotic block.
4. Chromosome suspension is dropped onto slides and air dried slides are kept in 70% ethanol at -20°C before being subjected to the painting procedure.
5. Before performing the three-color FISH assay, a slide pretreatment is carried out to remove cell debris and improve the quality of the metaphase spreads.
6. Denaturation of the slides is performed for 1 min at 72°C. The biotinylated and/or digoxigenated whole chromosome probes are denatured for 5 min and hybridized on the slides for two nights at 37°C in a humid chamber. After a postwashing series, detection of biotin with avidin-fluorescein-isothiocyanate (=FITC) and digoxigenin with anti-digoxigenin-rhodamine and a final counterstaining with DAPI finishes the painting procedure. Thus, chromosome 1 is painted in red, chromosome 2 in green and the mixture of both labels leads to a yellow color of chromosome 4, unpainted chromosomes appear in a blue color.
7. Evaluation is done on a fluorescence microscope with suitable filter combinations (FITC/Texas Red/DAPI). Chromosomal rearrangements (characterized mainly by a color transfer to other chromosomes – Fig. 1) are classified according to standard cytogenetic criteria from 1000 (unirradiated), 400 (0.7 Gy) and 200 (2.0 Gy) metaphases per case. Aberration frequencies are expressed as a number of breakpoints per mitosis (B/M). The frequencies of break events were estimated

as the minimum number regarded to be necessary for the formation of the observed aberrations.

Procedure

Irradiation procedures and lymphocyte cultures

1. Heparinized freshly drawn peripheral blood is irradiated with different doses (0.7 and 2.0 Gy) of 6 MV X-rays from a linear accelerator (Mevatron) at a dose rate of 2.2 Gy/min.
2. 1 ml of whole peripheral blood is stimulated in 9 ml RPMI 1640 culture medium supplemented with 15% foetal calf serum, 2.5% phytohemagglutinin standard solution, 1% penicillin-streptomycin (10,000 IU/ml).
3. 180 μl Colcemid (0.09 μl/ml) is added to each blood culture 4 h before harvesting to block mitoses.
4. Chromosome preparations are carried out according to standard procedures (see. e.g. Chapter 3 by Liehr and Claussen).
5. Air dried slides are kept in 70% ethanol at -20°C before being subjected to the three-color FISH painting.

Pretreatment of metaphase spreads on slides

Pretreatment is useful when the quality of chromosome preparation is suboptimal or the metaphase spreads are covered with cytoplasm. Pretreatment with RNase and pepsin can remove cell debris and significantly improve the quality of the metaphase spread. All washing procedures mentioned below are performed in a 100 ml Coplin jar with a maximum of 7 slides and shaken gently.

1. Take the slides from the 70% ethanol storage solution and air dry.
2. Pipette 100 μl of RNase A solution (50 mg/ml in 2×SSC) onto each slide and cover with a coverslip for 15-30 min at 37°C in a humid chamber. The incubation time is dependent on the cell density and cytoplasm present on the slide. The slides can be looked at by light microscopy and treatment can be repeated when the outcome is not satisfactory.

3. Put the slides in a Coplin jar, and wash three times with 2×SSC for 5 min at room temperature (RT).
4. Immerse the slides in 1×PBS for 5 min.
5. Incubate the slides for 5-10 min at 37°C with 0.005% pepsin solution (prewarm 99 ml of distilled water and 1 ml of 1 M HCl at 37°C in a water bath and dilute 50 µl of a 10% pepsin stock solution just before performing the pepsin step).
6. Add 5 ml 1 M $MgCl_2$ to 95 ml 1×PBS and wash the slides for 5 min at RT.
7. For post-fixation, discard $MgCl_2$/PBS and replace it with 1% formaldehyde $MgCl_2$/PBS solution for 10 min at RT.
8. Wash the slides with 1xPBS for 5 min, dehydrate the slides through an ethanol series (70 %, 90% and 100% ethanol for 5 min each) and air dry.

Three-color fluorescence in situ hybridization (FISH)

For the three-color FISH approach described here, different wcp probes for chromosomes 1, 2 and 4 can be used. When unlabeled probes are available (e.g. derived from American Type Culture Collection) the probes have to be biotinylated and/or digoxigenated according to the points 1 to 5 mentioned below. On the other hand, labeled probes are available from different commercial resources (e.g. ONCOR or Vysis) – when using such probes start with point 6.

Labeling reaction

1. Plasmid or phage DNA from the DNA libraries of these chromosomes are isolated according to manufacturers instructions (e.g. Qiagen). Depending on the quality of purified DNA, in summary about 7.5 µg DNA (2.5 µg DNA from chromosome 1, 3 µg DNA from chromosome 2 and 2 times 1 µg DNA from chromosome 4) allows "painting" of metaphase spreads on a whole slide (up to 1000 metaphases).
2. Specific labeling of the chromosomes 1, 2 and 4 is performed using a modified approach of Lichter et al. (1988). The nick translation procedure for each chromosome is performed separately and on ice. In a reaction volume of 24 µl the DNA of the chromosome library of chromosome 1 was digoxigenated, of chromosome 2 biotinylated and chromosome 4 was biotinylated and digoxigenated (1:1) by nick translation. In detail:

10-16 μl double distilled water, 1-2.5 μg DNA from the specific chromosome and 6 μl of either Biotin-Nick Translation Mix (Boehringer Mannheim) or DIG-Nick Translation Mix (Boehringer Mannheim) are put into a reaction tube. The components are mixed and incubated in a 15°C water bath for 90 min. The reaction is stopped by adding 24 μl of a stop solution (20mM EDTA, 20 mM Tris-HCL, 0.1 M NaCl, 0.5% dextran blue and 0.1% bromphenol blue in distilled water) and 52 μl TE 10/1 buffer.

3. Purification of the probe:
   The 100 μl reaction volume (from point 2) is pipetted onto a prepared Sephadex column (Sephadex G50) and centrifuged. The labeled DNA of chromosomes 1, 2 and 4 is collected in the same tube together with 16 μl Cot 1 DNA (Gibco) (=1 μg labeled DNA/2 μl Cot 1), 2 μl HTD (herring testis DNA) and the 2.5-fold volume of 100% ethanol.
4. Precipitation of the DNA is done overnight or up to several months at -20°C.
5. To pellet the DNA, centrifuge (13,000 rpm) at 4°C for 45 min. The supernatant is discarded and the pellet is washed with 70% ethanol. After centrifugation the pellet can be dried under vacuum. Dilute the probe DNA carefully in 35 μl hybridization buffer (hybridization buffer = incubate 1 g dextran sulfate in 2.5 ml deionized formamide, 0.5 ml 20×SSC, 0.5 ml 0.5 M sodium phosphate buffer and 1.5 ml distilled water at 70°C for 3 h). 35 μl hybridization buffer is the appropriate volume if a coverslip 24×50 mm is used.

Denaturation and hybridization

6. Denature the probe (as prepared in steps 1 to 5 or a commercially available probe combination, according to manufacturers instructions) at 75°C for 5 min, chill on ice and allow preannealing for 35 min at 37°C.
7. For denaturation of the chromosomes on the slides, add a denaturation solution of 70 μl deionized formamide, 10 μl 0.5M sodium phosphate buffer, 10 μl 20×SSC and 10 μl distilled water on each slide and cover with a coverslip. Incubate the slides on a heating block at 72°C for 1 min.
8. Remove the slides from the heating block, remove the coverslips and immerse the slides in prechilled 70% ethanol (-20°C).
9. Dehydrate in an ethanol series (70%, 90%, 100% on ice 5 min) and air dry.

10. Put the 35 μl of the denaturated and preannealed probe (point 6) onto the denatured slide, cover with a coverslip (24×50 mm) without making bubbles and seal with rubber cement.

11. Transfer the slides for hybridization in a humid chamber at 37°C for 2 nights.

12. Remove the rubber cement and place the slides in a Coplin jar filled with 4×SSC/0.2%Tween to remove the coverslips carefully.

13. Post-wash the slides in prewarmed 4×SSC solution containing 50% formamide (45°C) and in 0.1×SSC (60°C) 3×5 min each.

14. Immerse the slides in a Coplin jar with 4×SSC/0.2% Tween.

15. Add 150 μl 5% BSA blocking solution (bovine serum albumin (Sigma) in 4×SSC/0.2% Tween) onto each slide, cover with a coverslip 24×60 mm and incubate for 20 min at 37°C in a humid chamber.

16. Remove the coverslips and place the slides in a Coplin jar with 4×SSC/0.2% Tween.

17. Just before use, prepare detection solution I and pipette 100 μl onto each slide and incubate at 37°C for 60 min in a humid chamber (detection solution I: FITC-avidin (CAMON Vector Laboratories A2011) diluted 1:500 in 5%BSA solution).

18. Remove the coverslips and wash 3×5 min with 4×SSC/0.2% Tween at 37°C.

19. Add 100 μl of detection solution II to each slide, cover with a coverslip and incubate in a humid chamber at 37°C for 2 h (prepare fresh detection solution II: biotinylated antiavidin [CAMON Vector Laboratories BA0300] diluted 1:100 and Anti-digoxigenin-rhodamine diluted 1:35 [Boehringer Mannheim, 1207750] in 5%BSA solution).

20. Repeat step 16.

21. Add 100 μl detection solution I on each slide, cover with a coverslip and incubate in a humid chamber at 37°C for 60 min and repeat step 16.

22. Counterstaining is performed with 4,6-diamidinophenylindol (DAPI) [1 μg dissolved in 1 ml 4×SSC/Tween]; incubate 5 min in the dark.

23. Immerse the slides 3-4 times in distilled water, air dry, add 20 μl antifade Vectashild and cover with a 24×60 mm coverslip.

24. Analyze the slides by fluorescence microscopy. Chromosome 1 appears red, chromosome 2 green and chromosome 4 due to a mixture of green and red labels has a yellow color (Fig. 1).

Aberration scoring

Microscopy has to be performed on a fluorescence microscope equipped with suitable filter combinations (FITC/Texas Red/DAPI). For analysis, only metaphases clearly exhibiting all six painted chromosomes are evaluated. Aberration frequencies are expressed as breakpoints per mitosis (B/M). Thus, translocations, dicentrics and ring chromosomes are calculated as two break events, whereas open breaks and isolated acentric fragments of one color are registered as one break event. As far as available, 1000 metaphases have to be scored in unirradiated lymphocytes, while 400 and 200 metaphases are examined after irradiation of 0.7 and 2.0 Gy, respectively. The chromosomal reaction to in vitro irradiation is expressed

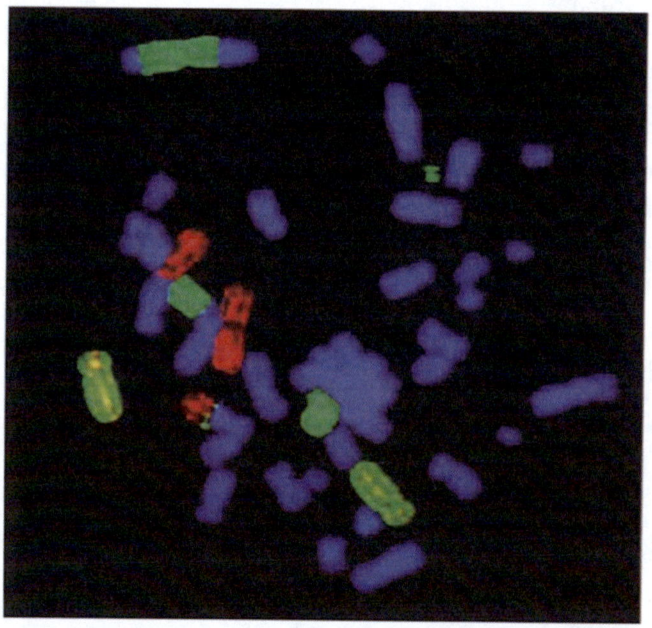

Fig. 1. Three-color FISH painting of chromosomes 1, 2 and 4 in a metaphase showing aberrations of chromosomes 1 and 2. Because of the bright labeling even small rearrangements were seen as color junctions. The picture was made using a Zeiss Axioplan microscope and the software of MetaSystems (Altlussheim, Germany)

on the one hand as the number of breaks per mitosis corrected for the breaks per mitosis in the unirradiated sample, or on the other hand as the number of complex aberrations. Complex aberrations are defined as exchange aberrations involving three or more breaks in two or more chromosomes (Savage and Simpson 1994).

Results

As mentioned in the protocol we do the three-color FISH studies for the detection of increased radiosensitivity using two different doses of X-rays (0.7 and 2.0Gy). As shown by Neubauer et al. (1996, 1997), intrinsic radiosensitivity can be detected and quantified by the three-color FISH technique (Fig. 1).

Genetic factors may determine increased radiosensitivity, as is well-known in patients suffering from Ataxia telangiectasia (AT), an autosomal recessively inherited disease characterized by a hypersensitivity to ionizing radiation. Thus cells from these patients might show the strongest sensitivity to ionizing radiation.

In Figure 2 the results of three-color-FISH studies on lymphocytes of 10 AT-homozygous, 10 AT-heterozygous and 10 normal control individuals are depicted.

As expected, a significantly increased incidence of chromosomal aberrations were observed in lymphocytes of AT homozygotes, in comparison to the controls. Lymphocytes from AT-heterozygotes show also a clear

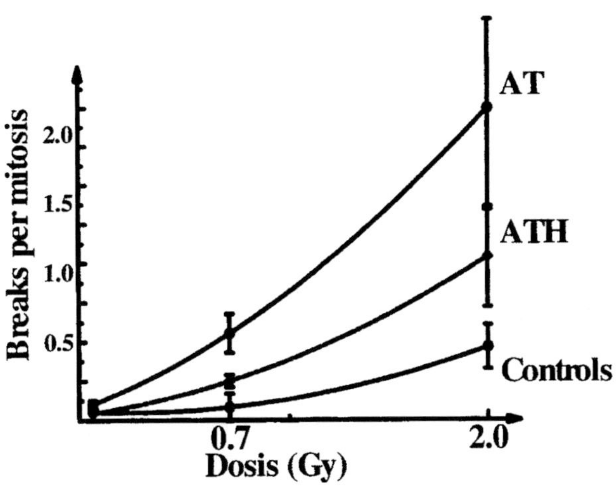

Fig. 2. Comparison of mean break frequencies in evaluated groups (AT-homozygotes, AT-heterozygotes and controls, 10 individuals, each). The curve was fitted according to the linear-quadratic equation

dose dependance of radiation induced aberrations. The avarage number of breaks indicates a moderate increase in in vitro radiosensitivity (see Figure 2). Occasional overlaps of aberration frequencies were observed only between AT-homozygotes and AT-heterozygotes. However, these two groups can be differentiated by the quality of aberrations, that is by the number of complex aberrations (data not shown). Complex aberrations are observed in a higher frequency in patients with increased radiosensitivity. AT-heterozygotes show less complex aberrations than AT-homozygote gene carriers.

Thus, using the three-color-FISH approach, two independent evaluation criteria are available to detect radiosensitivity in patients. Within a few days, an estimation of individual radiosensitivity can be achieved. In conclusion the three-color-FISH technique can be proposed as a predictive assay for detecting radiosensitive patients.

Acknowledgements. The author thanks Dr. M. Stumm, Magdeburg, and Dr. T. Dörk, Hannover, for providing AT homozygote and heterozygote lymphoblastoid cell lines.

References

Barrios, L, Miró, R, Caballín, MR, Fuster, C, Guedea, F, Subias, A, Egozcue, J (1989) Cytogenetic effects of Radiotherapy. Breakpoint Distribution in induced chromosome aberrations. Cancer Genet Cytogenet 41, 61-70

Cigarran, S, Barrios, L, Barquinero, J, F, Caballin, MR, Ribas, M, Egozcue, J (1998) Relationship between the DNA content of human chromosomes and their involvement in radiation-induced structural aberrations, analysed by painting. Int J Radiat Biol 74, 449-455.

Gebhart, E, Neubauer, S, Schmitt, G, Birkenhake, S, Dunst J (1996) Use of a three-colour chromosome-in situ-suppression (CISS) technique for the cytogenetic detection of past radiation exposure. Radiat Res 145, 47-52.

Greulich, KM, Kreja L, Heinze, B, Rhein, AP, Weier, HG, Bruckner, M, Fuchs P, Molls, M (2000) Rapid detection of radiation-induced chromosomal aberrations in lymphocytes and hematopoietic progenitor cells by mFISH. Mutat Res 452, 73-81.

Lichter, P, Cremer, T, Borden, J, Manuelidis, L, Ward, DC (1988) Delineation of individual human chromosomes in metaphase and interphase cells by in situ suppression hybridization using recombinant DNA libraries. Hum Genet 80, 224-234.

Natarajan, AT, Vyas, RC, Darroudi, F, Vermeulen, S (1992) Frequencies of X-ray-induced chromosome translocations in human peripheral lymphocytes as detected by in situ hybridization using chromosome-specific DNA libraries. Int J Radiat Biol 61, 199-203.

Neubauer, S, Gebhart, E, Schmitt, G, Birkenhake, S, Dunst J (1996) Is chromosome in situ suppression (CISS) hybridization suited as a predictive test for intrinsic radiosensitivity in cancer patients? Int J Oncol 8, 707-712.

Neubauer, S, Dunst, J, Gebhart, E (1997) The impact of complex chromosomal rearrangements on the detection of radiosensitivity in cancer patients. Radiother Oncol 43, 189-195.
Savage, JRK, Simpson, P (1994) FISH 'painting' patterns resulting from complex exchanges. Mutation Res 312, 51.60.
Schmid, E, Bauchinger M (1969) On the time relationship of radiation-induced chromosome aberrations in human. Mutation Res 8, 599-611.

Fluorescence in Situ Hybridisation of a Complete Set of 41 Telomere Specific Probes in a Single Simultaneous Hybridisation Experiment Using the Chromoprobe Multiprobe System-T

MARTIN LAWRIE, MARIA OCANA GIL, and DON CARDY

Introduction

It is estimated that around 3% of the population exhibit some degree of mental retardation where the pathogenesis of their disorder is poorly understood (1). The importance of subtelomeric chromosome rearrangements associated with this idiopathic mental retardation (IMR) has been shown to be as much as 8% in moderate to severe retardation and of 0.5% in moderate retardation (2, 3, 4). There are also reports of disruption of these regions of the chromosomes being implicated in recurrent miscarriages, haematological malignancies and polycystic kidney disease (5).

The detection of such subtle chromosome rearrangements in patients suspected of having a chromosome anomaly has up to now been limited to standard cytogenetic staining techniques with a limit of 850-1000 band resolution. This is not only time-consuming but also limited because of the similar banding patterns (G negative) found at the telomeres. Thus, cryptic deletions and rearrangements can prove to be impossible to detect by conventional methods.

The advent of Fluorescence in situ hybridisation (FISH) technology has made new approaches available for cytogenetic screening (6,7). Whole chromosome painting and reverse chromosome paints have not, however, been able to address the screening needs of the telomeres because these regions are underrepresented in the whole chromosome libraries so that

✉ Martin Lawrie, Cytocell Limited, Somerville Court, Banbury Business Park, Adderbury, Banbury, Oxfordshire, OX17 3SN, UK (*phone* +44-1295-810910; *fax* +44-1295-812333; *e-mail* m_lawrie@cytocell.co.uk)
Maria Ocana Gil, Cytocell Limited, Somerville Court, Banbury Business Park, Adderbury, Banbury, Oxfordshire, OX17 3SN, UK
Don Cardy, Cytocell Limited, Somerville Court, Banbury Business Park, Adderbury, Banbury, Oxfordshire, OX17 3SN, UK

any subtle deletion event within these regions of the chromosome cannot be detected (8, 9, 10). Chromosome-specific unique sequence probes are more appropriate but are often not sufficiently telomeric to exclude a rearrangement. The complex and repeated structure of the telomeric regions of human chromosomes (Fig. 1) has also served to complicate generation of suitable probes. The true telomere, capping the end of the chromosome, is constructed of between 3,000 and 20,000 base pairs of tandemly repeated TTAGGG sequences. This provides chromosome stability but displays complete homology to each end of every chromosome. Immediately proximal to this is a variable sized region of telomere associated repeats (TAR) which show shared homology between different groups of chromosome telomeres, whilst the DNA sequences unique to the chromosome end are situated proximal to the true telomere and the TAR.

A complete set of telomere specific FISH probes capable of detecting deletion, triplication and balanced translocation rearrangements of subtelomeric DNA in every chromosome was generated and reported three years ago (4, 11). In this first generation set, there were a total of 41 as

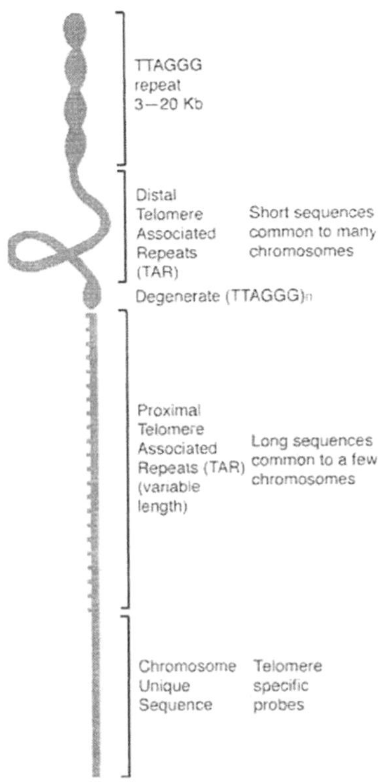

Fig. 1. Fine structure of the terminal portion of human chromosomes

opposed to 48 (the p-arms of the acrocentric chromosomes are excluded because they contain repetitive and ribosomal DNA only and the X and Y chromosomes share homologous sequences at both pseudoautosomal regions). These probes were produced as cosmid clones with size ranges from 20-40 kb, making the direct labelling of these for rapid FISH analysis difficult. In addition, this first generation set contained probes which were not as close to the end of the chromosome as would be preferred. More recently, a second generation of probes have been developed. These probes have been developed as PAC clones and are approximately 100 kb in size making the direct labelling of them possible. These new probes are of varying distance from the telomere, ranging from 30-450 kb as a result of the variable length of the TAR.

Standard FISH techniques would require the individual testing of every telomere probe for each different chromosome, making routine screening impractical. We have combined the complete set of specific telomere probes with our Chromoprobe Multiprobe technology to produce a novel FISH technique to allow the simultaneous hybridisation and analysis of all probes on one slide. The Multiprobe is a disposable glass device with 24 raised bosses to which each of the 41 telomere probes are reversibly applied following a patented process. Each square thus contains its p-arm probe in green and its q arm probe in red (except the acrocentric chromosomes which have q- arms only, and the sex chromosomes which share their telomere specific sequences). This offers cytogenetic diagnostic laboratories a screening method that is relatively quick, straightforward and cost effective. Using the Chromoprobe Multiprobe system, the FISH protocol is simplified for a number of reasons. Firstly, the p arm and q arm telomeres are differentially fluorescently labelled in green and red; both probe and target DNA are denaturated simultaneously in a single step; there are no detection or amplification steps; the device is hybridised overnight; and rapid, formamide-free post hybridisation stringent washes are used throughout.

Procedure

Sample slide preparation

1. Soak a template slide for 2 min in 100% methanol and polish dry with a clean soft tissue.

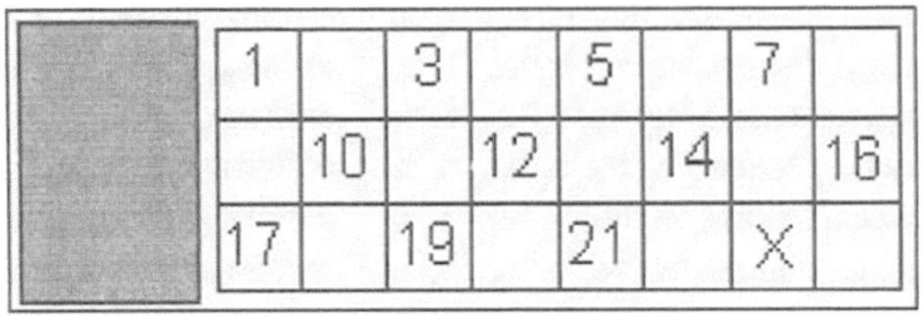

Fig. 2. Locations of first 12 spots of cell sample on template slide. Spotting in this manner will allow sample to spread correctly

2. It is important that the intended sample has a sufficiently high cell density for adequate analysis. For example, there should be at least 20 nuclei in a field of view using a $\times 10$ objective, and there should be at least 5 metaphase cells in each square of the template slide.
3. Pipette 2 µl of methanol/acetic acid fixed cell suspension (e.g. peripheral blood lymphocytes) onto all 24 areas in a series of alternating squares as shown below (Fig. 2). This will prevent the drops of suspension from interfering with each other as they spread.
4. Once the first group of drops has air dried, spot the remaining squares with 2 µl drops in the same manner. After the slide has dried, examination of the slide under phase contrast will reveal whether any squares have been missed.

Hybridisation

1. Pre-warm a Chromoprobe Multiprobe Hybridisation Chamber (or a plastic slide box measuring $130 \times 80 \times 30$ mm) to 37°C in a water bath.
2. Mix 30 µl of hybridisation solution per device by repeated pipetting and pre-warm to 37°C. Pre-warm Multiprobe to 37°C by placing device **label side down.** Avoid touching the raised boss surfaces.
3. Wash sample slide in $2\times$SSC for 2 min at room temperature.
4. Dehydrate sample slide through an ethanol series: 2 min each in 70%, 85% and absolute ethanol at room temperature. Air dry and place at 37°C to warm.
5. Pipette 1 µl of pre-warmed hybridisation solution on to each of the bosses on the pre-warmed device while it remains at 37°C.
6. Carefully invert the sample slide over the Multiprobe such that the number 1, is located over the top right-hand area of the Multiprobe (Fig. 3).

7. Carefully lower the slide over the Multiprobe so that the drops of hybridisation fluid make contact with the slide. Apply gentle, even pressure to ensure that the hybridisation fluid is spread to the edges of each of the raised areas on the Multiprobe.
8. Lift the slide/Multiprobe carefully and invert so that the slide is under the Multiprobe.
9. Incubate at 37°C for 10 min.
10. Transfer the slide/Multiprobe device to the 75°C hotplate taking particular care to hold it level (ensure the sample slide is in good contact with the hotplate). Incubate at 75°C (+/-1°C) for **2 min.**
11. Place slide/Multiprobe device in the Chromoprobe Multiprobe Hybridisation Chamber, replace, but do not seal, the lid and float the chamber in an uncovered 37°C water bath overnight. An incubator should not be used for the hybridisation procedure, as the multiprobe system is an open one and the correct levels of humidity are vital for good hybridisation. Damp tissues must not be used inside the chamber.

Post hybridisation washes

1. Remove the Multiprobe device carefully from the sample slide and incubate the slide in 0.4×SSC (pH 7.0) at 72°C for 2 min.
2. Remove the slide and incubate in 2×SSC, 0.05% Tween 20 for 30 s at room temperature.
3. Drain off excess solution by placing the slide on its edge on a tissue for around 20 s. Do not allow to dry.
4. Add one 20 µl drop of DAPI-antifade solution to both ends of the slide and apply a large (24×60 mm) coverslip slowly to avoid the formation of air bubbles.

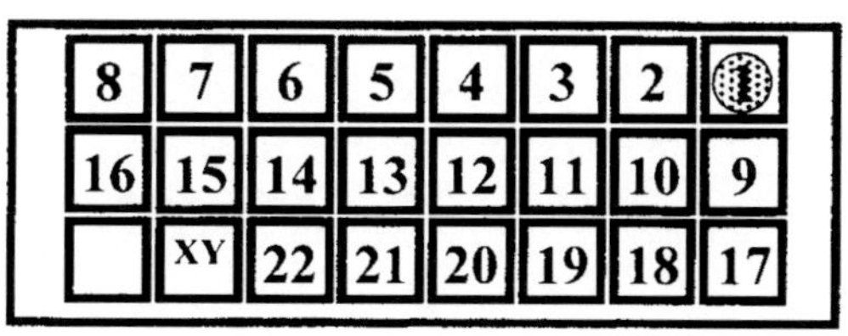

Fig. 3. Location of telomeric probes on Multiprobe bosses

5. Blot with a piece of filter paper/tissue.
6. Leave for 10 min in the dark before viewing by epifluorescence microscopy using a Texas Red, FITC and DAPI triple band pass filter.

References

Knight S, Regan R, Nicod A, Horsley S, Kearney L, Homfray T, Winter R, Bolton P, Flint J. Subtle Chromosome rearrangements in children with unexplained mental retardation. Lancet 1999. 354: 1676 – 1681

Flint J, Wilkie AOM, Buckle VJ, Winter RM, Holland AJ, McDermid HE: The detection of subtelomeric chromosomal rearrangements in idiopathic mental retardation. Nat. Genet. 1995. 9: 132- 139.

Giraudeau F, Aubert D, Young I, Horsley SW, Knight SJL, Kearney L, Vergnaud G, Flint J: Molecular-cytogenetic detection of a deletion of Ip36.3. J. Med. Genet. 1997. 34: 314 -17

Knight S, Horsley S, Regan R, Lawrie M, Maher E, Cardy D, Flint J, Kearney L. Development and clinical application of an innovative fluorescence in situ hybridisation technique which detects submicroscopic rearrangements involving telomeres. Eur. J. Hum. Genet. 1997 5: 1-8

European Polycystic Kidney Disease Consortium. Cell. 1994. 77:881-894

Lichter P, Cremer T, Tang C-JC, Watkins PC, Manuelidis L, Ward DC: Rapid detection of human chromosome 21 aberrations by in situ hybridization. Proc. Natl. Acad. Sci. USA 1988. 85: 9138-9142.

Pinkel D, Landegent J, Collins C, Fuscoe J, Segraves R, Lucas J, Gray J: Fluorescence in situ hybridization with human chromosome specific libraries: Detection of trisomy 21 and translocations of chromosome 4. Proc. Natl. Acad. Sci .USA 1988. 85: 9664-9668.

Collins C, Kuo WL, Segraves R, Fuscoe J, Pinkel D, Gray JW: Construction and characterization of plasmid libraries enriched in sequences from single human chromosomes. Genomics 199.1 11: 997-l006.

Schrock E, du Manoir S, Veldman T, Schoell B, Wienberg J, Ferguson-Smith MA, Ning Y, Ledbetter DH, Bar-Am I, Soenksen D, Garini Y, Ried T: Multicolor spectral karyotyping of human chromosomes. Science 1996. 273:494-497

Speicher MR, Ballard SG, Ward DC: Karyotyping human chromosomes by combinatorial multifluor FISH. Nat. Genet. 1996.1 2:368-372

National Institutes of Health and Institute of Molecular Medicine Collaboration: A complete set of human telomeric probes and their clinical application. Nat. Genet. 1996.14:86-89.

Part E

FISH in Zoology and Virology

Zoo-FISH

HARRY SCHERTHAN*

Introduction

Zoo-FISH or comparative chromosome painting between distantly related mammalian species (Scherthan et al. 1994) provides a means of rapidly generating crude comparative chromosome maps which disclose regions of synteny conserved among eutherian genomes (Comparative Genome Organization 1996). Such maps contribute particularly to comparative genomics in species combinations where comparative mapping data are limited or not available at all. Furthermore, they allow reconstruction of the karyotypic rearrangements which accompanied mammalian radiations (for review see O'Brien et al. 1997, Chowdhary et al. 1998).

Principles and applications

Comparative chromosome painting (also termed heterologous- or cross-species chromosome painting) denotes a molecular cytogenetic technique which utilizes chromosomal in situ suppression hybridization (Lichter et al. 1988, Pinkel et al. 1988, Bouffler 1994) with human whole chromosome paint probes (WCPs) to delineate regions of conserved synteny between the complements of species within a mammalian order (Wienberg et al. 1990). Due to the high degree of molecular homology among primate genomes (Kohne et al. 1972, Sibley et al. 1990), comparative chromosome painting results in readily detectable conserved homology segments (Wienberg et al. 1990, Jauch et al. 1992) and has proven to be a valuable

Harry Scherthan, University of Kaiserslautern, Dept. of Human Biology and Human Genetics, Erwin-Schroedingerstr. 14, Kaiserslautern, 67663, Germany (*phone* +49-631-205-2107; *fax* +49-631-205-2878; *e-mail* scherth@uni.de)
* present address: Max-Planck-Inst. for Molecular Genetics, Ihnestr. 73, Berlin, D-14195, Germany.

tool to elucidate karyotypic rearrangements in primate evolution (for review see Wienberg and Stanyon 1997). However, the application of comparative chromosome painting to species combinations spanning different mammalian orders failed initially. Since it was likely that the concentration of informative (conserved) DNA molecules contributing to a paint signal will diminish with increasing evolutionary distance of two species under investigation, it was supposed that only highly concentrated probes

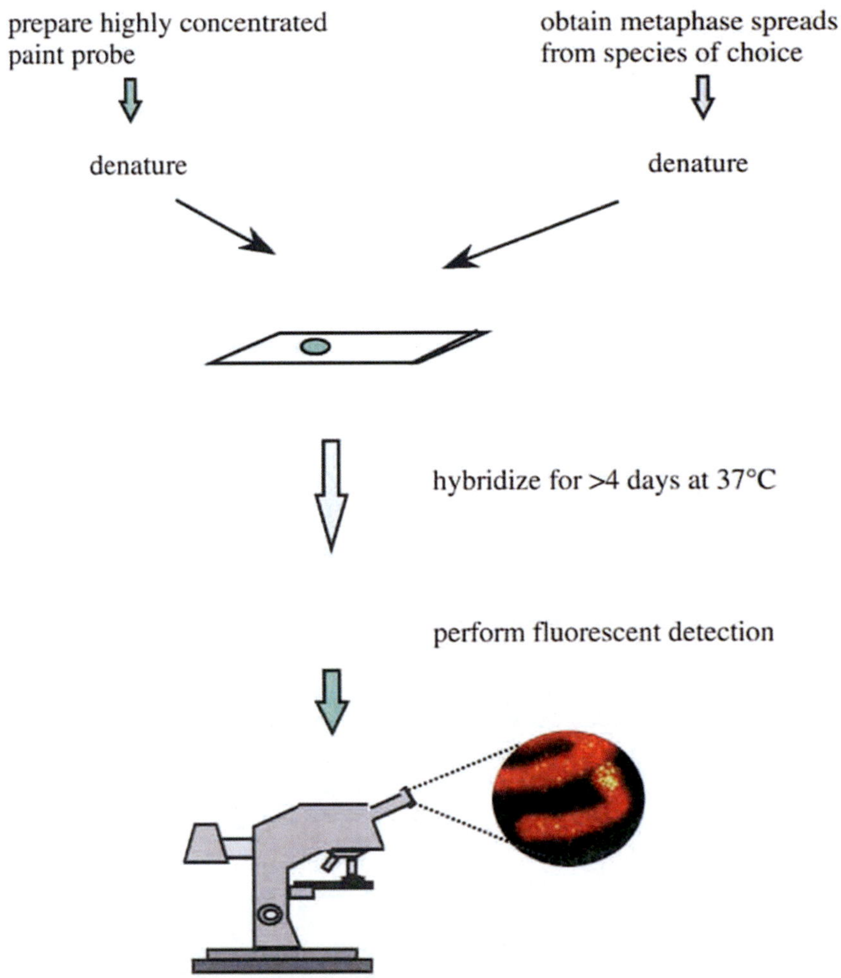

Scheme 1. Outline of the strategy applied in Zoo-FISH mapping. As compared to standard chromosome painting, extended hybridization time and highly concentrated DNA probes are requisite to achieve labeling at segments of conserved synteny

would allow for the extension of the comparative chromosome painting approach to more distantly related mammals. The generation of plasmid and PCR amplifiable human chromosome specific DNA libraries (Collins et al. 1991, Vooijs et al. 1993) eventually allowed such concentrated WCP probes to be generated. These, in combination with optimal chromosome preparations and extended hybridization times, which favor the hybridization of single copy sequences, finally allowed the extension of the comparative chromosome painting approach to phylogenetically more distant mammals (Scheme 1, Fig. 1). Consequently, this protocol was termed Zoo-FISH (Scherthan et al. 1994) and numerous studies in various labs have since shown that it is capable of establishing chromosomal homologies between the genomes of species which diverged more than 100 million years ago (for review see e.g. Chowdhary et al. 1998). Using this protocol it has been demonstrated that the human karyotype arrangement closely resembles that of the putative mammalian ancestor (Dixkens et al. 1998).

Most Zoo-FISH investigations used human chromosome specific plasmid or PCR-generated library DNA probes, because these were the first available. Moreover, the human genome is the most densely mapped of all mammalian species and therefore serves as the reference point for transfer of comparative map data between mammalian species (O'Brien et al. 1993). The generation of chromosome specific DNA libraries from numerous mammalian and other species (Ferguson-Smith 1997) has permitted

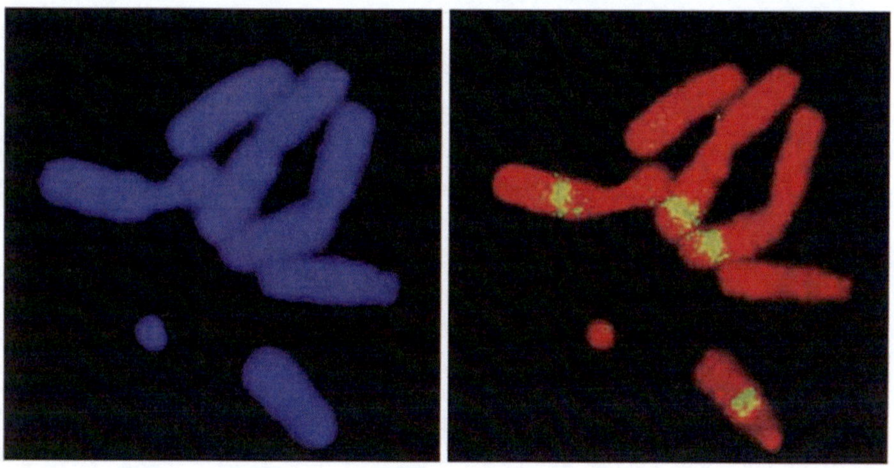

Fig. 1. Brightly blue (DAPI, *left*) stained Indian muntjac chromosomes (MMV) after Zoo-FISH with the human chromosome 1 pBS library probe (*right*). Two conserved syntenic segments (fluorescein, yellowish) are seen. One at MMV1q2 and another at Xq3/Y_1q3 (Frönicke and Scherthan 1997). Red color results from propidium iodide counter staining

the development of reverse and forward comparative chromosome painting protocols by which a refined resolution of comparative chromosome maps is achieved (Goureau et al. 1996, Wienberg et al. 1997, Nash et al. 1998). Presently, it remains to be seen whether the Zoo-FISH approach proves effective in other than eutherian species combinations. The Zoo-FISH protocol outlined below displays the general procedure for the preparation and hybridization of human plasmid WCPs (Collins et al. 1991) to the chromosomes of a mammalian species of choice.

Materials

Equipment

- Pasteur pipettes
- Plastic centrifuge tubes, 15 ml
- Glass slides. Remove impurities by submerging slides for several hours in 80% ethanol in a Coplin jar. Dry with a Kim wipe prior to use
- Coverslips of 22×60 mm and round coverslips with 10 mm diameter (Hecht Assistent)
- Coplin jars with covers
- Hot plate or heating block
- Moist chamber to prevent slides from drying out. This can be e.g. a plastic box and lid with wet paper and glass rods to raise slides above the moist surface.
- Water bath, preferably shaking. It is used to adjust the temperature of the solutions in Coplin jars. Always measure temperatures inside Coplin jars.
- Fluorescence microscope with appropriate filter sets and camera system

Reagents and solutions

- Methanol and glacial acetic acid (Merck)
- Fixative; make fresh each time by mixing 3 parts of ice-cold methanol with 1 part of glacial acetic acid, always keep on ice.
- 20×SSC (3 M NaCl, 0.3 M Na_3citrate, pH 7.0). Store at RT. Make up 0.05×SSC from this stock. Discard dilute solutions after use.

- Formamide, research grade (Merck). A small quantity of deionized formamide is required for the hybridization solution. Small aliquots are prepared by filling the tip of a 1.5 ml microfuge tube with ion exchange resin (20-50 mesh; Biorad). Add 1 ml of formamide, mix and store at -20°C.

  Note: Formamide is a harmful chemical and should be handled with care.

- 70% formamide/30% 2×SSC, pH 7.0. This denaturation solution can be stored for several weeks at 4°C.
- C_0t1 DNA from human and other species may be obtained from a commercial supplier. In case it is not available, C_0t1 DNA may be prepared according to published protocols (Kappler et al. 1998).
- Hybridization solution is composed of: 50% deionized formamide, 10% 20×SSC, 20% of a 50% (w/v) dextran sulfate (Pharmacia) solution, 10% 50 mM sodium phosphate pH 7.0, human C_0t1 DNA at a final conc. of 2 μg/μl, probe DNA at 300 ng/μl. Use deionized water to adjust volume.
- Tween-20 (Sigma)
- BT buffer (0.15 M $NaHCO_3$, 0.1% Tween-20, pH 8.3). After dissolution of $NaHCO_3$, pH will be in the appropriate range. Make up fresh BT for each experiment.
- BSA (fraction V, Serva)
- Avidin-FITC (e.g., ExtrAvidin-fluorescein, Sigma)
- Biotinylated goat-anti-avidin antiserum (Vector Labs)
- Propidium iodide (PI, Sigma). Dissolve at 1 mg/ml in sterile water, store stock at -20°C.
- DAPI (4,6-Diamidine-2-phenylindole dihydrochloride) (Sigma). Stock solution: 1 mg/ml in sterile water, store at -20°C.
- Antifade solution: Vectashield (Vector Labs) is recommended as a mounting medium for fluorescence microscopy, since it efficiently reduces fading of fluorochromes during microscopic analysis (Florijn et al. 1995). For counter staining of DNA, fluorescent dyes are added to the antifade solution in the following concentrations: DAPI 0.5 μg/ml and/or PI 1 μg/ml.

Procedure

Probe labeling

Nick translation, random priming and PCR protocols have proven to be effective methods for labeling probe DNAs. For ease of handling, commercially available labeling kits may be utilized. For a detailed description of nick translation, the reader is referred to step by step protocols published elsewhere (e.g., Cremer and Lichter 1992). A critical point in Zoo-FISH is the fragment size and concentration of the labeled DNA probe. Fragment sizes of 100-300 bp have yielded good signal-to-noise ratios. It is recommended to monitor the size and amount of labeled probe by running an aliquot ($^1/_{20}$) of the nick reaction on a 1% agarose mini gel.

1. Obtain purified WCP DNA probe by plasmid preparation or PCR. Make sure that the probe DNA is free of RNA.
2. Label 1 µg of pBS chromosome specific DNA library probe by nick translation using a commercial kit according to the recommendations of the supplier (e.g. BioNick Kit; Life Technologies).
3. After the appropriate incubation time, set 1 µl of each reaction aside and perform minigel analysis and a dot blot test to control the amount and quality of DNA labeling, respectively (Scherthan and Cremer 1994).
4. Precipitate labeled DNA probe (1 µg) by addition of 3 µg of human C_0t1 DNA and 3 vol of ethanol.
5. After incubation at -20°C for 30 min overnight, spin for 30 min at high speed.
6. Discard supernatant and wash pellet briefly with 70% ethanol.
7. Air dry pellet by placing the open tube for 5-10 min in a dry incubator or drying block at 65°C.
8. When remnants of liquid have evaporated, dissolve pellet in 3.3 µl hybridization solution.

Preparation and storage of metaphase chromosomes

Metaphase chromosomes of the species under investigation are preferably obtained from primary fibroblast cultures or cultured peripheral blood lymphocytes. Chromosomes are spread according to standard methanol/acetic acid (3/1) fixation protocols (Islam and Levan 1987, Frönicke and Scherthan 1997).

1. Grow fibroblasts to half confluency, arrest them at metaphase by colcemid treatment and swell cells for 30 min in 0.0375 M KCl.
2. Fix the cell pellet more than five times with always freshly prepared ice cold fixative.
3. Dissolve pellet in 0.5 ml fixative and drop ~50 µl of cell suspension on an ethanol-cleaned slide, tilt and drain excess fixative.
4. Place slide over a water bath at ≥60°C. Let slides sit until the upper side appears dry.
5. Remove preparations and inspect them by phase contrast microscopy on low power. Cytoplasm is evident as dark rings around metaphase plates and nuclei. Repeat fixation until cytoplasm is absent. In phase contrast chromosomes should have an evenly dark appearance and metaphase plates should be closely spaced.
6. Allow dried slides to sit at RT for up to 2 h.
7. Slides are sealed in a plastic container and stored at -20°C.

Pretreatment and denaturation of chromosome preparations

Prior to in situ hybridization, fresh or thawed chromosome preparations are placed for 50 min on a heating block or hot plate at 90°C to harden chromosomal chromatin, which limits swelling of the chromosomes during denaturation and extended hybridization time.

1. Remove slides with forceps and allow to cool to room temperature.
2. Cover preparations with 100 µl 70% formamide/30% 2×SSC, pH 7.0.
3. Mount a 22×60 mm coverslip and incubate for 2 min on a hot plate or heating block at 70°C to denature chromosomal chromatin.

4. Pick up slides with forceps, rinse off coverslips and denaturation solution with a jet of cold deionized water from a wash bottle and air dry.

   Note: Dehydration through an alcohol series has proven dispensable.

In situ hybridization

1. Denature hybridization solution for 5 min at 93°C.
2. Place microfuge tube for 1h at 37°C to preanneal any shared repetitive sequences.
3. Apply an appropriate volume (1 µl for a round coverslip with 10 mm is sufficient) to a region of the previously denatured slide with a sufficient density of metaphase plates.
4. Mount coverslip and seal with rubber cement.
5. Place slides for ›96 h in a sparsely moistened chamber or incubator at 37°C.
6. After hybridization peel off rubber cement with forceps.
7. Submerse slides in a Coplin jar with 0.05×SSC until coverslips float off.
8. Perform post-hybridization washes for 3×5min in 0.05×SSC at 37°C.

Signal detection

Two color fluorescent detection (Fig. 2) is achieved by incubation with avidin-FITC (Pinkel et al. 1986) and anti digoxigenin rhodamine Fab fragments. In single-color FISH experiments (Fig. 1) either label may be used.

1. Replace the 0.05×SSC (from step 8 above) by 0.5% BSA/BT buffer and incubate for 5 min at 37°C.
2. Remove slides from Coplin jar, briefly drain excess fluid and place them in a moist box.

   Note: Preparations should never dry up during this and subsequent steps.

3. Apply 100 µl BT buffer containing avidin-FITC (2.5 µg/ml) and anti-digoxigenin rhodamine Fab fragments at 0.5 µg/ml.

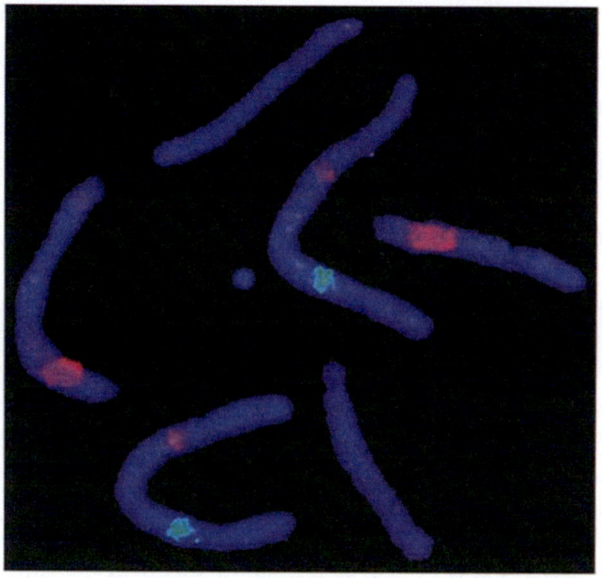

Fig. 2. Zoo-FISH with human chromosome (HSA) 17 and 4 WCPs to Indian muntjac chromosomes. HSA 17 detects a single region of conserved homology on MMV1p 1.7-p2.1 (fluorescein, green), while HSA4 paints two conserved syntenic segments (rhodamine, red): a small one at MMV1q 1.9-q2.1 and a large segment at Xq/Y_1q 4.2-q5.1 (note that the Xq and Y_1q are homologous chromosome arms) Frönicke and Scherthan 1997 (REF). The *small blue dot* in the center represents the Y_2 chromosome

4. Cover with large coverslip and incubate for 1 h at 37°C.
5. Gently slide off or submerse slides in a Coplin jar containing BT buffer to remove coverslips.
6. Wash slides 3×3 min in BT buffer at 37°C.
7. Apply 100 µl BT buffer containing biotinylated goat anti-avidin antibody (2 µg/ml).
8. Cover with coverslip and incubate for 30 min at 37°C in a moist chamber.
9. Wash 3×3 min in BT buffer at 37°C.
10. Repeat step 3 with avidin-FITC only and incubate for a final 30 min at 37°C.
11. Repeat step 6, remove slides from Coplin jar and briefly drain excess fluid.

12. Apply 18 μl antifade solution containing DAPI as DNA specific counterstain and mount a 22×60 mm coverslip.

13. Cover slides with a tissue and apply gentle pressure to remove excess liquid and trapped air bubbles. Inspect under fluorescence microscope equipped with appropriate filter sets for the visualization of green, red and blue fluorescence. Images may be recorded with conventional microphotography or digital image acquisition systems.

Troubleshooting

1. Since the quality of the chromosome preparations is crucial for success of a Zoo-FISH experiment, fixation of the cell pellet at least five times with freshly prepared ice cold fixative (methanol/acetic acid, 3/1) should be performed. This should produce metaphase spreads free of cytoplasm with a crisp appearance upon DAPI staining (Fig. 1). Thymidine synchronization and BrdU incorporation (Drouin et al. 1990, Hayes et al. 1991) may be applied to obtain long chromosomes which facilitate FISH mapping due to enhanced DAPI contrast (Carter 1994, Frönicke and Scherthan 1997).

2. A variable performance of human WCPs in Zoo-FISH experiments has been noted (Frönicke 1997). Therefore, probe concentrations in the range of 200-330 ng/μl have to be tested. In contrary to PBS libraries, which contain vector sequences, PCR generated probes contain negligible amounts of vector/adaptor sequences which allows them to be used at lower concentrations (150-200 ng/μl). In some cases, e.g. HSA 11, probes from other sources are the key to a successful experiment (Goureau et al. 1996, Frönicke et al. 1997, Dixkens et al. 1998). It should be noted that WCPs of the human Y chromosome generally fail to delineate corresponding segments of conserved synteny in Zoo-FISH experiments (Chowdhary et al. 1998).

3. When pre-labeled commercial WCP probes are used in Zoo-FISH, hybridization times have to be extended to >7 days (Rettenberger et al. 1995, Goureau et al. 1996).

4. Suppression DNA (usually the C_0t1 fraction of the genome from which the paint probe has been generated) is added to the Zoo-FISH hybridization solution, since some repetitive DNA sequences may be shared among genomes (e.g. telomere repeat sequence motifs have been found in satellite DNA of Rodents and Cetaceans, or may be located

at interstitial sites (Meyne et al. 1990). Such repeat loci may interfere with the specificity of the reaction. Preannealing of the probes generally contributes to a higher signal to noise ratio.

5. RNase digestion, which is routinely performed to prevent elevated levels of background fluorescence due to unspecific binding of DNA probes to RNA molecules in remaining cytoplasm or in nuclei, can be omitted for metaphase chromosome preparations free of cytoplasm. However, it is imperative that preparations of probe DNA are free of RNA.

6. In the initial Zoo-FISH protocol (Scherthan et al. 1994) probe and target DNAs were denatured simultaneously because this method usually creates stronger signals. However, separate denaturation is recommended (Rettenberger et al. 1995, Frönicke et al. 1996), since the shearing forces which are created during simultaneous denaturation by the dextran sulfate polymer of the hybridization solution may cause inferior chromosome morphology.

7. Dehydration of freshly denatured chromosome preparations through an ethanol series has proven dispensable. It is sufficient to wash down the coverslip and hybridization solution with a jet of cold deionized water from a wash bottle and air dry the preparations with an air jet.

8. Formamide containing wash solutions can be replaced effectively with dilute salt solutions, thereby reducing the handling of formamide to a few μl.

9. The high amounts of labeled probe required for Zoo-FISH may cause considerable background levels which impair the use of multi-color FISH schemes. Generally, it is recommended to perform single or two color detection since these produce reliable results (Figs. 1,2). This contrasts with the intra- and intergeneric approach to comparative molecular cytogenetics, where multi-color painting is feasible (Müller et al. 1998).

10. Comparison of the relative genome fraction delineated with a particular human WCP in a mammalian genome revealed that equivalent genomic fractions have been maintained during mammalian evolution. It should be noted that values obtained by such measurements have to be adjusted for the presence of intervening centromeres, heterochromatin blocks or NORs (Scherthan et al. 1994; Chowdhary et al. 1996, Frönicke et al. 1996,1997).

11. DAPI banding patterns, which contribute to chromosome identification after FISH (for review see Carter 1994), may be contrast enhanced by the addition of Actinomycin D (10 µg/ml) to the DAPI containing antifade solution. Alternatively, GTG banding prior to FISH may be performed to identify chromosomes (Scherthan et al. 1989, 1994). Although this approach is laborious, it may be followed when digital image analysis is not feasible.

Acknowledgements. I am grateful to J.W. Gray, UCSF, for generously providing the linker adaptor and pBS libraries, and to L. Frönicke, T. Raudsepp and B.P. Chowdhary for fruitful collaborations.

References

Bouffler SD (1994) Whole-chromosome hybridization. Intn Rev Cytology 153;171-232

Carter NP (1994) Cytogenetic analysis by chromosome painting. Cytometry 18;2-10.

Comparative Genome Organization; First International Workshop on Comparative Genome Organization. (1996) Mamm. Genome 7;717-734.

Collins C, Kuo WL, Segraves R, Pinkel D, Fuscoe J, Gray JW (1991) Construction and characterization of plasmid libraries enriched in sequences from single human chromosomes. Genomics 11;997-1006.

Chowdhary BP, Frönicke L, Gustavsson I and Scherthan H (1996). Comparative analysis of the cattle and human genomes: detection of ZOO-FISH and gene mapping based chromosomal homologies. Mamm Genome 7;297-302.

Chowdhary BP, Raudsepp T, Frönicke L and Scherthan H (1998) Emerging patterns of comparative genome organization in some mammalian species as revealed by Zoo-FISH. Genome Research 8;577-589.

Dixkens C, Klett C, Bruch J, Kollak A, Serov OL, Zhdanova N, Vogel W, Hameister H (1998) Zoo-FISH analysis in insectivores: "Evolution extols the virtue of the status quo". Cytogenet Cell Genet 80;61-67.

Drouin R, Lemieux N, Richer CL (1990) Analysis of DNA replication during S-phase by means of dynamic chromosome banding at high resolution. Chromosoma 99;273-80

Ferguson-Smith MA. (1997) Genetic analysis by chromosome sorting and painting: phylogenetic and diagnostic applications. Eur J Hum Genet.5;253-65.

Florijn RJ, Slats J, Tanke HJ, and Raap AK (1995) Analysis of antifading reagents for fluorescence microscopy. Cytometry 19;177-182.

Frönicke L (1997) Zoo-FISH-Untersuchungen zur Genomorganisation der Säugetiere. PhD Thesis, FB Biologie, Univ. Kaiserslautern, Abt. Humanbiologie.

Frönicke L, Chowdhary BP, Scherthan H and Gustavsson I (1996) A comparative map of the porcine and human genomes demonstrates ZOO-FISH and gene mapping based chromosomal homologies. Mamm Genome 7;285-290.

Frönicke L and Scherthan H (1997) Zoo-FISH analysis of human and Indian muntjac karyotypes (*Muntiacus muntjak vaginalis*) reveals satellite-DNA clusters at the margins of conserved syntenic segments. Chromosome Res. 5;254-261.

Frönicke L, Müller-Navia J, Romanakis K and Scherthan H (1997) Chromosomal homologies between human, harbor seal (*Phoca vitulina*) and the putative ancestral carnivore karyotype revealed by Zoo-FISH. Chromosoma 106;108-113.

Goureau AM, Yerle M, Schmitz A, Riquet J, Milan D, Pinton P, Frelat G and Gellin J (1996) Human and porcine correspondence of chromosome segments using bidirectional chromosome painting. Genomics 36;252-262.

Hayes H, Petit E, Dutrillaux B (1991) Comparison of RBG-banded karyotypes of cattle, sheep, and goats. Cytogenet Cell Genet 57;51-55.

Islam MQ and Levan G (1987) A new fixation procedure for improved quality G-bands in routine cytogenetic work. Hereditas 107; 127-130.

Jauch A, Wienberg J, Stanyon R, Arnold N, Tofanelli S, Ishida T, and Cremer T. (1992) Reconstruction of genomic rearrangements in great apes and gibbons by chromosome painting. Proc Natl Acad Sci USA 89: 8611-8615.

Kappler R, Schlegel J, Romanakis K, Mennel H-D and Scherthan H (1998) Comparative genomic in situ hybridization discloses chromosomal copy number changes in a transplanted brain tumor line of the rat (*Rattus norvegicus*). Mamm Genome 9:193-197.

Lichter P, Cremer T (1992) Chromosome analysis by non isotopic in situ hybridization. *In*: Human Cytogenetics - A practical Approach (2nd edition), Vol. I 2nd ed. Rooney D.E. and Czepulkowsky B.H., eds. IRL Press, Oxford pp; 157-192.

Lichter P, Cremer T, Borden J, Manuelidis L and Ward DC (1988) Delineation of individual chromosomes in metaphase and interphase cells by *in situ* suppression hybridization using recombinant DNA libraries. Hum Genet 80;224-234.

Kohne DE, Chiscon JA, Hoyer BH (1972) Evolution of mammalian DNA. In: LeCame LM, Neyman J, Scott EL (eds). 6th Berkeley Symp. Math. Stat. Prob., Vol.5, pp193-209. Univ. of California Press, LosAngeles.

Meyne J, Baker RJ, Hobart HH, Hsu TC, Ryder OA, Ward OG, Wiley JE, Wurster-Hill DH, Yates TL, Moyzis RK (1990) Distribution of non-telomeric sites of the (TTAGGG)n telomeric sequence in vertebrate chromosomes. Chromosoma 99: 3-10.

Müller S, O'Brien PC, Ferguson-Smith MA, Wienberg J (1998) Cross-species colour segmenting: a novel tool in human karyotype analysis. Cytometry 33;445-52.

Nash WG, Wienberg J, Ferguson-Smith MA, Menninger JC, O'Brien SJ (1998) Comparative genomics: tracking chromosome evolution in the family ursidae using reciprocal chromosome painting. Cytogenet Cell Genet 83;182-92.

O'Brien SJ, Womack JE, Lyons LA, Moore KJ, Jenkins NA and Copeland NG (1993) Anchored reference loci for comparative genome mapping in mammals. Nature Genet 3;103-112.

O'Brien SJ, Wienberg J, Lyons LA (1997) Comparative genomics: lessons from cats. Trends Genet 13;393-399

Pinkel D, Landegent J, Collins C, Fusco J, Segraves R, Lucas J, Gray JW, (1988) Fluorescence in situ hybridization with human chromosome specific libraries: detection of trisomy 21 and translocations of chromosome 4. Proc Natl Acad Sci USA 85, 9138-9142.

Rettenberger G, Klett C, Zechner U, Bruch J, Just W, Vogel W, Hameister H (1995) ZOO-FISH analysis: cat and human karyotypes closely resemble the putative ancestral mammalian karyotype. Chromosome Res 3;479-486.

Scherthan H, Lutz S, Metzdorf R and Blin N (1989) Interspecies relationship of a chromosome specific DNA. Hereditas 111;183-188.

Scherthan H, Cremer T, Arnason U, Lima-de-Faria A, Weier H-U and Frönicke L (1994) Comparative chromosome painting discloses homologous segments in distantly related mammals. Nature Genet 6;342-347.

Scherthan H and Cremer T (1994) Methodology of non isotopic in situ-hybridization in paraffin embedded tissue sections. in: *Methods in Molecular Genetics* (K.W. Adolph ed). Academic Press, San Diego, Vol. 5;223-238.

Sibley CG, Comstock JA, Ahlquist JE (1990) DNA hybridization evidence of hominoid phylogeny: a reanalysis of the data. J Mol Evol 30;202-236.

Vooijs M, Yu L-C, Tkachuk D, Pinkel D, Johnson D, Gray JW (1993) Libraries for each human chromosome, constructed from sorter-enriched chromosomes by using linker-adaptor PCR. Am J Hum Genet 52;586-597.

Wienberg J, Jauch A, Stanyon R, and Cremer T (1990) Molecular Cytotaxonomy of primates by chromosomal in situ suppression hybridization. Genomics 8;347-350

Wienberg J, Stanyon R (1997) Comparative painting of mammalian chromosomes. Curr Opin Genet Dev 7;784-791.

Wienberg J, Stanyon R, Nash WG, O'Brien PC, Yang F, O'Brien SJ, Ferguson-Smith MA (1997) Conservation of human vs. feline genome organization revealed by reciprocal chromosome painting. Cytogenet Cell Genet 77;211-217.

FISH on Insect Cells Transfected with Heterologous DNA

THOMAS LIEHR

Introduction

Insect cells transfected with heterologous DNAs (plasmids) can be used for the assessment of functional properties of genes that have been isolated or subjected to in vitro mutagenesis (Rubin and Spradling 1983, Snow et al. 1989). An example of a cell system used for such approaches are Schneider cells (Schneider 1972). These are derived from embryonic cells of *Drosophila melanogaster*, can be transfected with special plasmids (Bunch et al. 1988) and used for heterologous expression. Several, in general, reliable in vitro methods can be performed to control the transfection efficiency in such cells (i.e. PCR, Southern-, Northern- or Western-blot approaches). Another fast and reliable method to detect transfected Schneider cells, offering the possibility of assessing the transfection rate on a single cell level, has recently been described and is based on a modified FISH technique (Rautenstrauss et al. 1998; see Figure 1) and described in the following.

Procedure

Preparation of the Schneider cells for FISH

1. Schneider cells transfected for transient expression (e.g. with a derivate of the plasmid pRmHa-3 (Bunch et al. 1988)) and wild type Schneider cells (without any plasmid, used as negative controls) have to be cultivated according to Schneider (1972).

Thomas Liehr, Institut für Humangenetik, Kollegiengasse 10, Jena, 007743, Germany (*phone* +49-3641-935533; *fax* +49-3641-935502; *e-mail* i8lith@mti-n.mti.uni-jena.de)

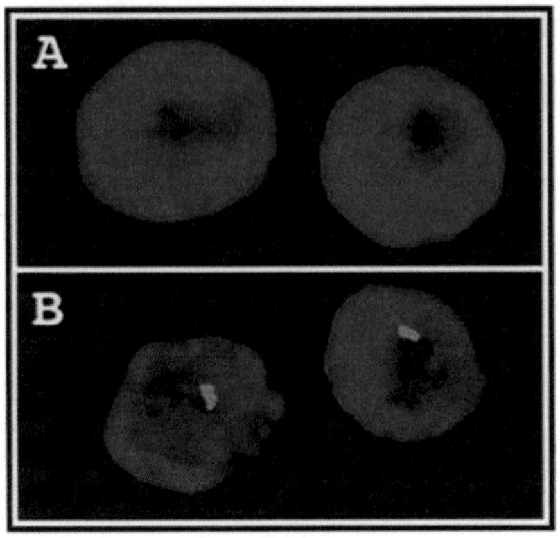

Fig. 1. Fluorescence in situ hybridization (FISH) of **A** non-transfected and **B** transfected. Schneider cells with plasmid pRmHa-3. The two nuclei in **A** represent the negative control (non-transfected Schneider cells); the two nuclei in **B** represent typical transfected cells. The plasmid pRmHa-3 was labeled with biotin (green). The nuclei are counterstained in blue with DAPI (= diaminophenylindol). Images were captured on a Zeiss Axioplan microscope using the ISIS digital FISH imaging system (MetaSystems, Altlussheim, Germany) using an XC77 CCD camera with on-chip integration (Sony)

2. Cells of one tissue culture flask are collected in a 15-ml tube and sedimented by centrifugation (1000 rpm, 5 min) (see comment 1).
3. The supernatant is carefully discarded and the pellet is resuspended in 5 ml 1% formaldehyde solution (in 1×PBS) and incubated at room temperature (RT) for 2-3 h (see comment 2).
4. Cells are sedimented by centrifugation (1000 rpm, 5 min) and the supernatant is discarded.
5. The resulting pellet is washed twice in 5 ml 1×PBS and resuspended in 0.5 ml of 1×PBS.
6. Approx. 70 µl of the fixed Schneider cells are transferred onto one clean dry slide for each cell line, and fixed by air drying overnight. Afterwards, slides can be stored at -20°C for several weeks before being used in FISH.

Labeling of the probes

1. The plasmid probe which has been used to transfect the Schneider cells is now used as the specific FISH probe (e.g. a derivative of the plasmid pRmHa-3 (Bunch et al. 1988)). Thus, it has to be labeled, e.g. with biotin by nick translation using the kit from Boehringer Mannheim (Cat. No. 1745816).
2. For this purpose 1 μg of the probe is diluted in 16 μl of double distilled water, and 4 μl of the nick translation solution is added. This solution is mixed carefully with a tip of a 20 μl Eppendorf pipette and the 1.5-μl microtube is incubated at 15°C for 90 min.
3. During this time, a Sephadex column can be prepared. First, a 1-ml syringe is filled up to the 0.1-ml line with silanized glass wool.
4. Then, Sephadex G80 solution is filled up to the 1.0-ml line (Sephadex G80 solution: 8 g of Sephadex G80 powder is dissolved in 100 ml 1×TE solution and incubated overnight at RT; after that it is ready for use and can be stored at RT for several months).
5. Put the syringe in a 15-ml tube and put it in a centrifuge at 3000 rpm for 1 min to separate the TE solution from the Sephadex.
6. Repeat steps 4 and 5.
7. Add 100 μl 1×TE solution on the prepared Sephadex column and repeat step 5 to wash the column.
8. Now the Sephadex column is ready for use and can be stored for 2-3 h at 4°C before use. Do not forget to discard the 1×TE solution collected in the 15-ml tube.
9. Add 20 μl of "stopmix" and 60 μl of 1×TE to the nick translation solution from step 2 ("Stopmix": 0.25% bromphenol blue, 0.25% dextran blue, 0.5 M EDTA).
10. The 100 μl solution from step 9 is put onto the sephadex column from step 8 and centrifuged at 3000 rpm for 3 min. The light blue solution, (due to the dextran blue of the "stopmix"), coming out of the column, is collected in the 15-ml tube. Usually, the resulting solution is approx. 100 μl. As 1 μg has been introduced into the nicktranslation reaction, the DNA concentration of the solution is around 10 ng/μl.

Fluorescence in situ hybridization (FISH)

1. Add 100 µl denaturation-buffer to the slides and cover with (24×50 mm coverslips). (Denaturation-buffer: 70% (v/v) deionized formamide, 10% (v/v) filtered double distilled water, 10% (v/v) 20×SSC, 10% (v/v) phosphate buffer; make fresh as required; phosphate buffer: prepare 0.5 M Na_2HPO_4 and 0.5 M NaH_2PO_4, mix these two solutions (1:1) to get pH 7.0, then aliquot and store at -20°C).
2. Incubate slides on a warming plate for 5 min at 75°C.
3. Remove the coverslips immediately with forceps and place slides in a Coplin jar filled with 70% ethanol (4°C) to conserve target DNA as single strands.
4. Dehydrate slides in ethanol (70%, 90%, 100%, 4°C, 3 min each) and air dry.
5. For each slide to be hybridized, precipitate 200 ng of the biotin labeled probe together with 4 µg Cot1-DNA and 20 µg herring sperm DNA with 2.5 vol 100% ethanol for 20 min at -80°C or 12-20 h at -20°C (see comment 3).
6. Pellet the DNA by centrifugation at 15,000 rpm for 15 min, discard the supernatant and dry the DNA pellet at RT or using a speed-vac.
7. Dissolve the pellet from step 6 in 20 µl of hybridization-buffer, vortex and spin down. (Hybridization-buffer: Dissolve 2 g dextran sulfate in 10 ml 50% deionized formamide/2×SSC/50 mM phosphate buffer for 3 hat 70°C. Aliquot and store at -20°C).
8. Denature the probe-solution at 75°C for 5 min and do a prehybridization step at 37°C for 30 min.
9. Add 20 µl of probe-solution onto each denatured slide, put 24×50 mm coverslips on the drops and seal with rubber cement (Fixogum, Marabu).
10. Incubate slides for three nights at 37°C in a humid chamber.
11. Take the slides out of the 37°C humid chamber, remove rubber cement with forceps and coverslips by letting them swim off in 4xSSC/ 0.2%Tween (RT, 100 ml Coplin jar) (see comment 4).
12. Postwash the slides 3 x 5 min in formamide-solution (45°C) followed by 3 x 5 min in 2×SSC (37°C) in a 100-ml Coplin jar, with gentle agitation.

13. Put the slides in 4×SSC/0.2% Tween (100 ml, RT), for a few seconds.
14. Add 50 μl of solution I to each slide, cover with 24×50 mm coverslips and incubate at 37°C for 30 min in a humid chamber. [Solution I: FITC-avidin (CAMON Vector Laboratories A2011)/4xSSC + 0.2% Tween + 5% BSA (1:300); make fresh as required].
15. Remove the coverslips and wash 3×3 min in 4×SSC/0.2% Tween (RT, with gentle agitation).
16. Add 50 μl of solution II to each slide, cover with 24×50 mm coverslips and incubate at 37°C for 45 min in a humid chamber. [Solution II: Biotinylated antiavidin (CAMON Vector Laboratories BA0300)/ 4×SSC + 0.2% Tween + 5% BSA (1:100); make fresh as required].
17. Repeat step 15.
18. Repeat steps 14 and 15.
19. Counterstain the slides with DAPI-solution (100 ml in a Coplin jar, RT) for 8 min. (DAPI-solution: Dissolve 5 μl of DAPI (4,6-diamidino-2-phenylindol.2HCl stock-solution; Serva 18860) in 100 ml 4xSSC/ 0.2% Tween; make fresh as required).
20. Wash slides several times in water for a few seconds and air dry.
21. Add 15 μl of antifade Vectashield (CAMON Vector Laboratories H1000), cover with coverslips and look at the results under a fluorescence microscope.

Comments

1. Remember to prepare the cells to be tested for transfection and the wild type cells as negative controls. The approach described here always needs a negative control.
2. Incubation in formalin-buffer can be done with or without agitation, it leads to no difference in the quality of prepared Schneider cells.
3. For the precipitation, sodium acetate does not need to be added; the dextran blue from the "stopmix" can replace it here.
4. During the washing steps, it is important to prevent the slide surface drying out, otherwise background problems may arise.

Acknowledgements. Supported by the Herbert Quandt Stiftung der VARTA AG. The continuos support of the Carl Zeiss GmbH (Jena, Germany) is gratefully acknowledged.

References

Bunch, TA, Grinblat, Y, Goldstein, LS (1988) Characterization and use of the Drosophila metallothionein promoter in cultured *Drosophila melanogaster* cells. Nucleic Acids Res, 16:1043-61

Rautenstrauss, B, Fuchs, C, Ekici, A, Nelis, E, Van Broeckhoven, C, Liehr, T (1998) Assay of transfection rate in insect cells on a single cell level. Genet Anal-Biomol Engineering, 14:103-104

Rubin, GM, Spradling, AC (1983) Vectors for P element-mediated gene transfer in Drosophila. Nucleic Acids Res 11:6341-51

Schneider I (1972) New cell line from embryonic cells of *Drosophila melanogaster.* J Embr Exp Morph 27:353-356

Snow, PM, Bieber, AJ, Goodman, CS (1989) Fasciclin III: a novel homophilic adhesion molecule in Drosophila. Cell 59:313-23

Yeast FISH: Delineation of Chromosomal Targets in Vegetative and Meiotic Yeast Cells

HARRY SCHERTHAN* and EDGAR TRELLES-STICKEN*

Introduction

Fluorescence in situ hybridization provides an effective means to delineate chromosomes and chromosomal regions of interest, during all stages of the cell cycle, irrespective of chromatin composition and compaction. This property makes FISH particularly useful for studying chromosome behavior in species with minute genomes and/or ill-defined metaphase chromosomes. Genetic model organisms like fission and budding yeast display such cytological characteristics. Consequently, chromosome painting and locus-specific yeast FISH (Scherthan et al. 1992, 1994, Uzawa et al. 1992, Guacci et al. 1994, Weiner and Kleckner 1994, Gotta et al. 1996) has become an indispensable tool in the analysis of chromosome dynamics and organization in wild-type and mutant yeast strains.

Principles and applications

As compared to human and other higher eukaryotes, the bakers yeast *Saccharomyces cerevisiae* (n=16) challenges classical cytology, since metaphase occurs without nuclear envelope breakdown, highly condensed chromosomes are absent and yeast nuclei lack a lamina which makes them vulnerable to distortions during cell wall removal and isolation procedures. Conventional cytogenetic analysis of yeast mitosis, therefore, only reveals a morphological change of the nuclear/chromosome mass which is seen to elongate, bifurcate and eventually to separate during ana-

✉ Harry Scherthan, Edgar Trelles-Sticken, University of Kaiserslautern, Dept. of Human Biology and Human Genetics, Erwin-Schroedingerstr., Kaiserslautern, 67663, Germany (*phone* +49-631-205-2107; *fax* +49-631-205-2878; *e-mail* scherth@uni.de)
* present address: Max-Planck-Inst. for Molecular Genetics, Ihnestr. 73, Berlin, D-14195, Germany.

phase B. Despite its powerful genetics, *S. cerevisiae* cytology is further hampered by the very small size of the object: diploid nuclei are only 2-4 μm in diameter and harbor 32 chromosomes with a total DNA content of 24 Mbp of predominantly unique DNA sequences (Goffeau et al. 1997). For these reasons, light and electron microscopic studies of yeast metaphase chromosomes at best revealed tiny chromatin lumps which are hardly reminiscent of the metaphase chromosomes of more complex eukaryotes (Kater 1927, Wintersberger et al. 1975). Unlike in mitosis, pachytene chromosome spreads (bivalents) can be obtained from meiotic yeast cultures when homologues are connected by their joint synaptonemal complex (SC; Kuroiwa et al. 1984, Dresser and Giroux 1988).

SCs can be visualized in intact nuclei by 3D electron microscopy or by silver staining in surface spread meiocyte nuclei (for details see Esposito et al. 1990, Loidl et al. 1998). Introduction of FISH to yeast cytology (Scherthan et al. 1992, Uzawa et al. 1992) provided the first means to delineate individual chromosomes and their subregions in spreads and undisrupted nuclei. Therefore, molecular yeast cytology has become a major tool to physically monitor chromosome behavior during wild-type and mutant mitotic and meiotic cell cycles (Funabiki et al. 1993, Loidl et al. 1994, Nag et al. 1995, Gotta et al. 1996, Guacci et al. 1997, Trelles-Sticken et al. 1999). Below, we outline a protocol for yeast FISH which is routinely used in our lab.

Materials

- Standard equipment used in the cytological laboratory including: Coplin jars with lids, glass slides, coverslips, micropipettes, heating block, plastic or metal boxes, microcentrifuge and incubator.
- Phase contrast and fluorescence microscope equipped with appropriate filter sets for excitation of the different fluorochromes applied. A conventional or preferably a digital image recording and processing system.

Reagents

- Antifade solution: Vectashield (Vector Labs) is recommended as mounting medium for fluorescence microscopy, since it efficiently reduces fading of fluorochromes during microscopic analysis (Florijn et al. 1995)
- Avidin-FITC (e.g., ExtrAvidin-fluorescein; Sigma)

- Bacto agar (Difco)
- Bacto peptone (Difco)
- Biotinylated goat-anti-avidin antiserum (Vector Labs)
- BSA (fraction V; Serva)
- BT buffer (0.15M $NaHCO_3$, 0.1% Tween-20, pH 8.3). After dissolution of $NaHCO_3$, pH will be in the appropriate range. Make up fresh BT for each experiment.
- Carrier DNA from salmon sperm or *E. coli* (Sigma). Dissolve at 10 mg/ml in distilled (d)H_2O and shear by sonication to 100 bp–1 kbp.
- DAPI (4,6-diamidine-2-phenylindole dihydrochloride; Sigma). Stock solution: 1 mg/ml in sterile dH_2O, store at -20°C.
- D-Glucose (Sigma)
- Dithiothreitol, make up 0.5 M stock with dH_2O and store DTT stock at -70°C.
- DNA counterstains: The fluorescent dyes DAPI and PI are added to the antifade solution in the following concentrations: DAPI 0.5 µg/ml and/or PI 1 µg/ml.
- Ethanol (p.a.), ethanol (70%)
- Fixative: 3.7% acid-free formaldehyde, 4% sucrose, pH 7.4
- Formaldehyde 37%, acid free (Merck, Darmstadt)
- Hybridization solution: 50% deionized formamide, 10% 20×SSC, 20% of a 50% (w/v) dextran sulfate solution, 10% of 50 mM sodium phosphate, pH 7.0, probe DNA at 30 ng/µl, sheared herring carrier DNA at a final conc. of 1 µg/µl. Use deionized water to adjust volume.
- Lipsol. Lab cleaning reagent from LIP Ltd., Shipley, UK
- PBS, 10×: 1.3 M NaCl, 0.07 M Na_2HPO_4, 0.03 M NaH_2PO_4, pH 7.4
- Pre-sporulation medium: 1% yeast extract, 2% Bacto peptone, 1% potassium acetate, pH 7.0. Make up fresh for each use.
- Propidium iodide (PI; Sigma). Dissolve at 1 mg/ml in sterile water, store stock at -20°C.
- RNaseA (Sigma). Stock solution: 10 mg/ml RNaseA in dH_2O, heat for 10 min to 95°C, store at -20°C.

- RNaseA working solution: Use RNaseA at 200 μg/ml in 2×SSC
- Rubber cement (Fixogum; Marabu). Bicycle repair fluid may serve as a substitute.
- Sodium N-lauroylsarcosine (Fluka Chemie), 2% in dH_2O, store at 4°C.
- Solution I (Sol I): 1 M Sorbitol, pH 7. May be stored at 4°C up to 2 weeks.
- Solution II (Sol II): 0.1 M MES [2-(N-morpholino)-ethane sulfonic acid]; 1 mM EDTA; 0.5 mM $MgCl_2$; 1 M sorbitol; adjust pH to 6.4 with NaOH.
- Sporulation medium: 2% Potassium acetate, pH 7.0. Make up fresh for each use.
- SSC, 20×: 3 M NaCl, 0.3 M sodium citrate, pH 7.0
- Sucrose
- Target unmasking fluid (TUF; Dianova)
- Tween-20 (Sigma)
- Yeast extract (Difco)
- YPD agar: YPD + 2% Bacto agar
- YPD medium: 1% (w/v) Yeast extract; 2% Bacto peptone; 2% D-glucose
- Zymolyase 100T (Kirin Brewery Ltd.) Make up a stock solution (10 mg Zymolyase 100T/ml dH_2O) and store at -70°C.

Subprotocol 1
Cell culture

Procedure

1.1 Culture of vegetative yeast cells

1. Streak out yeast strain of interest on a freshly prepared YPD-plate and incubate for 2 days at 30°C.

2. Transfer a colony from the plate into an appropriate volume (e.g. 20 ml) of liquid YPD medium and grow the culture at 30°C to an OD600=1.
3. Analyze budding rate (mitotic cells) by transferring an aliquot of cells to a phase contrast microscope. Logarithmic growth is indicated by a budding rate of 50-60%.
4. Transfer an aliquot of your culture (e.g. 9 ml) to a centrifuge tube on ice containing $^1/_{10}$ volume of acid-free formaldehyde (37%) (Pringle et al. 1991).
5. Incubate for 30 min on ice.
6. Sediment cells at 700 g for 3 min and wash them once in a large volume of dH_2O.
7. Sediment cells at 700 g for 3 min and resuspend them in 1/10 of the original volume of Solution I. Cells are now ready for spheroplasting (see procedure 1.3).

1.2 Preparation of meiotic yeast nuclei

Since the yeast cell cycle is regulated by the availability of nutrients, vegetative cells can be induced to switch from vegetative growth (budding) to meiosis (sporulation) by the deprivation of nitrogen and fermentable carbon sources. In fast sporulating strains like SK1 (Kane and Roth 1974) cells enter and progress through meiosis in a rather synchronous manner (Padmore et al. 1991) which ultimately yields haploid spores.

1. Streak out diploid a/α strain of interest on a YPD agar plate (see procedure 1.1., step 1) and transfer a colony to pre-sporulation medium (e.g. 20 ml). Grow cells at 30°C overnight to a density of OD_{600}=1-1.5. Control the budding rate which should be 40% (see troubleshooting).
2. Sediment cells at 3000 rpm for 5 min at RT.
3. Wash the cells twice with sporulation medium at 30°C.
4. Resuspend the cell pellet in half the initial volume of sporulation medium (here 10 ml) and return the culture back to 30°C. This time point represents t=0 in a meiotic time course experiment.
5. Remove aliquots from the culture at t=0 and subsequent time points.

6. To obtain extensively spread nuclei, put the cell suspension aliquots immediately on ice and, after spheroplasting (procedure 1.3), refer to procedure 1.5 for further processing. To prepare undisrupted or mildly spread nuclei, add the aliquots to $^1/_{10}$ volume of 37% formaldehyde on ice. After 30 min perform spheroplasting (see below) and continue as described under procedure 1.4 or 1.5.

1.3 Spheroplast preparation

1. Add 2% (v/v) of a 0.5 M DTT stock to the aliquots of the cell cultures obtained in procedure 1.1, step 7 or procedure 1.2, step 6.
2. Add 1% (v/v) of Zymolyase 100T Stock solution and vortex the tube briefly.
3. Incubate for 10 – 20 min at 37°C.
4. Check degree of cell wall degradation by placing an aliquot (5-10 μl) of the digested suspension on to a slide.
5. Mix with an equal volume of a 2% sodium N-lauroylsarcosine.
6. Immediately cover the mixture with a cover glass and observe the sample instantly under a phase contrast microscope (16× magnification).
7. After a few seconds the cells should be seen bursting (initially bright cells become dark and then lyse). The cell wall digest is complete when most of the cells (80%) rupture instantly. This is a critical indicator for the success of FISH (see Troubleshooting).
8. Stop the digest by adding an equal volume of ice cold Solution II; sediment cells as described in procedure 1.1, step 6, and resuspend pellet in $^1/_{20}$ volume Sol II as compared to initial aliquot volume.
9. The resulting pellet may be subjected either to preparation of three-dimensionally preserved nuclei (procedure 1.4) or to detergent spreading (procedure 1.5).

1.4 Preparation of structurally preserved yeast nuclei for FISH

1. Wash spheroplasts (see procedure 1.3, step 9) two times in 1×SSC at RT and fix once by resuspending the pellet in 70% ethanol.

2. Spin down cells and dissolve pellet in $^1/_{20}$ volume absolute ethanol with respect to the starting aliquot (see procedure 1.2).
3. Distribute 20 µl of ethanol-fixed nuclei uniformly on a clean glass slide by streaking out with one edge of a cover slip without scratching along the surface of the slide.
4. Dry down the preparation on a heating block at 60°C for 10-15 min.
5. Apply 18 µl of antifade solution containing DAPI as a DNA specific dye and examine the density of nuclei under a fluorescence microscope. If required, remove cover slip, wash in dH_2O, air dry, then add more ethanol fixed spheroplasts to the slide and allow them to adhere (see above).
6. To open up fixed chromatin, incubate preparations for 1 h-overnight in 4×SSC/0.5% Tween 20/pre-boiled RNaseA (1 µg/ml) at 60°C (Scherthan et al. 1993).
7. Slide down coverslip, wash slides at room temperature for 5 min in PBS.
8. Digest remaining RNA for ›60 min at 37°C by adding 100 µl of pre-boiled RNaseA at a concentration of 200 µg/ml 2×SSC under a large coverslip. Use a moist chamber for incubation.
9. Wash preparations once in PBS and drain off excess fluid.

1.5 Preparation of nuclear spreads

Since undisrupted yeast nuclei generally display a small diameter (2-4 µm) the technique of detergent spreading may be used to enhance cytological resolution in the yeast nucleus (Dresser and Giroux 1988, Loidl et al. 1991, Jin et al. 1998). Depending on the mode of pre-fixation (see procedure 1.2, step 6), it is possible to control the degree of nuclear spreading and to preserve the relative morphology of nuclei during spreading (Trelles-Sticken et al. 1999).

Limited detergent spreading of nuclei

Limited nuclear spreading may be applied to obtain moderately enlarged nuclei with relatively preserved nuclear architecture.

1. Subject cells to pre-fixation as described under procedure 1.1 and spheroplast them as far as procedure 1.2, step 7.

2. Stop cell wall digest by adding ice cold Sol II.
3. Sediment cells and resuspend them in a volume of Sol II corresponding to a 1/20 of the starting volume.
4. Place an aliquot of spheroplasts (e.g. 20 µl) on a clean glass slide.
5. Add 1% Lipsol and fixative in the ratio: 1 vol spheroplasts: 2 vol Lipsol: 4 vol fixative.
6. Briefly tilt the slide after each step to mix the solutions evenly within the resulting drop.
7. After addition of the fixative, distribute the suspension evenly by streaking with the side of a pipette tip over the slide without touching the surface.
8. The degree of nuclear spreading may be altered in two ways: Changing the given ratios in favor of Lipsol results in nucleoids with a bigger diameter, while the resulting diameter will be smaller when the relative volume of detergent is reduced in the mixture.

Preparation of extensively spread nuclei

Extensively spread nuclei are obtained when the aliquots removed from the culture are directly transferred to ice and the pre-fixation step is omitted. Spheroplasting and detergent spreading of unfixed cells is then performed as described above, with the exception that fixation is performed according to the following scheme:

- Successively mix: 1 vol spheroplasts/2 vol fixative/4 vol Lipsol/4 vol fixative.
- Tilt and air dry.

Subprotocol 2
Probe labeling and FISH

Nick translation, random priming and long range PCR protocols have proven effective methods to label probe DNAs in our hands. For ease of handling, we recommend the use of commercially available labeling kits. For a detailed description of the set up of a nick translation reaction, the reader may refer to detailed protocols published elsewhere (Lichter and Cremer 1992). A critical aspect for successful yeast FISH is that all probes used are completely free of RNA. With the availability of the complete DNA se-

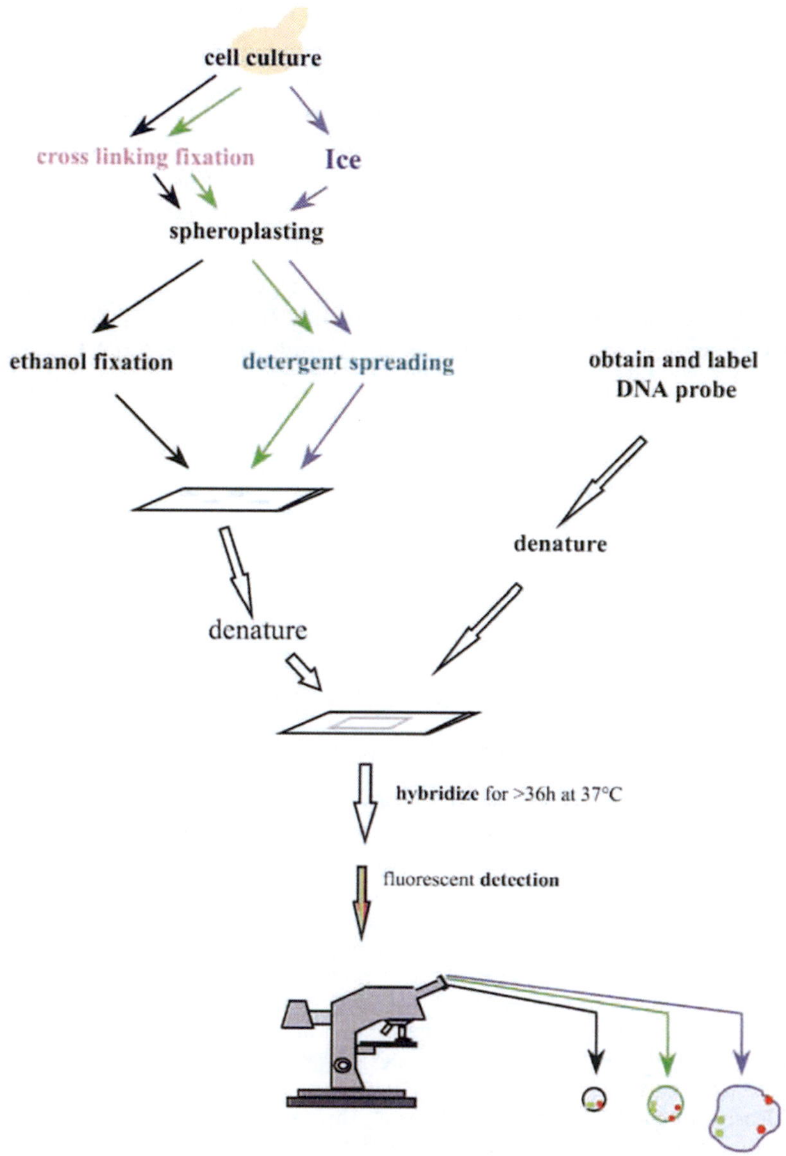

Scheme 1. Flowchart of the preparation procedure for undisrupted (*black arrowed* route) and spread nuclei (*green* and *blue arrowed* routes) for yeast FISH. Depending on the mode of fixation prior to detergent spreading, either moderately spread (green arrows) or extensively spread nuclei (blue arrows) may be obtained. Nuclear spreads offer a higher resolution and thereby facilitate visual analysis of FISH signals. It should be noted that particular three-dimensional chromosome arrangements are disrupted by extensive nuclear spreading (see Fig. 1d,e)

quence of the yeast genome (http://genome-www.stanford.edu/Saccharomyces) primers covering the region of interest may be designed and long range PCR can be used to produce potent FISH probes. Generally, DNA probes larger than 6 kb and probe concentrations of 20-50 ng/µl are required to generate distinctly visible hybridization signals (Scherthan et al. 1992, 1994). It is recommended to determine the size and amount of labeled probe by running an aliquot ($^1/_{20}$) of the nick translation reaction in a 1% agarose mini gel.

DNA-Probes are preferably labeled with haptens like biotin, digoxigenin or estradiol. Haptens are detected with appropriate fluorochrome-conjugated antibodies which yield fluorescent signals which are readily seen upon visual inspection (see subprotocol 3.). The two color FISH scheme presented below (Fig. 1) also allows three colors to be simultaneously observed under the microscope, where the third color is induced by 1:1 ratio mixing of biotin and digoxigenin labels (Fig. 1c, h). We also used direct fluorochrome labeling but this usually results in weak hybridization sig-

Fig. 1. FISH on spread and undisrupted *S. cerevisiae* nuclei. **a** Two-color chromosome painting reveals paired and separated chromosomes I (red, rhodamine) and III (green, fluorescein) in spread meiocyte nuclei (DAPI, blue) obtained 2 h after induction of sporulation. Conventional fluorescence microscopy. *Bars* in **a-f** represent 5 µm. **b** Spread pachytene nucleus showing delineated chromosome I (red) and III (green) bivalents. Conventional fluorescence microscopy. **c** Multi-color FISH delineates bivalents of chromosomes *I* (red), *III* (green) and *V* (orange) in an extensively spread pachytene nucleus (DAPI, blue) (conventional fluorescence microscopy). **d** FISH with a composite probe for all telomeres (green) and centromeres (red) to cells obtained during mitotic growth shows the presence of a distinct centromere cluster and peripheral telomere aggregates in (*i*) a structurally preserved nucleus, (*ii*) a mildly spread nucleus, and in (*iii*) an extensively spread nucleus. Note that telomere signals disperse upon enhanced spreading, while centromeres don't. (**e**) Undisrupted and spread meiocyte at the bouquet stage. Telomeres (green) form a single cluster, while centromeres (red) are dispersed. This is more obvious in the larger spread meiocyte to the *right*. **f** Telomere (green) and cosmid FISH (red, tagging a chr. *XI* region) to undisrupted nuclei. Telomere clusters are peripherally distributed in a vegetative nucleus with separated cosmid signals (left). The meiocyte nucleus to the right displays paired cosmids (one signal) and a single telomere cluster (green; bouquet stage). **g-i** FISH to spreads of diploid *Schizosaccharomyces pombe* cells (note that wild-type *S. pombe* is haploid, n=3). **g** Metaphase figure with differentially marked telomeres of chromosomes 2 by FISH with two cosmid probes (red, green). **h** Horse tail nucleus obtained at meiotic prophase (see Kohli 1994). Telomeres (red) and adjacent rDNA (orange, by red/green labeling) mark the leading edge of this bouquet nucleus, while centromeres (green) locate near the opposite end. **i** Painting of chromosomes 1 (green) and 2 (red) in a mildly spread diploid nucleus. Homologue association leads to one signal which delineates a joint territory of each chromosome pair. *Bar* in **i**: 2 µm, it also applies to **g** and **h**

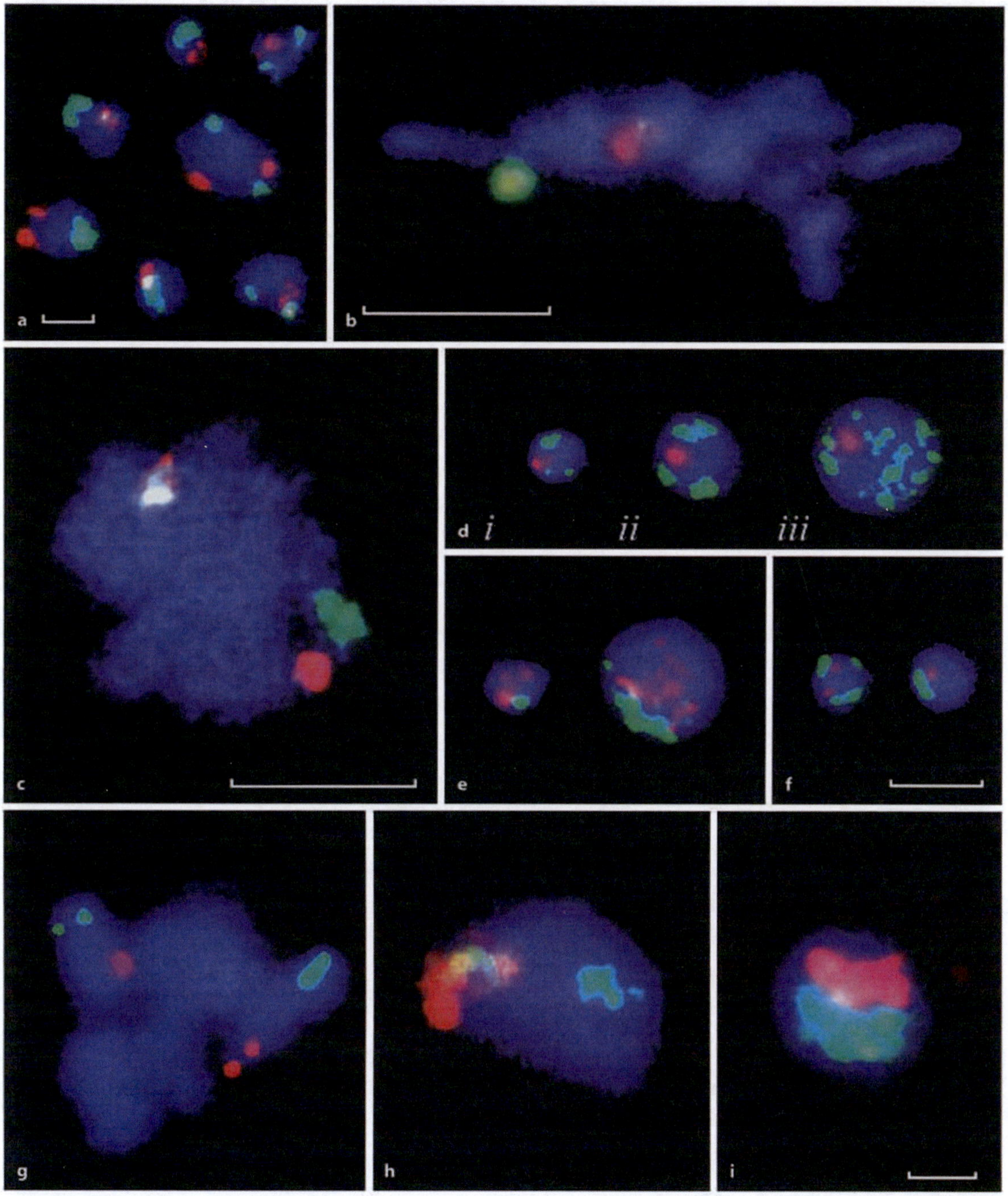
a
b
c
d i
ii
iii
e
f
g
h
i

nals (due to the small amount of fluorochromes deposited at the target region[s]), hence analysis of such FISH experiments requires image recording with a charge-coupled device camera and computer-based image analysis system.

Procedure

1. Label 1 μg of probe DNA by the method of your choice (see above). Remove $^{1}/_{20}$ of the reaction volume to perform a minigel analysis.
2. To the labeling reaction, add $^{1}/_{10}$ volume of 300mM EDTA, 1 μl carrier DNA and precipitate the DNA by adding 3 volumes of ethanol.
3. 4. Incubate for 30 min at -20°C, centrifuge for 30 min at 12,000 *g* and resuspend pellet in hybridization solution to result in a final probe DNA concentration of 30-50 ng/μl.
4. Place tube for 30 min at 60°C to completely dissolve the pellet in the hybridization solution.
5. Denature hybridization solution for 3 min at 95°C, cool on ice.
6. On the preparations, determine the region with highest cell density on the slide and mark it from underneath with a diamond pen.
7. Denature preparations by applying 100 μl of 70% formamid/2×SSC under a large coverslip and place slide for 5 min at 74°C. Wash down coverslip and denaturation solution with a jet of ice cold water and air dry.
8. Add required amount of hybridization solution (1 μl for a round coverslip of 10 mm diameter is sufficient) to the marked region, add cover slip and seal with rubber cement. Slides are placed in a dry box and renaturation is allowed to proceed at 37°C for 2 days in a moist incubator (Scherthan et al. 1992).
9. After in situ hybridization, rubber cement is removed with forceps and coverslips are floated off in a Coplin jar containing 0.05×SSC.
10. Perform 3×5 min stringency washes in 0.05×SSC at 42°C to remove excess and unspecifically bound probe molecules.

Subprotocol 3
Detection of hybrid molecules

Procedure

We present an indirect detection system for two colors, since only these render readily distinguishable signals (Fig. 1) which are requisite for the visual analysis of a large number of nuclei and hybridization patterns in e.g. time course experiments.

1. Preparations are blocked against unspecific binding by placing them 5 min in a Coplin jar containing 0.5% BSA/BT buffer at 42°C.
2. Remove slides from Coplin jar, briefly drain excess fluid and place them in a moist box. (Preparations should never dry up during this and subsequent steps).
3. Apply 100 µl BT buffer containing avidin-FITC (2.5 µg/ml) and anti-dig-rhodamine Fab fragments at 0.5 µg/ml.
4. Cover with large cover slip and incubate for 1 h at 42°C.
5. Gently slide down or float off coverslips in BT buffer.
6. Wash slides 3×3 min in BT buffer at 42°C.
7. Apply 100 µl BT buffer containing biotinylated goat anti-avidin antibody (2 µg/ml).
8. Cover with cover slip and incubate for 30 min at 42°C in a moist chamber.
9. Wash 3×3 min in BT buffer at 42°C.
10. Repeat step 3 with avidin-FITC only and incubate for a final 30 min at 42°C.
11. Repeat step 6, remove slides from Coplin jar and briefly drain excess fluid.
12. Apply 18 µl antifade solution containing DAPI as a DNA-specific counterstain and mount a 24×60 mm coverslip.
13. Place slides in a drying block or under a paper towel. Apply gentle pressure to remove excess liquid and trapped air bubbles.

14. Inspect with a fluorescence microscope equipped with appropriate filter sets for the visualization of green, red and blue fluorescence. Images may be recorded with conventional microphotography or digital image acquisition systems. Double band pass filters for green and red excitation are recommended for visual analysis.

Troubleshooting

- Store cell culture media at 4°C for no more than 2 weeks.
- Since the success of the experiments, in particular meiotic time courses, depends on the culture conditions and since the quality of the various cell culture components varies depending on the supplier, it is recommended to always test the performance of the components used.
- Set up new experiments when wild-type budding rates do not exceed 40%.
- In meiotic time courses, sporulation rates of more than 80% should be achieved to allow for acquisition and meaningful interpretation of quantitative FISH data.
- Complete cell wall removal is imperative for a successful FISH experiment, since incomplete cell wall degradation results in high background fluorescence. This is particularly the case in experiments with structurally preserved nuclei, which are fixed prior to spheroplasting. One should always ensure that the cell wall is maximally degraded but at the same time, too much cell lysis has to be avoided.
- Another source for high background in yeast FISH is remnants of RNA in the probe and/or in the cells.
- Cells which have been grown to stationary phase exhibit a thickened cell wall. Therefore, prolonged spheroplasting times as compared to the ones given in the protocol above, are required to obtain nuclei suitable for FISH.
- Successful FISH detection of chromosome regions depends on the access of the DNA probe to its target sequence. In preparations fixed with cross-linking agents like formaldehyde, access is generally achieved by limited proteolytic treatment and/or treatment with chaotropic reagents, strong detergents or pre-incubation with 4xSSC/Tween

(Scherthan et al. 1993, 1994, Guacci et al. 1994,1997). Proteases can be used to effectively remove cytoplasm and nuclear proteins, but in our hands proteolytic treatment, like harsh detergent treatment, resulted in detrimental effects on yeast nuclear structure. We found that repeated denaturation can be used as an alternative to proteolytic pre-treatment.

- Composite DNA probes for the delineation of whole chromosomes or chromosomal subregions like telomeres or centromeres may be produced by combining several chromosome specific phage or plasmid clones (Scherthan et al. 1992, Gotta et al. 1996, Jin et al. 1998). Note that probe DNAs which are situated in duplicated regions of the yeast genome (Goffeau et al. 1997) or contain t-RNA genes or TY elements, may cause supernumerary signals which generally impair interpretation of the FISH results.
- The protocols given above work in our hands on vegetative and meiotic cells and may be combined with immunostaining of nuclear proteins by performing FISH detection and immunodecoration of nuclear antigens simultaneously (Trelles-Sticken et al. 1999). Other protocols which use coagulating fixatives instead of cross linking with formaldehyde are available (Gotta et al. 1999). Hence, protocols which use different fixation regimens should be tested and compared for optimal performance in a particular application.
- Images may be recorded with conventional microphotography, digital image acquisition systems like cooled CCD cameras or laser scanning microscopes. Deconvolution may be used to reduce noise in the gray scale images.

Acknowledgements. We would like to thank Drs J. Bähler, B. Dujon, D. Kaback, J. Loidl and E. Louis for help with the materials and J. Loidl for an introduction to the spreading technique. This work was supported by a grant of the Deutsche Forschungsgemeinschaft (SCH350/8-3).

References

Dresser ME, Giroux CN (1988) Meiotic chromosome behavior in spread preparations of yeast. J Cell Biol 106:567-573.

Esposito RS, Dresser ME, Breitenbach M (1990) Identifying sporulation genes, Visualizing Synaptonemal Complexes, and Large Scale Spore and Spore Wall Purification. Guide to Yeast Genetics and Molecular Biology. 194:110-131.

Florijn RJ, Slats J, Tanke HJ, and Raap AK (1995) Analysis of antifading reagents for fluorescence microscopy. Cytometry 19:177-182.

Funabiki H, Hagan I, Uzawa S, Yanagida M (1993) Cell cycle-dependent specific positioning and clustering of centromeres and telomeres in fission yeast. J Cell Biol 121:961-976.

Goffeau A, et al (1997) The Yeast Genome Directory. Nature 387(Suppl): 1-105.

Gotta M, Laroche T, Formenton A, Maillet L, Scherthan H, Gasser S.M, (1996) Colocalization of telomeric sequences with Sir3, Sir4 and Rap1 detected by combined immunofluorescence and FISH in *S.cerevisiae.* J Cell Biol 134:1349-1363.

Gotta M, Laroche T, Gasser SM (1999) Analysis of nuclear organization in *Saccharomyces cerevisiae*. Methods Enzymol 304:663-672.

Guacci V, Hogan E, Koshland D (1994) Chromosome condensation and sister chromatid pairing in budding yeast. J Cell Biol 125:517-530.

Guacci V, Hogan E, Koshland D (1997) Centromere position in budding yeast: evidence for anaphase A. Mol Biol Cell 8:957-972.

Jin Q-W, Trelles-Sticken E, Scherthan H, Loidl J (1998) Yeast nuclei display prominent centromere clustering that is reduced in nondividing cells and in meiotic prophase. J Cell Biol 141:21-29.

Kane SM, Roth R (1974). Carbohydrate metabolism during ascospore development in yeast. J. Bacteriol. 118:8-14.

Kater J McA (1927) Cytology of *Saccharomyces cerevisiae* with special reference to nuclear division. Biol Bulletin 52:436-449.

Kohli J (1994) Meiosis. Telomeres lead chromosome movement. Curr Biol 4:724-727.

Kuroiwa T, Kojima H, Miyakawa I, Sando N (1984) Meiotic karyotype of the yeast *Saccharomyces cerevisiae*. Exp Cell Res 153:259-265.

Lichter P and Cremer T (1992) Chromosome analysis by non isotopic in situ hybridization, in Human Cytogenetics – A practical Approach (2nd edition), Vol. I (Rooney D.E. and Czepulkowsky B.H., eds.) IRL Press, Oxford, pp. 157-192.

Loidl J, Nairz K, Klein F (1991) Meiotic chromosome synapsis in a haploid yeast. Chromosoma 100:221-228.

Loidl J, Scherthan H, Kaback D.B (1994) Physical association between nonhomologous chromosomes precedes distributive disjunction in yeast. Proc Natl Acad Sci USA 91:331-334

Loidl J, F Klein, Engebrecht J (1998). Genetic and morphological approaches for the analysis of meiotic chromosomes in yeast. In: Berrios, M (ed.) Nuclear structure and function. pp.257-285. Academic Press, San Diego.

Nag DK, Scherthan H, Rockmill B, Bhargava J, Roeder GS (1995) Heteroduplex DNA formation and homolog pairing in yeast meiotic mutants. Genetics 141:75-86.

Padmore R, Cao L, Kleckner N (1991) Temporal comparison of recombination and synaptonemal complex formation during meiosis in *S. cerevisiae*. Cell 66:1239-1256.

Pringle JR, Adams AE, Drubin DG, Haarer BK (1991) Immunofluorescence methods for yeast. Methods Enzymol 194:565-602

Scherthan H, Loidl J, Schuster T Schweizer D (1992) Meiotic chromosome condensation and pairing in *Saccharomyces cerevisiae* studied by chromosome painting. Chromosoma 101:590-595.

Scherthan H, Schweizer D. and Loidl J (1993) Delineation of individual chromosomes of *Saccharomyces cerevisiae* by two colour in situ hybridisation. Trends Genetics 9:41.

Scherthan H, Bähler J, Kohli J (1994) Dynamics of chromosome organization and pairing during meiotic prophase in fission yeast. J. Cell Biol. 127:273-285.
Trelles-Sticken E, Loidl J, Scherthan H (1999) Bouquet formation in budding yeast: initiation of recombination is not required for meiotic telomere clustering, J Cell Sci 112:651-658.
Uzawa S, Yanagida M (1992) Visualization of centromeric and nucleolar DNA in fission yeast by fluorescence in situ hybridization. J Cell Sci 101:267-275.
Weiner BM, Kleckner N (1994) Chromosome Pairing via Multiple Interstitial Intractions before and during Meiosis in Yeast. Cell 77: 977-991.
Wintersberger U, Binder M, Fischer P (1975) Cytogenetic demonstration of mitotic chromosomes in the yeast *Saccharomyces cerevisiae.* Mol Gen Genet 142:13-17.

In Situ Detection of Viral Nucleic Acids by Chemiluminescence

Monica Musiani, Patrizia Pasini, Aldo Roda, and Marialuisa Zerbini

Introduction

Chemiluminescence is the release of light produced by a chemical reaction from particular chemically excited molecules as a result of the transition from an electronically excited state to a lower energy state, usually the ground state. This phenomenon has provided the biologist with a unique analytical tool in various analytical techniques in which small amounts of analytes can be detected and quantified by measurement of the light emitted by chemiluminescent reactions (Martin et al. 1995, Lorimier et al. 1996, Erhardt et al. 1996, Girotti et al. 1996, Kricka LJ 1998, Ross et al. 1999; Guerasimova et al. 1999; Roda et al. 1999). In the last few years, it has been demonstrated that chemiluminescent detection systems which combine the use of an enzyme and a substrate that is converted to a light-emitting product, have the highest potential sensitivity; moreover chemiluminescent systems can give precise and accurate quantitative results since the luminescent signal intensity is proportional to enzyme concentration present in the reaction (Lamarcq et al. 1995).

Different chemiluminescent substrates for alkaline phosphatase (AP), based on adamantyl 1,2-dioxetane phenyl phosphate derivatives, are now commercially available [e.g. CSPD (PE-Tropix, Bedford, MA, USA) and Lumi-Phos Plus (Lumigen, Southfield, MI, USA)]. These substrates allow

✉ Monica Musiani, University of Bologna, Department of Clinical and Experimental Medicine, Division of Microbiology, Via Massarenti 9, Bologna, 40138, Italy (*phone* +39051-302435; *fax* +39051-341632; *e-mail* musiani@med.unibo.it)
Patrizia Pasini, University of Bologna, Department of Pharmaceutical Sciences, Via Belmeloro 6, Bologna, 40126, Italy
Aldo Roda, University of Bologna, Department of Pharmaceutical Sciences, Via Belmeloro 6, Bologna, 40126, Italy
Marialuisa Zerbini, University of Bologna, Department of Clinical and Experimental Medicine, Division of Microbiology, Via Massarenti 9, Bologna, 40138, Italy

the detection of a few zeptomoles of AP using a photomultiplier tube (Bronstein et al. 1990) or attomoles of enzyme using a video camera (Roda et al. 1996) and have a glowing kinetics with a steady-state light emission lasting several minutes, which permits easy handling and analysis of the samples with an increased detection sensitivity in the analysis of nucleic acids.

For horseradish peroxidase (HRP) chemiluminescent detection, luminol-based reagents are also commercially available [e.g. the enhanced chemiluminescent reagent ECL (luminol/enhancer/H_2O_2) (Amersham, Amersham, UK) and SuperSignal Ultra (luminol/enhancer/stable peroxide) (Pierce, Rockford, IL, USA)]. Both reagents contain substances, such as p-iodophenol, which enhance the light production deriving from the HRP-catalyzed oxidation of luminol by hydrogen peroxide, thus permitting detection of HRP with very high efficiency. The kinetics is of a glow-type with a steady-state light emission maintained for at least 10 min (Matthews et al. 1985, Thorpe and Kricka 1986, Thorpe and Kricka, 1987).

Continuing improvements in chemiluminescent substrates have recently been matched by new developments in photon imaging instrumentation such as high performance luminographs based on a CCD video camera or high dynamic range pick up tube (Saticon) combined with a video amplifier. These instruments not only allow a quantification of emitted light at a single photon level, but also permit localization of the chemiluminescent emission on a target surface (Scott and Inaba 1989, Wick 1989, Hooper and Ansorge 1991, Bräuer et al. 1993, Roda et al. 1996). Moreover, connecting the luminograph to an optical microscope, it is possible to localize the light emission on a target surface or inside tissues or cells (Hiraoka et al. 1987, Mueller-Klieser et al. 1988, Hawkins and Cumming 1990, Lorimier et al. 1993, Mueller-Klieser and Walenta 1993, Musiani et al. 1996a, Roda et al. 1998, Pasini et al. 1998). The evolution of photon counting imaging together with the continuous development of chemiluminescent substrates for AP and HRP has recently led to the development of chemiluminescence in situ hybridization assays for the detection of viral genomes having AP and HRP as reporter molecules (Musiani et al. 1998).

The chemiluminescent in situ hybridization assay for viral genome detection has the advantage of combining the specificity of labeled probes, the sensitivity of enzymatic chemiluminescent substrates and the spatial morphological resolution and localization of the signal of the in situ hybridization.

Materials

Equipment The basic instrumentation for chemiluminescence in situ hybridization includes:

- a conventional light microscope whose optics must be selected in order to obtain a magnification factor of at least 360 with a 40× objective and a minimum spatial resolution of 1 μm. The lens coupling system should be as simple as possible and objectives should have the highest numerical aperture compatible with focal aberration and depth of field, thus reducing the potential loss of light derived from the chemiluminescent reaction and achieving the maximum analytical detectability. The microscope should also be provided with a micrometric system to automatically adjust the position of the sample, allowing exactly the same field of view to be found when the slide has been replaced after washing and handling procedures. The microscope has to be enclosed in a dark box to prevent contact with the external light during the measurements;
- an ultra-sensitive video camera specifically designed to pick up the static chemiluminescent signal. The video camera must be characterized by high sensitivity, very low instrument noise, wide dynamic range, and suitable spatial resolution. In the last few years, high-performance CCD cameras have been proposed as an alternative to older devices based on vidicon tubes. Cooled, back-illuminated CCD cameras provide very high detection sensitivity, due to their high quantum efficiency and low background noise, allowing a quantification of emitted light at a single photon level. In addition, often these devices do not require an image intensification step, which could negatively affect the image quality and the signal-to-noise ratio. The camera should be calibrated with a light standard such as a liquid scintillation radioactive isotope (^{14}C, ^{3}H) or an electronically controlled light emitting diode (LED);
- a personal computer (PC) provided with software for the camera management and the image processing and analysis. Concerning the camera management, the mode of integration of light should be chosen to have a wide dynamic range without saturation of the light sensitive elements. It should be possible to acquire the light signal for different intervals of time with the possibility of having a multi-frame time-dependent image. Besides chemiluminescent images, the camera has to be able to acquire and store light-transmitted images in order to prop-

erly identify the morphological structure of the sample. A series of corrections should be performed after the acquisition of the images, such as background, gamma, geometric, flat-field, defect corrections. A series of mathematical functions should also be available to improve the quality of the chemiluminescent and live images regarding image processing and analysis. Pseudocolor and overlay functions are necessary to superimpose chemiluminescent image to light-transmitted image in order to appropriately localize the light emission from the target. A quantitative analysis of the light emitted in a given area of the sample (pixel) should be possible. It should also be possible to standardize the above mentioned parameters to permit a precise, accurate and reliable comparison of specimens analyzed in different sets of experiments. High quality color prints of the images with appropriate comments, according to the guidelines of good laboratory practice and method standardization should be carried out.

In our studies, the luminescent signal from the hybrid formation was detected and analyzed using a high-performance, low light-level imaging apparatus (Luminograph LB 980, EG&G Berthold, Bad Wilbad, Germany), which permits emitted light measurement at the single-photon level. The video system consists of a 1″Saticon, high dynamic range pick-up tube (which is a Vidicon-type tube with Se-As-Tl light target photoconductor) linked to an image intensifier, by high transmission lenses, and also to a videoamplifier. This system is connected to a model BH-2 optical microscope (Olympus Optical, Tokyo, Japan), and to a PC provided with software for quantitative image analysis. The microscope is enclosed in a dark box to prevent contact with the external light.

Alternatively, when a higher degree of detectability was required, a Luminograph LB 981 (EG&G Berthold, Bad Wilbad, Germany), based on a slow-scan, back-illuminated, cooled CCD was used. The instrument set-up and CL imaging processing are quite similar to those of the LB 980 luminograph.

Detection of luminescence

Both instrument systems for the detection of chemiluminescence in in situ hybridization operate in three steps:

1. firstly, tissue structures and cells are recorded in transmitted light;
2. the luminescent signal is then measured with an optimized photon accumulation lasting 1 min;
3. after a computer elaboration of the luminescent signal with pseudocolors corresponding to the light intensity, an overlay of the two

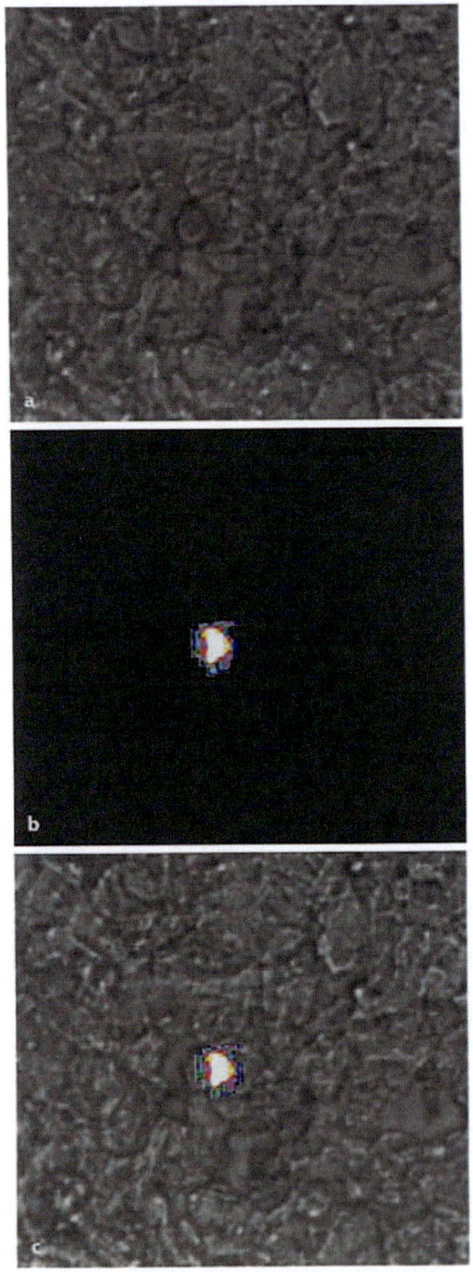

Fig. 1. Detection of luminescence

images on the screen provided by the transmitted light and by the luminescent signal allows the spatial distribution of the target analyte to be localized and evaluated.

Digital images are stored in the computer and quantitatively analyzed. The light emission from each cell can be quantified by defining a fixed area and summing the total number of photon fluxes from within this area. The light emission is then expressed as photons/second/area (Fig. 1).

Samples

- Sections (3-5 μm thin) of tissue samples, either fixed in formalin and embedded in paraffin or frozen in liquid nitrogen and stored at -80°C until use, are cut and placed on pretreated or silanated slides. Paraffin embedded tissue sections are dewaxed by two 5-min incubations in xylene and then washed in absolute ethanol. Frozen sections are fixed with 4% paraformaldehyde in PBS for 30 min. After fixation, the slides are washed three times in PBS for 10 min each and then the samples are dehydrated with ethanol washes (30%, 60%, 80%, 95% and 100%) for 5 min each. The slides are air dried and stored at 4°C. Samples can be stored for several months with no decrease in their interaction with the hybridization probes.
- Cellular specimens from body fluids are centrifuged at 1000 rpm for 10 min at room temperature, resuspended in 0.15 M PBS, pH 7.4, and smeared or cytospun onto pretreated or silanated glass slides. For blood-nucleated cells, a previous separation from peripheral blood by Ficoll gradient is recommended. Cellular samples are fixed and dehydrated as described above for frozen sections.

Probes

The probes used for chemiluminescence in situ hybridization (ISH) generally consist of double stranded DNA probes obtained from cloned DNA fragments but they can also consist of synthetic oligonucleotides, single stranded DNA probes generated by PCR, or single-stranded antisense RNA probes. Synthetic oligonucleotide probes, because of their small size, have good penetration properties, and the fact that they are single-stranded, rules out the possibility of annealing with complementary strands in the hybridization solution. Cloned probes, which are usually longer, have the advantage of having more sites available for labeling and of recognizing more of the target DNA, resulting in a stronger signal.

The probes used in chemiluminescence in situ hybridization are nonisotopically labeled probes, with indirect labels such as biotin, fluorescein or digoxigenin. These indirect labels are generally revealed by streptavidin, anti-fluorescein antibody and anti-digoxigenin antibody, respectively, with a covalently attached signaling group that is usually AP or HRP. Digoxigenin labeled probes have proven to be very sensitive and specific for the detection of viral genomes, and nowadays are the most used probes.

Hybridization mixture

- 50% deionized formamide
- 10% dextran sulfate
- 250 μg/ml of carrier calf thymus DNA
- 2 ng/μl of the labeled probe in 2×SSC buffer (0.3 M NaCl, 0.03 M sodium citrate, pH 7.0)

Procedure

Several protocols for the hybridization reaction have been described with minor modifications, generally involving a cell pretreatment followed by the hybridization reaction with the labeled probe, followed by chemiluminescent detection.

Pre-treatment of cells/sections

1. Hydrate cells/sections briefly in PBS.
2. Place cells/sections in 0.02 N HCl for 10 min.
3. Wash three times with PBS.
4. Treat cells/sections with 0.01% Triton X-100 in PBS for 2 min.
5. Wash three times with PBS.
6. Treat with pronase (0.5 mg/ml) in 0.05 M Tris HCl buffer, pH 7.6, containing 5 mM EDTA for 5 min.
7. Wash twice with PBS containing 2 mg/ml of glycine.
8. Postfix with 4% paraformaldehyde in PBS for 5 min.
9. Wash twice with PBS containing 2 mg/ml of glycine.
10. Dehydrate cells/sections by ethanol washes (30%, 60%, 80%, 95% and 100%).

Hybridization reaction

Incubate cells/sections with 10 μl of the hybridization mixture (see above). A clean glass coverslip is applied and the edges are sealed with nail polish to prevent loss of the mixture during denaturation and hybridization. Cells/sections and the hybridization mixture containing the labeled probe are denatured together by heating in an 85°C water bath for 6 min, and are then hybridized at 37°C for 3 h.

The coverslips are carefully removed and cells/sections are washed at stringent conditions in order to minimize non-specific hybridization (three times at 34°C, with 50% formamide and 2×SSC for 3 min each, followed by two 3 min washes in 2×SSC at 34°C and finally one 3 min wash in 2×SSC at room temperature).

Chemiluminescent detection

Wash cells/sections briefly in a 150 mM Tris HCl buffer, pH 7.5, containing 150 mM NaCl. Incubate with the appropriate HRP- or AP-labeled detector systems. In detail:

a) the samples hybridized with fluorescein labeled probes are incubated for 30 min at 37°C with HRP- or AP-conjugated Fab fragments to fluorescein isothiocyanate (e.g. HPV probes Dako A/S, Glostrup, Denmark) used at the dilution suggested by manufacturer's instructions or determined by preliminary block titration;

b) the samples hybridized with digoxigenin labeled probe are incubated for 30 min at 37°C with HRP- or AP-conjugated anti-digoxigenin Fab fragments (e.g. HPV probes Kreatech, Amsterdam, The Netherlands) used at the dilution suggested by the manufacturer's instructions or determined by preliminary block titration;

c) the samples hybridized with biotinylated DNA probe are incubated for 30 min at 37°C with HRP- or AP-conjugated to streptavidin (e.g. HPV probes Biohit, Helsinki, Finland and HSV probes Enzo Biochem, New York, USA) used at the dilution suggested by the manufacturer's instructions or determined by preliminary block titration.

Wash cells/sections three times for 5 min with a 150 mM Tris HCl buffer, pH 7.5 containing 150 mM NaCl.

To detect HRP labeled hybridized probes: Add a 20-40 μl drop of luminol-based chemiluminescent substrate (e.g. ECL system: luminol/H_2O_2/

enhancer) prepared following the manufacturer's instructions over the sample on the glass slide, and put the slide under the microscope connected with the video camera.

Acquire the light signal in a few (between two and five) minutes. In this case, the light emission kinetics is of a glowing type with a plateau stable for at least 10 min and whose intensity is proportional to the enzyme activity.

To detect AP-labeled hybridized probes: Wash cells/sections for 2 min with equilibration buffer (EB) (100 mM Tris HCl, 100 mM NaCl, 50 mM $MgCl_2$, pH 9.5). Add a 20-40 μl drop of dioxetane phosphate-based reagents (i.e. CSPD, CDP-star, Lumi-Phos Plus) over the sample.

Acquire the light emission after 20 min of incubation. The kinetics in this case, is also of a glowing type, but it is slower and about 15 min is the necessary time to reach a steady-state light emission.

Quantitative analysis

A control of uninfected cells and a negative biopsy specimen provide the threshold background levels. Threshold values are calculated analysing a mean of 50 cells from control negative samples when cellular smears have to be tested, and 10 microscopic fields from negative reference biopsies when a biopsy specimen has to be analyzed. The average SD values of the background light emission (expressed as photons/second/area) is then calculated. The average value of the background signal plus five-fold its standard deviation is considered the threshold value above which the chemiluminescent signal resulting from the hybridized nucleic acid can be determined as positive. The net light signal of the sample is obtained automatically on the screen after subtraction of the threshold values using the appropriate software.

Specificity controls for chemiluminescence in situ hybridization

The following control experiments are necessary to prove that the chemiluminescence hybridization reaction is detecting target sequences specifically.

A specific positive signal is to be detected when reference samples of target positive cells are treated with labeled probes.

No chemiluminescent signal should be observed when target positive cells/sections are hybridized with vector control DNA labeled probe and treated with anti digoxigenin Fab fragment conjugated with the enzyme and with the chemiluminescent substrate.

No luminescent signal should be detectable in target positive cells/sections after hybridization with unlabelled probes, followed by the immunoenzymatic chemiluminescent treatment.

Target positive cells/sections must be completely negative after hybridization with the labeled probes when the primary incubation with anti-digoxigenin antibody was either omitted or replaced by an incubation with non-immune sheep serum.

Double hybridization reaction

Chemiluminescence ISH can also be used for the simultaneous detection of different DNA sequences in the same sample, utilizing both HRP and AP as reporter enzymes (Gentilomi et al., 1997). In the double hybridization reaction, different probes to different targets (e.g. cytomegalovirus (CMV) DNA and Herpes simplex DNA) have to be labeled with indirect labels individually detectable by systems with covalently attached AP or HRP signaling groups.

The double hybridization reaction for the simultaneous detection of HSV DNA and CMV DNA in the same sample is performed as follows:

Overlay dehydrated cells/sections with 20 μl of the hybridization mixture (see above), containing both 2 ng/μl of biotin labeled HSV DNA probe and 2 ng/μl of digoxigenin-labeled CMV DNA probe.

Put cells/sections to hybridize and then wash them under stringent conditions as described above for single hybridization.

Wash cells/sections briefly in a 150 mM Tris HCl buffer, pH 7.5, containing 150 mM NaCl.

Incubate cells/sections for 30 min at room temperature with a mixture of streptavidin-HRP complex (NEN-DuPont, Boston, MA,USA) diluted 1/500 in blocking buffer (Boehringer) and polyclonal anti-digoxigenin Fab fragments conjugated to AP (diluted 1/500 in blocking buffer) (Boehringer).

Wash 3 times for 5 min each in TNT buffer (0.1 MTris HCl, pH 7.5; 0.15 M NaCl; 0.05% Tween 20). Perform the chemiluminescent detections of the two probes sequentially:

for the HSV DNA probe which utilized HRP as a reporter enzyme, add to the cells/sections 20-40 μl of luminol-based chemiluminescent to substrate (e.g. ECL system: luminol/H_2O_2/enhancer) prepared according to the manufacturer's instructions and measure the light output after 2-4 min incubation;

for the CMV DNA probe which utilized AP as a reporter enzyme, wash three times for 2 min each in equilibration buffer (100 mM Tris HCl, 100 mM NaCl, 50 mM $MgCl_2$, pH 9.5) and then incubate for 20 min with 20-40 µl of a dioxetane phosphate-based reagents (e.g. CSPD, CDP-star, Lumi-Phos Plus) prepared according to the manufacturer's instructions and acquire the light emission after having removed the chemiluminescent substrate.

In the double chemiluminescence reaction, the enzymatic reaction for HRP is usually performed before that of AP. This is done because the HRP/luminol reaction rapidly reaches a steady state light output which is maintained for a short time, then the signal decreases while the AP/dioxetane reaction slowly reaches the steady-state light emission which then lasts for a relatively long time. Short washes to remove the HRP substrate are thus sufficient to avoid any interfering photon emission during the AP reaction. Even if, for reasons of rapidity of the assay, the enzymatic reaction for HRP is usually performed before that of AP, consistent results can also be obtained even when the substrate reaction sequence is reversed.

Results

The advantages of chemiluminescence in situ hybridization are mainly: improved sensitivity, quantification of the data, objectivity of the evaluation and digital imaging of the results.

Improved sensitivity

Chemiluminescence in situ hybridization proved to be very sensitive, being able to detect as few as 10 to 50 viral genome copies in infected cells using both biotynilated probes and digoxigenin labeled probes with HRP and AP chemiluminescent detections respectively. In fact, positive chemiluminescent signals were obtained in HeLa cells which are known to contain about 10 to 50 integrated genome copies of human papillomavirus (HPV) 18 (Lorimier et al. 1996, Musiani et al. 1997). Chemiluminescence in situ hybridization has proved more sensitive than in situ hybridization followed by colorimetric detection and almost as sensitive as in situ hybridization followed by ^{35}S autoradiography; in fact, the chemiluminescence positivity of the results obtained from the detection of HPV DNA in HeLa cells was not obtained with colorimetric detection (Lorimier et al. 1993, Musiani et al. 1997) while chemiluminescent signal resolution was comparable to that provided by ^{35}S autoradiography (Lorimier et al. 1996). Moreover with chemiluminescence in situ hybridization to detect B19 parvovirus DNA in several samples of bone marrow cells, all the positive specimens from patients with a diagnosed B19 infection proved positive with a higher number of positive cells/specimen in comparison with colorimetric detection, thus permitting an easier evaluation of the sample (Musiani et al. 1996b).

Quantification of the data

Chemiluminescence in situ hybridization allows the quantification of the results; in fact results on smears of Caski and HeLa cells (which are known to contain 500-600 copies of HPV 16 DNA and 10-50 copies of HPV 18 DNA respectively) demonstrated that the luminescent signal changed in proportion to the number of known viral genome copies per cell (Musiani et al. 1997). In addition, studying the presence of cytomegalovirus DNA in infected cells, increasing values of emitted photons/cell corresponding to the presence of hybridized cytomegalovirus DNA, could be found in cells fixed at 48, 60, 72 and 96 h after infection, following the cytomegalovirus replication cycle (Musiani et al. 1996a).

Objectivity of the evaluation

With chemiluminescence in situ hybridization, since the positive signal is considered the one above threshold values, an objective evaluation of the results could be achieved without any training on the microscope to read the slides and so, doubts about positive or negative results are minimized.

Digital imaging of the results

Chemiluminescence in situ hybridization offers a permanent record of the reactions as all the images of the samples are stored in the computer and these images can be printed or sent for an evaluation in other laboratories using floppy disks or other computer networks.

In conclusion, the chemiluminescent in situ hybridization assay can be a useful tool for a sensitive detection and study of specific genetic sequences inside the cells. The use of the chemiluminescent in situ hybridization assay, which can be applied to detect one or two target nucleic acids in cell smears, archival frozen and paraffin embedded tissue samples, may also be promising for an estimation and quantification of nucleic acids present in tissue samples or cellular smears and for imaging gene expression in cells, providing a strong standardization of the methods, reagents and samples.

Acknowledgements. The research cited in this review has been partly supported by CNR Target Project on "Biotechnology", MURST (Ministero della Università e della Ricerca Scientifica e Tecnologica) and University of Bologna Funds for Selected Research Topics.

References

Bräuer R, Lübbe B, Ochs R (1993) Measuring luminescence with a low light-level imaging system using electronic light standards. In: Szalay AA, Kricka LJ, Stanley PE (eds) Bioluminescence and Chemiluminescence: status report. Wiley, Chichester. pp 13-17

Bronstein I, Edwards B, Voyta JC (1990) 1,2-Dioxetanes: novel chemiluminescent enzyme substrates. J Biolumin Chemilumin 4: 99-111

Erhardt A, Schaefer S, Athanassiou N, Kann M, Gerlich WH (1996) Quantitative assay of PCR-amplified hepatitis B virus DNA using a peroxidase-labeled DNA probe and enhanced chemiluminescence. J Clin Microbiol 34:1885-1891

Gentilomi G, Musiani M, Roda A, Pasini P, Zerbini M, Gallinella G, Baraldini M, Venturoli S, Manaresi E (1997) Co-localization of two different viral genomes in the same sample by double-chemiluminescence in situ hybridization. BioTechniques 23:1076-1083

Girotti S, Musiani M, Ferri E, Gallinella G, Zerbini M, Roda A, Gentilomi G, Venturoli S (1996) Chemiluminescent immunoperoxidase assay for the dot blot hybridization detection of Parvovirus B19 DNA using a low light imaging device. Anal Biochem 236: 290-295

Guerasimova A, Ivanov I, Lehrach H (1999) A method of one-step enzyme labelling of short oligonucleotide probes for filter hybridisation. Nucleic Acids Res 27:703-705

Hawkins E, Cumming R (1990) Enhanced chemiluminescence for tissue antigen and cellular viral DNA detection. J Histochem Cytochem 38:415-419

Hiraoka Y, Sedat JW, Agard DA (1987) The use of a charge-coupled device for quantitative optical microscopy of biological structures. Science 238: 36-41

Hooper CE, Ansorge RE (1991) Quantitative photon imaging in the life sciences using intensified CCD cameras. In: Stanley PE, Kricka LJ (eds). Bioluminescence and Chemiluminescence: current status. Wiley, Chichester. pp 337-344

Kricka LJ (1998) Prospects for chemiluminescent and bioluminescent immunoassay and nucleic acid assays in food testing and the pharmaceutical industry. J Biolumin Chemilumin 13:189-193

Lamarcq L, Lorimier P, Negoescu A, Labat-Moleur F, Durrant I, Brambilla E (1995) Comparison of seven bio- and chemiluminescent reagents for in situ detection of antigens and nucleic acids. J Biolumin Chemilumin 10: 247-256

Lorimier P, Lamarcq L, Labat-Moleur F, Guillermet C, Bethier R, Stoebner P (1993) Enhanced chemiluminescence: a high-sensitivity detection system for in situ hybridization and immunohistochemistry. J Histochem Cytochem 41:1591-1597

Lorimier P, Lamarcq L, Negoescu A, Robert C, Labat-Moleur F, Gras-Chappuis F, Durrant I, Brambilla E (1996) Comparison of ^{35}S and chemiluminescence for HPV in situ hybridization in carcinoma cell lines and on human cervical intraepithelial neoplasia. J Histochem Cytochem 44:665-671

Martin CS, Butler L, Bronstein I (1995) Quantitation of PCR products with chemiluminescence. BioTechniques 18: 908-912

Matthews JA, Batki A, Hynds C, Kricka LJ (1985) Enhanced chemiluminescent method for the detection of DNA dot-hybridization assays. Anal Biochem 151:205-209

Muller-Klieser W, Walenta S (1993) Geographical mapping of metabolites in biological tissue with quantitative bioluminescence and single photon imaging. Histochem J 25:407-420

Muller-Klieser W, Walenta S, Paschen W, Kallinowski F, Vaupel P (1988) Metabolic imaging in microregions of tumors and normal tissues with bioluminescence and photon counting. J Natl Canc Inst 80:842-848

Musiani M, Pasini P, Zerbini M, Roda A, Gentilomi G, Gallinella G, Venturoli S, Manaresi E (1998) Chemiluminescence: a sensitive detection system in in situ hybridization. Histol Histopathol 13:243-248

Musiani M, Roda A, Zerbini M, Pasini P, Gentilomi G, Gallinella G, Venturoli S (1996a) Chemiluminescent in situ hybridization for the detection of cytomegalovirus DNA. Am J Pathol 148:1105-1112

Musiani M, Roda A, Zerbini M, Gentilomi G, Pasini P, Gallinella G, Venturoli S (1996b) Detection of parvovirus B19 DNA in bone marrow cells by chemiluminescence in situ hybridization. J Clin Microbiol 34:1313-1316

Musiani M, Zerbini M, Venturoli S, Gentilomi G, Gallinella G, Manaresi E, La Placa MJr, D'Antuono A, Roda A, Pasini P (1997) Sensitive chemiluminescence in situ hybridization for the detection of human papillomavirus genomes in biopsy specimens. J Histochem Cytochem. 45:729-735

Pasini P, Musiani M, Russo C, Valenti P, Aicardi G, Crabtree JE, Baraldini M, Roda A (1998) Chemiluminescence imaging in bioanalysis. J Pharm Biomed Anal 18:555-564

Roda A, Pasini P, Baraldini M, Musiani M, Gentilomi G, Roberts C (1998) Chemiluminescent imaging of enzyme-labeled probes using an optical microscope-videocamera luminograph. Anal Biochem 257:53-62

Roda A, Pasini P, Musiani M, Girotti S, Baraldini M, Carrea G, Suozzi A (1996) Chemiluminescent low-light imaging of biospecific reactions on macro-and microsamples using a videocamera-based luminograph. Anal Chem 68:1073-1080

Roda A, Pasini P, Musiani M, Baraldini M (2000) Chemiluminescence imaging systems for the analysis of macrosamples: microtiter format, blot membrane, whole organs. Methods Enzymol 305:120-132

Ross R, Ross XL, Rueger B, Laengin T, Reske-Kunz AB (1999) Nonradioactive detection of differentially expressed genes using complex RNA or DNA hybridization probes. BioTechniques 26:150-155

Scott RQ, Inaba H (1989) Single photon counting imagery. J Biolumin Chemilumin 4:507-511

Thorpe GH, Kricka LJ (1986) Enhanced chemiluminescent reactions catalyzed by horseradish peroxidase. Methods Enzymol 133:331-354

Thorpe GH, Kricka LJ (1987) Enhanced chemiluminescence assays for horseradish peroxidase: characteristics and applications. In: Schölmerich J, Andreesen R, Kapp A, Ernst M, Woods WG (eds). Bioluminescence and Chemiluminescence: new perspectives. Wiley, Chichester, pp 199-208

Wick RA (1989) Photon counting imaging: applications in biomedical research. BioTechniques 7:262-268

Part F

Multicolor FISH Applications

Spectral Imaging Analysis of Metaphase and Interphase Cells

JINGLY FUNG, HEINZ-ULLI G. WEIER,
ROGER A. PEDERSEN, and HORST F. ZITZELSBERGER

Introduction

Spectral imaging is a novel technology combining spectroscopy and imaging for the analysis of fluorescence in situ hybridization (FISH). This system combines the techniques of fluorescence microscopy, charge-coupled device (CCD) camera, Fourier spectroscopy, and powerful computer software. It had initially been developed to screen human metaphase chromosome spreads after simultaneously hybridizing the chromosomes with 24 chromosome-specific whole chromosome painting probes. This particular application, termed Spectral Karyotyping (SKY) (Garini 1996), is mostly used in karyotype analysis of constitutional chromosome abnormalities, particularly in tumor studies.

In some circumstances, the target cells are few and metaphase spreads are very difficult to obtain. Blastomeres from human embryos for preimplantation genetic diagnosis (PGD), uncultured amniocytes from amniocentesis, and fetal cells from maternal blood for prenatal diagnosis are examples. Because aneuploidy of any of the 24 chromosomes may have

✉ Jingly Fung, University of California, Department of Obstetrics, Gynecology and Reproductive Sciences, Reproductive Genetics Unit, Box 0720, 533 Parnassus, San Francisco, CA, 94143-0720, USA (*phone* +415-476-8517; *fax* +415-476-6145; *e-mail* jlfung@itsa.ucsf.edu)
University of California, Department of Subcellular Structures, Life Sciences Division, E. O. Lawrence Berkeley National Laboratory, Berkeley, CA, 94720, USA
Heinz-Ulli G. Weier, University of California, Department of Subcellular Structures, Life Sciences Division, E. O. Lawrence Berkeley National Laboratory, Berkeley, CA, 94720, USA
Roger A. Pedersen, University of California, Department of Obstetrics, Gynecology and Reproductive Sciences, Reproductive Genetics Unit, Box 0720, 533 Parnassus, San Francisco, CA, 94143-0720, USA
Horst F. Zitzelsberger, gsf Forschungszentrum für Umwelt- und Gesundheit, Arbeitsgruppe Zytogenetik, Oberschleissheim, 85758, Germany

a negative impact on human development, the preimplantation and prenatal genetic diagnosis tests should score as many of the full complement of chromosomes as possible. This chapter describes a procedure for the enumeration of multiple chromosomes in interphase cells using Spectral Imaging (SIm)(Fung 1998). In interphase analysis, the use of whole chromosome painting probes is inappropriate because the chromatin domains of chromosomes overlap each other in the less condensed nucleus. Consequently, chromosome- or locus-specific probes for interphase analysis must be developed. Moreover, careful cell preparation and pretreatment are essential to ensure high hybridization efficiency and minimal signal overlap. This chapter describes in detail the techniques for the preparation of metaphase and interphase cell slides, the preparation of locus-specific probes, hybridization and detection, imaging acquisition and analysis using the Spectral Imaging system.

Outline

- Grow, swell and fix cells
- Prepare cell spreads on microscope slides
- Select probe set
- Pretreat slides
- Denature and hybridize probe set to cellular targets
- Remove unbound probe
- Detect bound probes with antibodies
- Acquire images with a Spectral Imaging system
- Analyze images with a library of pre-recorded spectra
- Display results in comprehensive and informative format

Materials

Equipment

- CCD camera (Hamamatzu)
- CDS-5 Cytogenetic Drying Chamber (Thermotron Industries)
- Centrifuge (MP4R, IEC)

- Dry bath (heat block) (Model 2001, Labline Instruments)
- Fluorescence microscope (Nikon E600, Nikon) equipped with 40× and 60× oil immersion lenses
- Incubator oven (FISHER, set to 37°C)
- Phase microscope (Nikon E800, Nikon)
- Pulsed Field Gel Electrophoresis (PFGE) system (BIO-RAD)
- SD200 SpectraCube spectral imaging system (Applied Spectral Imaging, Inc.)
- Shaking Incubators (New Brunswick): 30°C for yeast cell culture, 37°C for the culture of *E. coli*
- Thermal cycler for in vitro DNA amplification (model 4800, Perkin Elmer)
- Mini-Fluorometer (TKO 100, Pharmacia)
- Water bath (VWR)

Chemicals

- β-Agarase and buffer (New England Biolabs). Store at -20°C.
- 4 dNTP mix: dATP, dCTP, dGTP, and dTTP. Make 10 mM of each dNTP solution by combining 10 μl of a 100 mM solution (Roche Molecular Biochemical) with 60 μl of double distilled water. Store in 100 μl aliquots at -20°C.
- Agarose (Life Technologies).
- Bovine serum albumin (Sigma). Store at 4°C.
- Colcemid (Life Technologies), 10 mg/ml. Store at 4°C.
- Cy3-dUTP (Amersham), 1 mM. Store at -20°C.
- Digoxigenin-11-dUTP (Roche Molecular Biochemical), 1 mM. Store at -20°C.
- EDTA (ethylene diamine tetraacetic acid) (Life Technologies), 0.5 M, pH 8.0.
- Ethidium bromide (Life Technologies), 10 mg/ml.
- Fetal bovine serum (Sigma). Store at -20°C.

- Fluorescein-12-dUTP (Roche Molecular Biochemical), 1 mM. Store at -20°C.
- Formamide (Fisher). Store at 4°C.
- L-Glutamine (Sigma). Store at -20°C.
- Glycogen (Roche Molecular Biochemical), 20 mg/ml. Store at -20°C.
- Heparin (Life Technologies).
- Human COT1™ DNA (Life Technologies), 1 mg/ml. Store at -20°C.
- Low melting point agarose (Bio Rad). Store at room temperature.
- Methotrexate (Sigma). 10^{-4} M in Hank's BSS. Frozen in aliquot. Dilute to 10^{-5} M with Hanks BSS before use.
- Oligonucleotides used in in vitro DNA amplifications. Store at -20°C
  - JUN1: 5-CCCAAGCTTGCATGCGAATTCNNNNCAGG -3
  - JUN15: 5- CCCAAGCTTGCATGCGAATTC -3
- Penicillin/streptomycin (Life Technologies). Store at -20°C.
- Pepsin (Sigma), 10% stock solution (100 mg/ml) in sterile water. Store at -20°C.
- Phytohemagglutinin (Life Technologies), 10 mg/ml. Store at 4°C.
- Pronase (Sigma), 0.5% in FHM (see below). Store at -80°C.
- Proteinase K (Roche Molecular Biochemical): 20 mg/ml in 10 mM TrisHCl, pH 7.5. Store at -20°C.
- Random-Priming Kit: BioPrime kit (Life Technologies). Store at -20°C.
- RNase (Roche Molecular Biochemical), DNase-free: 20 mg/ml in sterile water, boil at 100°C for 10 min. Aliquot and store at -20°C.
- RPMI (Life Technologies). Store at 4°C.
- Salmon sperm DNA (3-5), 20 mg/ml. Store at -20°C.
- SKY kit (Applied Spectral Imaging, Inc.). Store at 4°C.
- T7 Sequenase II and Sequenase reaction buffer (Amersham Life Science Inc.). Store at -20°C.
- Thermus aquaticus (Taq) DNA polymerase (Perkin Elmer), 5 U/µl. Store at -20°C.

- Thymidine (Life Technologies), 10^{-5} M.
- Tween 20 (Sigma).
- Yeast artificial chromosome (YAC) library (Research Genetics). Store at -80°C.
- Zymolase (Sigma), 70,000 U/g, 10 mg/ml in 50 mM KH_2PO_4 (pH 7.8), 50% glycerol. Store at -20°C.

Buffers and other solutions

- AHC medium (BIO 101): Add 36.7 g/l of purified water. Autoclave at 121°C for 15 min.
- AHC agar medium (BIO 101): Add 53.7 g/l of purified water. Mix well to dissolve, then autoclave at 121°C for 15 min. Cool to 50°C, and pour plates. Store plates at 4°C.
- Fixative: acetic acid/methanol: 1/3 (vol/vol). Make fresh just before use.
- Denaturing solution: add 35 ml formamide, 10 ml distilled water, 5 ml 20×SSC. Store at 4°C. Prepare fresh every 2-4 weeks.
- ES Buffer: 0.5 M EDTA (pH 8.0), 1% Sarcosyl. Store at room temperature.
- Ethanol series: prepare 70%, 80%, and 100% ethanol in distilled water.
- 1% Formaldehyde: add 1.35 ml of 37% Formaldehyde to 50 ml of 1×PBS/$MgCl_2$. Store at 4°C.
- FHM: 1xFHM HEPES Buffered Media (Speciality Media). Store at -20°C.
- Formamide/SSC wash solution: add 15 ml 20×SSC, 60 ml distilled water, 75 ml formamide. pH 7.0. Store at 4°C.
- Hank's BSS (Life Technologies).
- Hybridization Master Mix (MM 2.1): 14.3% w/vol dextran sulfate, 78.6% formamide, 2.9×SSC, pH 7.0. Prepare by combining 1.45 ml of 20×SSC, 0.7 ml ultrapure water, 1.43 g dextran sulfate (Calbiochem), and 7.86 ml formamide. Heat to 70°C to dissolve dextran sulfate. Aliquot in 1.5-ml tubes and store at -20°C.
- Modified nucleotide mix (10x A4) for labeling in combination with 1 mM digoxigenin-11-dUTP, 1 mM fluorescein-12-dUTP or 1 mM Cy3-dUTP: combine 5 μl each of 100 mM dATP, 100 mM dGTP

and 100 mM dCTP with 2.5 µl of 1 M TrisHCl, pH 7.5, 0.5 µl of 0.5 M EDTA, pH 8.0 and 232 µl ultrapure water for a total of 250 µl. Store at -20°C. The concentration of nucleoside triphosphates is 2 mM each.

- 1×PBS (Phosphate buffered saline). 0.1 g/l $CaCl_2$, 0.1 g/l $MgCl_2 \times 6H_2O$,0.2 g/l KCl, 0.2 g/l KH_2PO_4, 8.0 g/l NaCl, 2.16 g/l $Na_2HPO_4 \times 7H_2O$.
- 1×PBS/$MgCl_2$, add 50 ml of 1 M $MgCl_2$ to 950 ml 1×PBS.
- SCE: 1 M sorbitol, 0.1 M Na citrate, 10 mM EDTA, pH 7.8.
- SSC: 20×SSC is 3 M NaCl, 0.3 M Na_3citrate × $2H_2O$, pH 7.0.
- 4×SSC/0.1% Tween 20: add 100 ml 20×SSC, 400 ml distilled water and 500 µl Tween 20. Mix well.
- 10× Taq buffer: 500 mM KCl, 100 mM TrisHCl, pH 8.3, 15 mM $MgCl_2$. Store at -20°C.
- TBE (Tris/borate/EDTA) buffer: 10× TBE is 890 mM Tris base, 890 mM boric acid, 20 mM EDTA.
- TE (Tris/EDTA) buffer: 1×TE is 10 mM TrisHCl, 1 mM EDTA, pH 7.4, 7.5 or 8.0.
- TE 50 buffer: 10 mM TrisHCl, 50 mM EDTA, pH 7.8.
- TrisHCl [tris (hydroxymethyl) aminomethane]: 1 M, pH 7.5 or pH 8.0.

Subprotocol 1
Preparation of slides and slide pretreatment

CAUTION: The key elements to consistent hybridization results seem to depend on (1) conditions during cell preparation and fixation, (2) relative humidity, temperature and air flow during preparation of metaphase spreads, and (3) slide pretreatment with proteolytic agents to enhance the accessibility of DNA targets to probes without causing undesirable loss of DNA. We will describe our slide preparation and pretreatment procedures in great detail. The DNA denaturation, hybridization and probe detection parameters appear to be less important and, consequently, control requirements are critical stringent during those steps.

Procedure

Metaphase spreads

Caution: Slide cleaning is an essential step prior to dropping the metaphase cells. To remove the most adherent residues, the glass surface can be wiped with Kim wipes which have been soaked with ethanol or fixative. Watch the surface. When smears appear, it is time to change the paper.

A. Metaphase spreads are made from short-term cultures of normal male lymphocytes
(For this part, refer to Chap. 3, FISH on Chromosomes Preparations of Peripheral Blood by Thomas Liehr and Uwe Claussen).

1. Cultivate normal male lymphocytes for 72 h at 37°C in RPMI 1640 supplemented with 20% fetal bovine serum, 1% penicillin/streptomycin, 1% L-glutamine, 1.32% heparin, and 4% phytohemagglutinin.
2. Block cultures for 17 h with methotrexate (10^{-5} M).
3. Incubate cultures in RPMI containing 1% thymidine (final concentration 10^{-5} M) for 5 h.
4. Block cells in mitosis with a 10 min treatment with colcemid.
5. Harvest and incubate cells in 75 mM KCl for 20 min at 37°C.
6. Pellet the cells and fix them in freshly made fixative (methanol/acetic acid, 3/1, vol/vol).
7. Drop cells on ethanol-cleaned slides inside a CDS-5 Cytogenetic Drying Chamber at 25°C and 45-50% relative humidity.
8. Check chromosome metaphase spreads using a phase contrast microscope. If there is any cytoplasmic residue visible around the metaphase spread, wash the remaining cells with fixative several times again before dropping them. Ideally, the chromosome metaphase spread appeared dark black. If metaphase spreads appear glassy, it suggests the cells have dried too slowly. Decrease the chamber's humidity by 2-5%. If metaphase spreads appear gray, it suggests the cells have dried too fast. Increase the chamber's humidity by 2-5%.
9. Store slides in slide boxes for at least two weeks at room temperature, then place in sealed plastic bags filled with nitrogen gas at -20°C until used.

B. Metaphase spread slides are pretreated following the instructions of the SKY kit

1. Look at the slide under phase contrast and select a region of interest containing at least five metaphase spreads and mark area with a diamond tip pen.
2. Incubate slides in the pepsin solution (15 μl stock pepsin solution in 50 ml of 0.01 M HCl solution) at 37°C for 3-5 min.
3. Wash slides twice in 1×PBS at room temperature for 5 min each.
4. Wash slides in 1×PBS/$MgCl_2$ at room temperature for 5 min.
5. Incubate slides in solution of 1% formaldehyde in PBS/$MgCl_2$ at room temperature for 10 min.
6. Wash slides in 1×PBS at room temperature for 5 min.
7. Dehydrate slides in 70%, 80%, 100% ethanol for 3 min each at room temperature.
8. Air dry slides at room temperature.
9. Look at the slide under phase-contrast, and note any remaining cytoplasm, debris, etc.

Interphase cells

Human blastomeres

All procedures followed protocols approved by the UCSF Committee on Human Research for using embryos for research. Prior written consent was obtained from all donors.

Caution: throughout the entire procedure, the embryos and blastomeres are observed under a microscope.

1. Incubate the donated embryos in 0.5% pronase/FHM to remove the zona pellucida.
2. Transfer the embryos into Ca^{2+}/Mg^{2+}-free PBS.
3. Separate the blastomeres by repeated pipetting of the embryo.
4. Transfer blastomeres to 1×PBS.
5. Incubate individual blastomeres in a hypotonic solution of 1% Na-citrate, 6 mg/ml bovine serum albumin in water for 5 min.

6. Mark a small circle on the back of an ethanol-cleaned slide using a diamond tip pen.
7. Place hypotonic treated blastomeres on slides.
8. Fix the blastomeres with one drop of methanol/acetic acid (3/1 or 1/1, vol/vol).
9. Check the nucleus under a phase contrast microscope and mark its position.
10. Dehydrate slides in three consecutive baths of 70%, 80% and 100% ethanol for 2 min each.
11. Slides are now ready for FISH or can be stored at -20°C.

Uncultured human amniocytes

Amniocytes were donated for research under an approved protocol in accordance with guidelines set by the UCSF Committee on Human Research. Uncultured amniocytes were obtained from 1-2 ml of amniotic fluid and fixed on slides within a few hours of amniocentesis.

1. Pellet the cells and resuspend the cells in a hypotonic solution consisting of 0.3% KCl for 30 min at 37°C.
2. Add several drops of freshly made ice-cold fixative (methanol/acetic acid, 3/1, vol/vol) and gently mix.
3. Pellet the cells and resuspend the cells with ice-cold fixative several times.
4. Drop the cells on fixative-cleaned slides above a boiling water bath.
5. Air dry the slides.
6. Check the nuclei on slides under a phase contrast microscope.
7. Incubate slides in 2×SSC at 37°C for 1 hr.
8. Pretreat slides with pepsin (50 µg/ml pepsin in 0.01 M HCl) at 37°C for 13 min.
9. Wash slides in 1×PBS at room temperature for 5 min.
10. Incubate slides in 1% formaldehyde in 1×PBS/$MgCl_2$ at room temperature for 5 min.
11. Wash slides in 1xPBS at room temperature for 5 min.
12. Dehydrate slides in 70%, 80%, and 100% ethanol for 2 min each.
13. Slides are now ready for FISH or can be stored at -20°C.

Subprotocol 2
Preparation of probes and hybridization mixture

Caution: The preparation of probes is best performed with kits which can be purchased from any one of several major companies (such as Roche Molecular Biochemicals, Life Technologies, VYSIS, Applied Spectral Imaging, Inc. and others). The DNA should be free of contaminants which might inhibit DNA polymerases, such as metals, detergents etc. In many cases, an alkaline lysis DNA isolation protocol followed by phenol-chloroform extraction and isopropanol precipitation should be sufficient. The PFGE agarose gel purification for YAC clones described below, as well as most cesium chloride gradient DNA isolation protocols will also purify DNA sufficiently for random priming.

Procedure

Whole chromosome painting probes for metaphase analysis

The chromosome painting probe for metaphase cell analysis is generated by flow sorting of human chromosomes and DNA amplification using degenerate oligonucleotide primed PCR (DOP-PCR) (Telenius 1992). The chromosome specific PCR products are labeled by directly incorporating haptenized (biotin and digoxigenin) or fluorochrome (Spectrum Green, Spectrum Orange, Texas Red) conjugated nucleotides, incorporated by PCR (Telenius 1992). Probes labeled with biotin or digoxigenin are detected with avidin-Cy5 and Cy5.5-conjugated antibodies against digoxin. Kits including all 24 chromosomes painting probes (vial 1 of SKY kit) can be purchased from Applied Spectral Imaging, Inc.

Locus-specific probes for interphase analysis: A. Preparation of DNA from yeast artificial chromosome (YAC)

The DNAs from YACs are isolated using pulsed field gel electrophoresis (PFGE) and amplified by degenerate oligonucleotide-primed PCR (DOP-PCR).

1. Grow the selected yeast clone (containing the YAC) on AHC agar for 2-3 days at 30°C. Pick colonies from the plates and culture the colonies in 5-35 ml AHC media at 30°C for 2-3 days in a shaking incubator.

Pulsed field gel electrophoresis (PFGE

2. Pellet the yeast cells from 5 ml AHC medium at 400 rpm for 6 min. Resuspend cells in 0.5 ml EDTA (0.125 M), pH 7.8. Spin down again and remove the supernatant.

3. Add 500 μl of SCE to a 70 μl pellet and resuspend. Mix with an equal volume of 1.5% LMP agarose at 43°C, quickly pipette up and down and vortex for 1-2 s to resuspend. Pipette into molds (BioRAD) and allow to solidify.

4. Remove plugs from molds and incubate samples in 2 ml SCE with 100 μl of zymolase (10 mg/ml) and shake at 150 rpm at 30°C for 2.5 h to overnight.

5. Remove SCE and add 2 ml of ES with 100 μl of proteinase K (20 mg/ml). Shake 5 h to overnight at 50°C.

6. Remove ES and rinse 5 times with 6 ml of TE50 for 30 min each rinse. Plugs are now ready for running PFGE or to be stored at 4°C.

7. PFGE running conditions: voltage gradient, 6 V/cm; switching time ramp, 79-94 s; running time, 38 h; agarose concentration, 1.0% LMP agarose; running temperature, 14°C; running buffer, 0.5×TBE.

8. Stain the gel with ethidium bromide (0.5 μg/ml in H_2O) and cut out the target band. Transfer gel slice to a clean tube.

9. Wash gel slices with water for 30 min, and then wash with 1× β-agarose buffer for 30 min.

10. Melt the gel completely by heating for 10 min at 85°C. Transfer the molten agarose to 43°C water bath. Add 1 μl β-agarase for every 25 μl molten agarose and continue incubation for 2 h. The sample can now be used for the DOP-PCR reactions or stored at 4°C.

Degenerate oligonucleotide-primed PCR (DOP-PCR)

There are two different DNA polymerase chain reactions, T7 DNA polymerase and Taq polymerase. The initial seven cycles of PCR use primer JUN1 annealed at 37°C and extended by T7 DNA polymerase. The PCR products are then further amplified at higher annealing temperature (50°C) with primer JUN15 and Taq polymerase for 35 cycles.

T7 DNA polymerase (7 cycles) PCR:

1. Prepare the T7 Mix (total 27 μl): 19.0 μl water, 5.2 μl 5× Sequenase reaction buffer, 0.8 μl 4 dNTP, 0.6 μl primer JUN1, and 0.4 μl Sequenase II (T7 DNA polymerase).

2. Denature 4 μl DNA (template) at 92°C for 6 min.
3. Add 8 μl T7 Mix. Place at room temperature for 2 min, and then at 37°C for 6 min.
4. Denature at 92°C for 3 min.
5. Add 2 μl T7 Mix. Place at room temperature for 2 min, and then at 37°C for 6 min.
6. Repeat steps (4)-(5) 5 times in total.
7. Denature DNA at 92°C for 3 min.
8. Add 8 μl T7 Mix. Place at room temperature for 2 min, and then at 37°C for 10 min.
9. The products at this stage can be used immediately for the Taq polymerase amplification reaction or stored at -20°C for up to several years.
10. Taq polymerase (35 cycles) PCR:
    In a 0.5-ml tubes mix in this order (for a total of 200 μl)
    - 152 μl water
    - 20 μl 10× Taq buffer
    - 4 μl 4dNTPs
    - 4 μl primer JUN15
    - 20 μl Sequenase reaction product
    - 0.67 μl Taq
11. Run 35 cycles of the following PCR reaction (Perkin Elmer thermocycle file).

Temperature (°C)	Time (s)
94	30
94	60
52	60
52	60
72	120
72	120 (plus 5-s extension per cycle)
72	600

12. The PCR products are precipitated in 1.2 vol isopropanol and resuspended in 1×TE buffer, and can be labeled in a random primer reaction.

B. Preparation of locus-specific probes via random primer reaction

1. Add 400 ng of DNA to water to a final volume of 14 µl in a 0.5-ml tube.
2. Boil tube at 100°C for 5 min, then quickly chill on ice. Add:

5 µl	10×A4 mix
6.5 µl	1 mM dTTP
3.5 µl	dig-11-dUTP, FITC-12-dUTP, or Cy3-dUTP(1 mM)
20 µl	2.5× random primers
For biotin labeling add 5 µl 10×dNTP (containing biotin-14-dCTP), 10 µl water, and 20 µl 2.5× random primers.	

3. Mix well and add 1 µl DNA polymerase I Klenow fragment (40 Units/µl).
4. Incubate tube at 37°C for 2 h.
5. Add 5 µl of 10x stop buffer (EDTA).
6. Store at -20°C.

C. Preparation of hybridization mixture for interphase cells analysis

Table 1 is an example of the 8-chromosomes probe set. Probes specific for repeated DNA on chromosome 15 (CEP15, satellite III, D15Z1, Vysis, Inc.), chromosome X (CEPX, alpha satellite, DXZ1, Vysis, Inc.), and chromosome Y (CEPY, satellite III, DYZ1, Vysis, Inc.) are labeled with either a green or red fluorochrome (Spectrum Green or Spectrum Orange, Vysis). The probe specific for repeated DNA of chromosome 16 was prepared from clone pHUR195. The locus specific DNA probes for chromosome 13 (YAC 900g6), chromosome 18 (YAC 945b6), chromosome 21 (YAC 141g6), and chromosome 22 (YAC 849e9) were obtained from yeast artificial chromosome (YAC) clones. The DNAs from YACs were isolated using PFGE and DOP-PCR. The DNAs from clones specific for chromo-

Table 1. Fluorochrome labeling scheme for chromosome-specific DNA probes[a]

Chromosome	Spectrum Green	FITC	Spectrum Orange	Cy3	Biotin (Cy5)	Digoxigenin (Cy5.5)
13		+				+
15			+			
16				+		+
18				+	+	
21						+
22					+	
X	+		+			
Y	+					

[a]Probes labeled with biotin or digoxigenin are detected with avidin-Cy5 and Cy5.5-conjugated antibodies against digoxin, respectively.

somes 13, 16, 18, 21, and 22 are labeled by random primer reaction incorporating biotin-14-dCTP, digoxigenin-11-dUTP, fluorescein-12-dUTP, or Cy3-dUTP.

Caution: when probes such as the chromosome 22-specific probes (YAC 849e9) cause cross-hybridization, we adjust the hybridization stringency and blocking protocol before evaluating another chromosome-specific probe. To suppress cross-hybridization signals generated by YAC 849e9, for example, we add an increased amount of human COT-1 DNA to the probe mixture. The ideal probe set should be comprised of bright single-copy locus-specific probes rather than DNA repeat probes to avoid cross-hybridization and domain clustering.

1. Mix 20-40 ng (0.5 µl-3 µl) of each probe, 1.3 µl human COT1 DNA, 1 µl salmon sperm DNA, 0.5 µl glycogen, 1/10 vol. 3 M sodium acetate, and 2 vol of 2-propanol.
2. Incubate at -20°C for 30 min.
3. Centrifuge at high speed (10,000 *g*) for 20 min.
4. Decant supernatant and wash pellet with cold 70% ethanol.
5. Decant the 70% ethanol, air dry the pellet.
6. Resuspend the pellet in 3 µl water.

7. Add 7 µl of hybridization master mix after the pellet is totally dissolved. This gives a total hybridization mixture of 10 µl.
8. Store the hybridization mixture at -20°C until use.

Caution: Often one fluorochrome shows a stronger signal compared to another fluorochrome when bound to the same DNA. For example, the intensity of probes detected with Cy5.5 is usually much stronger than the intensity of Cy3 labeled probes. This effect is probably due to a combination of different quantum efficiencies, probe labeling efficiencies, energy transfer and photon detection sensitivity. To avoid masking the Cy3 fluorescence with Cy5 or Cy5.5 emission, Cy3 probes are used at a higher concentration. Therefore, the composition of probes is very important for the interphase cell analysis by FISH when using the Spectral Imaging system.

Subprotocol 3
Hybridization and detection of bound probes

Procedure

These procedures follow the protocol provided in the SKY kit.

1. The hybridization mixture (vial-1 of SKY kit or locus-specific probes) is denatured at 76°C for 7 min, then allowed to pre-anneal at 37°C for 60 min.
2. Slides are denatured in denaturing solution at 76°C for 2–5 min. The denaturing time for slides depends on the different types of slides.
3. Then, slides are dehydrated in 70%, 80%, and 100% ethanol for 2 min each step, and allowed to air dry.
4. The hybridization mixture is applied to the slides under a glass coverslip, which is sealed with rubber cement. Finally, the hybridization proceeds at 37°C for 40-48 h in a moisture chamber.
5. Following hybridization, wash slides at 43°C three times in 50% FA/ 2×SSC for 10 min each, followed by two washes in 2×SSC for 10 min each and one wash in 0.4×SSC for 5 min.

6. Wash slides in 0.1% Tween 20/4×SSC at room temperature for 2 min. Then, 80 µl of blocking reagent (vial 2, SKY kit) are applied to each slide. Each slide is covered with a plastic coverslip, and incubated at 37°C for 45 min.
7. After careful removal of the coverslip, 80 µl of buffer I (vial 3, SKY kit, containing avidin Cy5 and mouse-anti-digoxin) are added to each slide and slides are recovered with a plastic coverslip. Incubate the slides at 37°C for 45 min.
8. After careful removal of the coverslip, wash the slides three times in 0.1% Tween 20/4×SSC at room temperature for 5 min on the shaker. Then, 80 µl of buffer II (vial 4, SKY kit, containing goat anti-mouse-antibody conjugated to Cy5.5) are applied to each slide and slides are recovered with a plastic coverslip. Slides are incubated at 37°C for 45 min.
9. After careful removal of the coverslip, slides are washed three time in 0.1% Tween 20/4×SSC at room temperature for 5 min each on a shaking platform, and finally, the slides are mounted in 10 µl of 4, 6-diamidino-2-phenylindole (DAPI, vial 5, SKY kit).

Subprotocol 4
Image acquisition and analysis

Images are acquired with the SD200 SpectraCube Spectral Imaging system (ASI, Inc.). The SD200 imaging system attached to a Nikon E600 microscope consists of an optical head (Sagnac interferometer) coupled to a multi-line CCD camera (Hamamatsu). The image data is stored in a Pentium 586/300 MHz computer and analyzed by the system software. The multiple band pass filter set used for fluorochrome excitation is custom-designed (SKY-1, Chroma Technology) to provide broad emission bands (giving a fractional spectral reading from ~450 to ~850 nm). Using a Xenon light source, the spectral image is generated by acquiring 80-130 interferometric frames per object. The sample emission spectra are measured in the visible and near infrared spectrum simultaneously at all points in the microscopic image. The CCD camera captures images at discrete interferometric steps that are stored as an image stack in the computer. Next, each interferogram is Fourier transformed resulting in a fluorescence spectrum for each pixel of the image. DAPI images are recorded using a DAPI specific optical filter set.

Procedure

Build combinatorial table (CTB) file

Building the combinatorial table file is the first step of the spectral karyotyping. Acquire the fluorescencence spectra of pure dyes using the SKY-1 filter set and build the combinatorial table file (CTB) (also refered to as 'reference spectral library') to be used in all subsequent measurements.

1. For spectral karyotyping on metaphase spreads, five pure fluorochromes (Spectrum Green, Spectrum Orange, Texas Red, Cy5, Cy5.5) have been acquired by the SD200 SpectraCube spectral imaging system and the CTB files for human and mouse karyotyping have been set up in the software.
2. For interphase cell analysis, acquire the spectrum of pure fluorochromes (Spectrum Green, FITC, Spectrum Orange, Cy3, Cy5, and Cy5.5).
3. The CTB file for interphase cell analysis is built according to the fluorochrome-labeling scheme shown in Table 1.
4. A pseudo-color for each target in the CTB table is assigned to all points in the image.

Acquiring the RGB color and DAPI images

Select a target of interest, acquire the image through the SKY-1 filter (saved as a raw file) and DAPI filter (saved as a tif file).

Data analysis

1. In the Sky View software, open the spectrum image file, dapi file, and CTB file.
2. The image acquired through the SKY-1 filter is displayed by assigning red, green or blue colors (RGB pseudo-color image) to three ranges of interest in the spectrum (usually 450-550, 550-650, 650-850 nm).
3. In the RGB pseudo-color image, delineate the signal domain by drawing contour lines around each signal (interphase cells analysis) or each chromosome (metaphase cells analysis).
4. Based on the measurement of the spectrum for each signal or metaphase, a spectral classification algorithm is applied that allows the assignment of a classification pseudo-color to all points in the image that had the same spectrum. This algorithm forms the basis for chromosome identification by spectral karyotyping.

5. A classification pseudo-color image and a karyotype table are generated.

   Note: Often the shape and the size of the signal domains drawn by the contour line will affect the karyotyping results. Careful drawing and editing of the contour lines is necessary.

Results

Metaphase spread analysis

Complex rearrangements in tumor cells are mapped rapidly by SKY using tumor metaphase cell preparations. Figure 1 demonstrates this by showing the SKY analysis of a metaphase spread prepared from the human prostate cancer cell line TSUPR1. Here, the DAPI image (Fig. 1A) helped to identify chromosomes. Following chromosome classification, individual chromosomes were assigned 'classification colors' which linked the fluorescence spectra to chromosome-specific probe mixtures (Fig. 1B, C). SKY analysis of TSUPR1 metaphases revealed a number of previously unknown translocations (t(2;8),t(3;19),t(6;7),t(6;15),t(15;18)) in the presence of several marker chromosomes (Fig. 1D). In summary, Fig. 1A shows the inverted image of the chromosome spread acquired through the DAPI filter. Figure 1B shows the same chromosomes in an RGB pseudo-color image acquired through the SKY filter. Based on the measurement of the complete spectrum for each point in this metaphase image, a spectral classification algorithm was applied that allowed the assignment of a pseudo-color to all points in the image that had the same spectrum. A karyotype table was obtained as shown in Fig. 1C. There were 84 chromosomes in this spread, with six of them classified as marker chromosomes. We changed the format of this karyotype table as shown in Fig. 1D, demonstrating chromosomes in classification colors to the left and normal idiograms to the right. Clearly, translocations in the cell of this prostate cancer cell line involved the chromosomes 3, 7, 8, 15, 18, 19.

Chromosome enumeration in interphase cells

Spectral Imaging can be applied to routinely score eight or more chromosomes in single interphase cell nuclei. The following examples show SIm results using three different cell types hybridized with a 8-chromosome

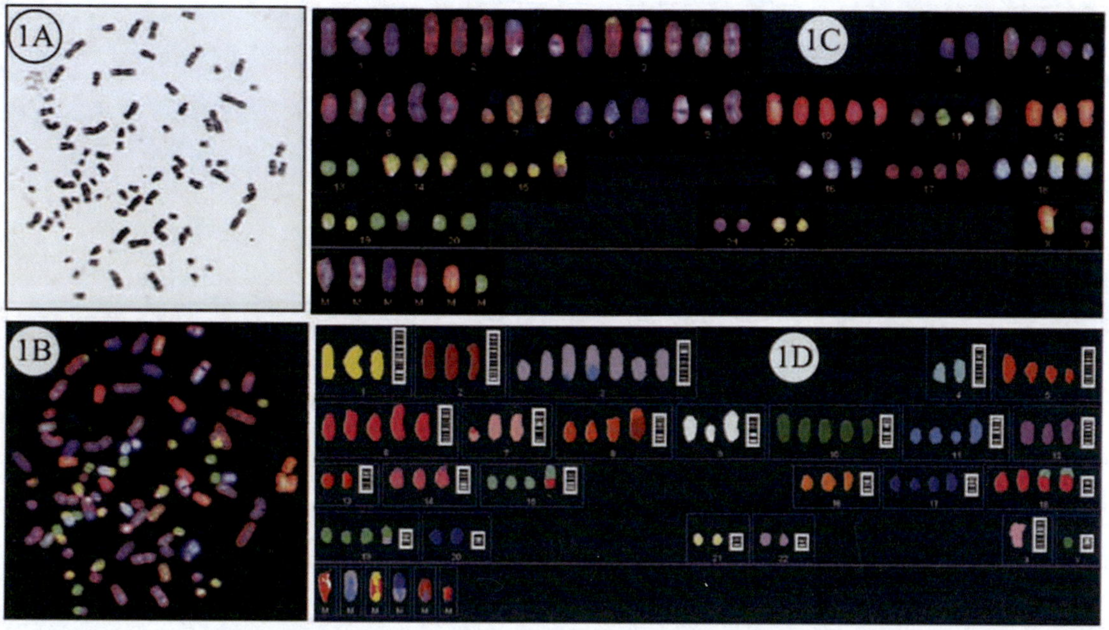

Fig. 1. SKY analysis of metaphase spreads from the prostate cancer cell line TSUPR1. **A** Inverted DAPI image. **B** RGB pseudo-color image. **C** Karyotype table. **D** Spectral karyotype showing chromosomes in classification colors to the *left* and normal idiograms to the *right*

probe set: lymphocytes, amniocytes, and blastomeres. Figure 2A shows an RGB pseudo-color image of an interphase lymphocyte nucleus with contour lines drawn around each signal domain. The corresponding classification color image of this nucleus is shown Fig. 2B. In its karyotype table shown in Fig. 2C, signals are displayed as pairs with the left member of each pair taken from the RGB pseudo-color image (Fig. 2A), and right member from the classification color image (Fig. 2B). This particular type of display allows the rapid detection of aneuploidy and, to some extent, helps to identify classification artifacts. The normal karyotype of this cell showed two copies each of chromosome 13, chromosome 15, chromosome 16, chromosome 18, chromosome 21, and chromosome 22 in addition of one copy each of the X and Y chromosomes (Fig. 2C). The size and shape of the signals in the classification color image (Fig. 2B) and the karyotype table (Fig. 2C) are the same as the size and shape of each contour (red, Fig. 2A).

Figure 2D is an RGB pseudo-color image of an uncultured interphase amniocyte cell. Its classification color image is shown in Fig. 2E. This ap-

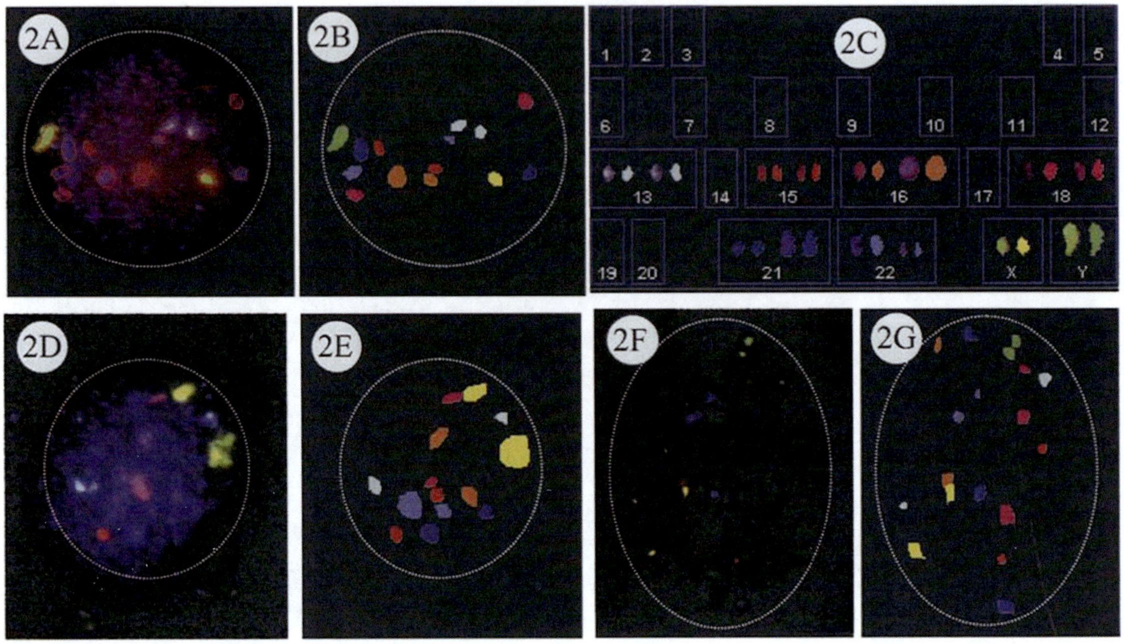

Fig. 2. SIm analysis of three different interphase nuclei. **A** Hybridization to a lymphocyte interphase nucleus. The RGB color image with contour lines along each signal shows 14 domains. **B** Domains in the classification color image can be arranged in **C** a karyotype table. Autosomal signals are grouped into two pairs with the left member of each pair from **A** and the right member from **B. D-E** Analysis of an amniocyte interphase nucleus; shown are the RGB color image (**D**) and the classification color image (**E**). **F-G** Analysis of a blastomere interphase nucleus; shown are the RGB color image (**F**) and the classification color image (**G**)

parently normal female cell shows a total of 14 signal domains representing two copies each of the six autosomal targets and two copies of chromosome X (karyotype not shown). This example also demonstrates the remarkable power of Spectral Imaging to discriminate true signals from high levels of background or autofluorescence.

Figure 2F shows an RGB pseudo-color image of a blastomere cell. Its classification color image is shown in Fig. 2G. This abnormal male cell showed a total of 18 signals representing two chromosomes 13, two chromosomes 15, two chromosomes 16, three chromosomes 18, three chromosomes 21, two chromosome 22, one chromosome X and two chromosome Y (karyotype not shown).

Comments

Purchasing the whole chromosome painting probes for spectral karyotyping or m-fish from one of the commercial suppliers might be more efficient than attempting to synthesize painting probes inhouse. If metaphase or other target cells are dense enough, a small coverslip can be used to conserve valuable probe material. For a 12 mm circular cover glass, for example, 2 μl of probe is often sufficient.

Applications

SKY on metaphase spreads

SKY allows the chromosomes to be "painted" in different colors and, thus, is an excellent tool to analyze complex rearrangements as well as to detect hidden chromosomal abnormalities that might be present in tumor tissues. Often chromosome analysis in tumors is complicated by polyploid genomes. SKY permits a detailed analysis of marker chromosomes in the complex rearrangements and tri- or tetraploidy cases (Fig. 1). SKY has been applied in several cancer studies (Schröck 1996, Zitzelsberger 1999), prenatal diagnosis (Ning 1999), and for analysis of human oocytes, polar bodies biopsied from human oocytes (Márquez 1998), artificially condensed chromatin of blastomeres (Willadsen 1999), and mouse karyotyping (Liyanage 1996).

SIm on interphase cells

Interphase FISH has widely been used to analyze chromosome or gene copy numbers in tumor tissues, interphase lymphocytes or for the rapid prenatal genetic diagnosis of aneuploidy. Commercially available kits presently allow enumeration of just five chromosomes in interphase cells using filter-based fluorescence microscope systems [X, Y, 13, 18, 21 (AneuVysion assay kit, Vysis) or 13, 16, 18, 21, 22 (MultiVysion PB multicolor probe panel, Vysis)]. Repeated hybridizations on single interphase cells can be applied to score additional chromosomes. Signal overlap, however, may cause severe problems in domain scoring and diagnosis of interphase cells. Therefore, SIm techniques were developed to recognize the majority of numerical chromosome aberrations in the few available fetal or embryonic interphase cells. Using SIm, it is now possible to simultaneously score at least 8 chromosomes in a single nuclei. The probe set described in this chapter was chosen to allow the simultaneous enumeration of 8 chromosome types. In fact, the probe set described here, i.e., human chromosomes 13, 15, 16, 18, 21, 22, X, and Y, is expected to detect more than 70% of all numerical chromosome aberrations responsible for spontaneous abortions in humans. Thus, SIm has become a powerful tool for multilocus genetic analysis of single interphase cells such as blastomeres from preimplantation embryos, uncultured amniocytes from amniocentesis, fetal cells from maternal blood, and tissue samples.

Acknowledgements. We thank Drs. S.H. Kim and R. Jensen, UCSF Cancer Center, for providing metaphase spreads from TSUPR1. This work was supported in part by the European Community, grant F14PCT950001L, by the US Department of Energy under contract DE-AC-03-76SF00098 and a grant from the National Institute of Health (1R21 CA88258). J.F. was supported in part by a NIEHS training grant 5-T32-ES07106-17. This publication was made possible by funds received from the Cancer Research Fund, under Interagency Agreement #97-12013 (University of California, Davis contract #00-00801V-20056) with the Department of Health Services, Cancer Research Section.

References

Fung J, Hyun W, Dandekar P, Pedersen RA, Weier, H.UG (1998) Spectral imaging in preconception / preimplantation genetic diagnosis (PGD) of aneuploidy: multi-colour, multi-chromosome screening of single cells. J Assist Reprod Genet 15:322-329

Garini Y, Macville M, du Manoir S, Buckwald RA, Lavi M, Katzir N, Wine D, Bar-Am I, Schröck E, Cabib D, Ried T (1996) Spectral Karyotyping. Bioimaging 4:65-72

Liyanage M, Coleman A, du Manoir S, Veldman T, McCormack S, Dickson RB, Barlow C, Wynshaw-Boris A, Janz S, Wienberg J, Ferguson-Smith MA, Schröck E, Ried T (1996) Multicolour spectral karyotyping of mouse chromosomes. Nat Genet 14:312-315

Márquez C, Cohen J, Munné S (1998) Chromosome identification in human oocytes and polar bodies by spectral karyotyping. Cytogenet Cell Genet 81:254-258

Ning Y, Laundon CH, Schröck E, Buchanan P, Ried T (1999) Prenatal diagnosis of a mosaic extra structurally abnormal chromosome by spectral karyotyping. Prenatal Diagnosis 19:480-482

Schröck E, du Manoir S, Veldman T, Schoell B, Wienberg J, Fergueson-Smith M, Ning Y, Ledbetter D, Bar-Am I, Soenksen D, Garini Y, Ried T (1996) Multicolor spectral karyotyping of human chromosomes. Science 273:494-497

Telenius H, Pelmear AH, Tunnacliffe A, Carter NP, Behmel A, Ferguson-Smith MA, Nordenskjold M, Pfragner R (1992) Cytogenetic analysis by chromosome painting using DOP-PCR amplified flow-sorted chromosomes. Genes Chromosomes Cancer 4:257-263

Willadsen S, Levronm J, Munné S, Schimmel T, Márquez C, Scott R, Cohen J (1999) Rapid visualization of metaphase chromosomes in single human blastomeres after fusion with in vitro matured bovine eggs. Hum Reprod 2:470-475

Zitzelsberger H, Lehmann L, Hieber L, Weier HUG, Janish C, Fung J, Negele T, Spelsberg F, Lengfelder E, Demidchik E, Salassidis K, Kellerer, AM, Werner M, Bauchinger M (1999) Cytogenetic changes in radiation-induced tumors of the thyroid. Cancer Research 59:135-140

Suppliers

5 Prime -3 Prime, 5603 Arapahoe, Boulder, CO 80303, USA

Amersham Pharmacia Biotech, Amersham Pharmacia Biotech, Inc., 800 Centennial Avenue, P.O. Box 1327, Piscataway, NJ 08855-1327, USA; phone 1-732-457-8000, fax: 1-732-457-0557; http://www.amersham.com.au/

Applied Spectral Imaging Inc., 2120 Las Palmas Drive, Suite D, Carlsbad, CA 92009, USA; phone (800) 611-3466, fax (760)929-2842

BIO 101: 1070 Joshua way, Vista, CA 92083, USA; phone (800)424-6101; fax (760) 598-0116; order@bio101.com

BIO-RAD Laboratories, 2000 Alfred Nobel Drive, Hercules, CA 94547-9980, USA; phone (510) 741-1000, fax (510) 741-5800, http://www.bio-rad.-com/index1.html

Calbiochem, 10394 Pacific Center Court, San Diego, CA 92121 USA; phone (800) 854-3417, fax (800)776-0999

Fisher Scientific, 9999 Veterans Memorial Drive, Houston, TX 77038-2499, USA; phone (800) 766-7000, fax (800) 926-1166

Hamamatsu Corp./Hamamatsu Photonic System Division, PO Box 6910, Bridgewater, NJ 08807, USA; phone (908) 231-1116, fax (908) 231-0852, http://www.hamamatsu.com

International Equipment Company (IEC), 300 Second Avenue, Needham Heights, MA 02194, USA, phone (617) 449-8060, fax (617) 444-6743

Labline Instruments, 15th and Bloomingdale Ave., Melrose Park, IL 60160-1491, USA; phone (800) 522-5463, fax (708) 450-0943, http://www.labline.-com, email: value@labline.com

Life Technologies, P.O. Box 9418, Gaithersburg, MD 20898, USA; phone (800) 828-6686, fax (800)331-2286, http://lifetech.com

New Brunswick Co., Inc., 44 Talmadge Rd., Edison, NJ 08818-4005, phone (908) 287 1200, fax (908) 287 4222, email bioinfo@nbsc.com, http://www.nbsc.com

New England Biolabs, Inc., 32 Tozer Road, Beverly, MA 01915-5599, USA; phone (800) 632-5227, fax (978) 921-1350, http://www.neb.com

Nikon, Inc., 1181 Chess Drive, Suite H, Foster City, CA 94404-1109, USA; phone (650) 513-0542, fax (650)513-0540

Perkin Elmer, 850 Lincoln Center Drive, Foster City, CA 94404, USA; phone (800) 345-5224, fax (415)572-2743

Pharmacia Biotech Inc., 800 Centennial Ave., P.O. Box 1327 Piscataway, NJ 08855-1327, USA; phone (800) 526 3593, fax (800) 329 3593, http://www.biotech.pharmacia.com

Research Genetics, Inc., 2130 Memorial Pkwy, S.W., Huntsville, AL 35801, USA; phone (205) 533 4363, fax (205) 536 9016, http://www.resgen.com/

Roche Molecular Biochemicals, 9115 Hague Rd., P.O. Box 50414, Indianapolis, IN 46250-0414, USA; phone (800) 428-5433; fax (800) 428-2883

Sigma, P.O. Box 14508, St. Louis, MO 63178, USA; phone (800) 325-3010, fax (800)325-5052

Specialty Media, 580 Marshall Street, Phillipsburg, NJ 08865, USA; phone (800) 543-6029, fax (908) 387-1670, http://www.specialtymedia.com/

Thermatron Industries, Inc., 291 Kollen Park Drive, Holland, MI49423, USA; phone (616) 393-4580, fax (616)392-5643

Vector Laboratories, Inc., 30 Ingold Road, Burlingame, CA 94010, USA; phone (415) 697-3600, fax (415)697-0339

VWR Scientific, 1310 Goshen Pkwy., West Chester, PA 19380, USA; phone (800)932-5000, fax (610) 436-1761; email solutions@vwrsp.com; http://www.vwrsp.com

Vysis, Inc., 3100Woodcreek Drive, Downers Grove, IL, USA: phone (800) 553-7042, fax (630) 271-7138

Abbreviations

SKY Spectral karyotyping
SIm Spectral Imaging

Glossary

SKY Spectral karyotyping
SIm Spectral Imaging

Microdissection of Chromosomes and Reverse FISH

Ilse Chudoba and Gabriele Senger

Introduction

Microdissection of chromosomes or chromosomal regions was first published by Scalenghe et al. in 1981. They dissected fragments from the polytene X chromosome of *Drosophila melanogaster*. The DNA was extracted and ligated to a lambda vector. Already in this early work, it was demonstrated that the resulting clones can be hybridized back to the original section of the X chromosome, where the dissection was carried out. However, application of this method to mammalian chromosomes was hampered due to the fact that normal metaphase chromosomes contain only two double helices of DNA and hence a large number of fragments (more than 100) had to be collected in order to obtain sufficient DNA for successful cloning in a lambda vector. A significant improvement of the original technique was achieved by the application of G-banding prior to the dissection procedure (Senger et al. 1990) and by establishing an efficient micro-cloning system (Lüdecke et al. 1989). However, the real breakthrough for the microdissection technique was the invention of the sequence-independent DNA amplification. Different protocols have been shown to be efficient in amplifying a DNA pool with no specific priming sites (Bohlander et al. 1992, Telenius et al. 1992). However, it turns out that the polymerase chain reaction using a degenerate oligonucleotide primer (DOP-PCR) (Telenius et al. 1992) is the preferred amplification technique which is applied in most experiments. The original technique of Telenius was adapted in different ways (Meltzer et al. 1992, Viersbach et al. 1994, Müller-Navia et al. 1995, Rubtsov et al. 1996, Chudoba et al., 1996). Regardless of the follow-up applications, such as positional cloning or fluor-

✉ Ilse Chudoba, MetaSystems GmbH, Robert-Bosch-Str. 6, Altlussheim, 68804, Germany (*phone* +49-6205-39610; *fax* +49-6205-32270; *e-mail* ic@metasystems.de)
Gabriele Senger, Praxis für Medizinische Genetik, Roritzerstraße 2, Regensburg, 93047, Germany

escence in situ hybridization (FISH), the initial step after collecting the microdissected fragments is always the amplification by DOP-PCR.

Materials

Equipment

- Inverted microscope (e.g. Zeiss, Axiovert 135)
  We recommend the use of an inverted microscope which is equipped with a long distance condenser. This makes the handling of the dissection needle before and after the dissection procedure as well as the positioning of the needle much easier. The microscope is equipped with two objectives: 10× (Plan-Neofluar) for searching for adequate metaphase spreads and 100x (Plan-Neofluar, oil-immersion), which is used to identify chromosomes and to observe and control the actual microdissection process. In order to align the chromosomes perpendicular to the microdissection needle, a freely rotating working stage is used.
- Micro-manipulator (Zeiss, Maerzhaeuser, Eppendorf)
  We recommend an electronically controlled micro-manipulator for guiding the microdissection needle.
- Pipette puller (e.g. Narishige PC-10)
  The actual tools for the microdissection are extended solid glass needles. For their preparation solid glass rods (1–2 mm in diameter) are processed with a standard pipette puller. Preferably a two-step vertical puller is used. In the first step the glass rod is reduced in diameter and the second step produces the needles. Usually, both of the needles that are thus generated can be used.
- Thermocycler
  Any thermocycler with or without heated lid can be used. However, as annealing of the DOP-primer takes place at a relatively low temperature, the cycler must be able to actively chill down the heating block.
- PCR working station
  Since the amplification system is initially independent of the priming sites, any DNA introduced into the reaction is amplified. Hence, maximum care should be taken in order to avoid any DNA contamination. A PCR working station provides a defined space and is equipped with a short wave (254 nm) UV-lamp. The UV-lamp can be used to decontaminate the working space itself as well as the tools used such as racks, tweezers and most importantly, the glass needles.

- Microliter pipettes
  Again, due to the high risk of DNA contamination, we reserve one set of microliter pipettes (0.01–2 µl, 0.1–20 µl, 10–200 µl) for the manipulation prior to the DOP-PCR and for the DOP-PCR itself, respectively. For all manipulations after the DNA fragments have been tagged with the DOP-primers, different pipettes are used.
- Needle holder
  A refillable lead pencil is used to hold the needles.

Consumables

- Coverslips (24×60 mm)
  Due to the arrangement of an inverted microscope and the oil immersion objective, coverslips are used for spreading metaphases instead of regular microscope glass slides.
- Filter tips (suitable for the microliter pipettes)
  For all manipulations of liquids, filter tips are used.
- PCR tubes
  Depending on the thermocycler 500- or 200-µl tubes are used.
- Plastic centrifuge tubes, sterile, (50 ml) (Greiner, #227 261)

Reagents

- DOP-Primer (=6MW) (Telenius et al. 1992) 5' ccg act cga gn$_6$ atg tgg 3
- Sodium dodecyl sulfate (SDS) (Sigma, L 4509)
- Bacto Trypsin (Difco, #0153-61-1)
- Phosphate-Buffer, pH 6.88 (Merck, #7294)
- Giemsa-solution (Merck, #9204)
- Proteinase K (Boehringer, #1413 783)
- Sequenase, Version 2.0, (Amersham, #US70775)
- Taq DNA Polymerase (Appligene Oncor, #120186AX)
- dNTP set, PCR grade (Boehringer Mannheim, #1969 068)
- Chromatide Texas Red-12dUTP (Molecular Probes, #C-7631)
- 20×SSC (Eppendorf, #0032 006.701)
- *E. coli* tRNA (Boehringer, #109 541)
- Deionized Formamide (Appligene Oncor, #S4117)

- Dextran sulfate (Serva, #18705)
- Tween 20 (Sigma, #P-1379)
- Human Cot-1 DNA (Life Technologies, #15279-011)
- NaOH
- DAPI (4',6-Diamino-2-Phenylindole, Sigma, #D 9542)
- Vectashield mounting medium (Vector laboratories, #H-1000)
- PBS (Life Technologies, #14190-094)
- Fixogum (Marabu, #2901 10 000)

Note: In order to reduce the risk of DNA and nuclease contamination, it is essential to wear gloves when handling any chemicals and liquids. All chemicals should be restricted for use in microdissection experiments only.

Chromosome preparation

Any chromosome preparations harvested following routine protocols can be used. Cell suspensions stored in Carnoy's fixative at -20°C for several years have been successfully used. However, treatment with intercalating substances (e.g. ethidium bromide) should be avoided, because they seem to reduce the PCR efficiency significantly.

Procedure

Preparation of metaphase spreads

In order to obtain adequate spreading of metaphases, the coverslips have to be cleaned. For this purpose they are soaked in a 2% SDS solution, at least overnight. Prior to use, they are rinsed thoroughly under running tap water and thereafter twice in sterile double distilled H_2O (dd H_2O). Dissection of the chromosomes is preferably carried out on freshly spread metaphases. They become harder with age and excision of defined chromosomal fragments becomes inexact. Moreover, excised fragments become fragile and tend to break into several small pieces, which makes them more difficult to collect.

The cell suspension is dropped onto the wet coverslips and air dried.

GTG-banding

One ampoule of the lyophilized Bacto trypsin (Difco) is dissolved in 10 ml sterile distilled water. Store 100 µl aliquots at -20°C.

Banding procedure is carried out in 50 ml sterile plastic centrifuge tubes. Altogether 4 tubes are necessary which should be prepared as follows:

- 35 ml phosphate buffer, pH 6.88 (Merck)
- 35 ml phosphate buffer, pH 6.88 (Merck) and 100 µl Bacto trypsin
- 35 ml phosphate buffer, pH 6.88 (Merck) and 1.5 ml Giemsa (Merck)
- 50 ml sterile distilled water

1. 1-min incubation in tube 1
2. 1-min incubation in tube 2 (trypsin digestion)
3. 2-min incubation in tube 3 (staining of the chromosomes)
4. rinse (tube 4) coverslips briefly and air dry

Pre-incubation in tube 1 is optional, but was observed to improve GTG-banding of freshly spread metaphases. The incubation times of trypsin digestion (tube 2) and Giemsa staining (tube 3) is variable and have to be adjusted to laboratory standards.

Microdissection

To facilitate handling of the fragile coverslips, a metal plate with a rectangular hole in the middle (~5.5×5.5 cm) is used as a carrier.

Prepare PCR-tubes containing 4 µl of reaction mix A.

Reaction mix A	Conc. stock	Actual conc.	Final conc.	Volume required
H2O				11.4 µl
Sequenase buffer	5×	1.25	1×	5.0 µl
Proteinase K	14 mg/ml[a]	2.5 mg/ml	2 mg/ml	3.6 µl

[a] Varies with lot no.

conc. stock: concentration of the stock solution

actual conc.: concentration resulting for the actual reaction mix A

final conc.: concentration resulting after adding reaction mix B

The volumes (reaction mix A–D) are calculated for four tubes. Two of them are used for collecting the chromosomal fragments, one serves as a negative control and the last one as a positive control by adding 5 pg of genomic DNA.

1. The coverslip is placed across the cut out of the metal plate with the metaphase spreads facing upwards. The plate is positioned on the microscope working stage with the cut out above the objective.
2. The dissection needle is introduced into the refillable lead pencil and attached to the micro-manipulator. Under low magnification (objective 10x) the tip of the needle is centered into the field of vision but "parked" in a position sufficiently distant above the coverslip.
3. An appropriate metaphase spread is chosen and centered in the field of vision. The objective is changed to 100× and the particular chromosome to be dissected is orientated perpendicular to the needle.
4. Now the needle is carefully lowered until the tip becomes visible. Still above the surface of the coverslip the tip of the needle is positioned right in front of the chromosomal region to be dissected or the marker chromosome to be isolated. The needle has to be lowered until it touches the surface of the coverslip. Any further lowering of the needle will result in a forward movement of the needle tip and excision of a chromosomal fragment (Fig. 1).
5. By raising the needle and placing the tip right on the isolated fragment it usually adheres to the needle and can be taken up.
6. The fragment is transferred to the PCR-tube by breaking off the tip in the tube.
7. Several fragments are excised in the same manner by repeating step 3 through step 7.
8. Overlay with mineral oil (if a thermocycler without heated lid is used).

Before starting the PCR, a protein digestion is carried out by incubating for about 60 min at 60°C. This can be done in the thermocycler.

DNA amplification by DOP-PCR

In the original DOP-PCR protocol published by Telenius et al. in 1992, Taq Polymerase was used for the initial 8 cycles, sometimes referred to as a

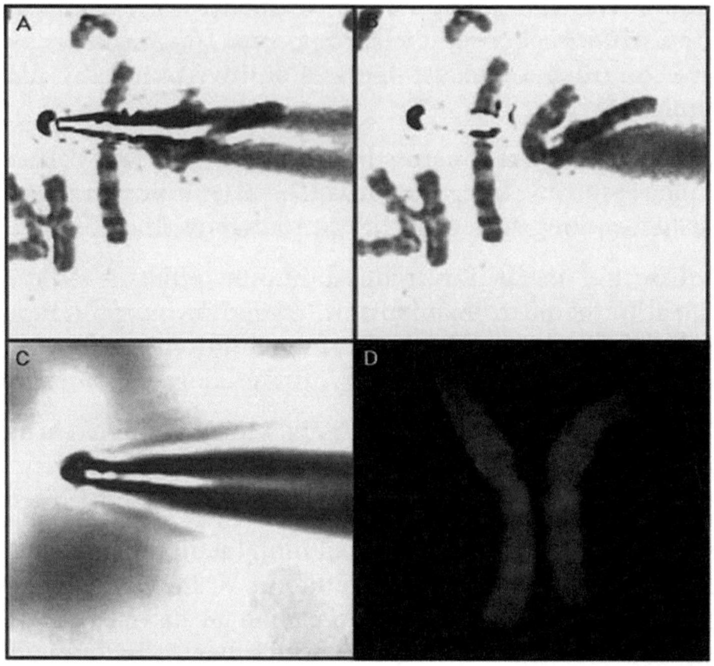

Fig. 1. Microdissection of the chromosomal region 1p22p31 is shown (A, B). In C the chromosomal fragment is taken up and adheres to the tip of the glass needle. In D hybridization of the resulting DNA library to normal chromosomes is displayed

"pre-amplification" step (Meltzer et al. 1995). However, in our hands the much more laborious protocol using T7 polymerase (Sequenase V 2.0) for pre-amplification proved to be more reliable and efficient especially in microdissection experiments where only single fragments were amplified (Zhang et al. 1993).

1. Prepare reaction mix B and add 1 µl to each tube containing the microdissected chromosomes or chromosomal fragments.

Reaction mix B	Conc. Stock	Actual conc.	Final conc.	Volume required
DOP	40 µM	24 µM	5 µM	3.0 µl
dNTPs	2.5 mM each	1 mM	200 µM	2.0 µl

2. Prepare reaction Mix C. Approximately 2 μl are needed for each reaction. If 4 tubes are to be amplified prepare 10 μl.

Reaction mix C	Conc. stock	Final conc.	Volume required
Sequenase dilution buffer			8.75 μl
Sequenase Version 2.0	8×	1×	1.25 μl

3. Insert tubes into the thermocycler and start the program (see below). The first step is a thorough denaturation of the DNA, which at the same time inactivates the Proteinase K.
4. During the initial 8 cycles 0.24 μl (about 0.3–0.4 units Sequenase) of reaction mix C is added at each annealing step (step 2). This is necessary, because the Sequenase is not thermostable and becomes inactivated during each denaturation step.
5. Prepare reaction mix D.
6. Add 50 μl of reaction mix D to each tube during step 5 (see below). If necessary increase the amount of mineral oil in the reaction tube.

Reaction mix D	Conc. stock	Final conc.	Volume required
H_2O			142.48 μl
Incubation mix	10×	1×	21.0 μl
dNTPs	2.5 mM each	220 μM	18.48 μl
Primer (DOP)	40 μM	1.1 μM	5.78 μl
MgCl2	25 mM	2.5 mM	21.0 μl
Taq DNA polymerase	5 U/μl	0.03 U/μl	1.26 μl
Final volume			210.0 μl

PCR-profile

	Block temperature	Time	Comments
Step 1	95 °C	5 min	Initial denaturation
Step 2	30 °C	2 min	Preamplification annealing
Step 3	37 °C	2 min	Preamplification extension
Step 4	95 °C	1 min	Preamplification denaturation
Repeat steps 2–4 8 times			
Step 5	30 °C	2 min	Transition
Step 6	95 °C	1 min	Denaturation
Step 7	56 °C	1 min	Annealing
Step 8	72 °C	2 min	Extension
Repeat steps 6–8 30 times			
Step 9	72 °C	5 min	Final extension

In order to protect the DNA from nuclease degradation, EDTA is added to a final concentration of 5mM. For long term storage, the tubes are frozen at -20°C.

Labeling of DOP-amplified DNA libraries

The DNA fragments resulting from the initial preamplification are all tagged with the DOP-primers. Therefore, the same primers can be used in order to amplify the DNA products and incorporate labeled nucleotides. Either Biotin- or Digoxigenin-labeled nucleotides as well as nucleotides directly coupled with fluorochromes can be used. By incorporation of directly labeled nucleotides, any time consuming post-hybridization detection steps are avoided and hence, preferred. In the following protocol we use Texas Red labeled dUTPs but Biotin- or Digoxigenin-dUTP as well as other directly labeled nucleotides can be incorporated. For each microdissection library 20 µl of labeling mixture is prepared. Add 2 µl of the PCR product from the first PCR.

Labeling mixture (calculated for 4 libraries)	Conc. stock	Final conc.	Volume required
H_2O			53.04 µl
Incubation mix	10×	1×	8.50 µl
dNTPs	2.5 mM each	200 µM	6.80 µl
Texas Red-dUTP	1 mM	40 µM	3.40 µl
Primer (DOP)	40 µM	2.0 µM	4.25 µl
MgCl2	25 mM	2.5 mM	8.50 µl
Taq DNA polymerase	5 U/µl	0.03 U/µl	0.51 µl
Final volume			85 µl

Profile of the labeling PCR

	Block temperature	Time	Comments
Step 1	95 °C	1 min	Denaturation
Step 2	56 °C	1 min	Annealing
Step 3	72 °C	2 min	Extension
Repeat steps 1–3 20 times			
Step 4	72 °C	5 min	Final extension

Purification

In order to separate the labeled DNA fragments from an excess of primers and labeled nucleotides, the DNA is precipitated.

1. Add:
   - 100 µg *E. coli* tRNA
   - 1/10 vol 3 M sodium acetate pH 5.6
   - 2 vol 100% ethanol (-20°C)
2. Mix and incubate at -20°C, 30 min
3. Cool down the centrifuge to 4°C.
4. Centrifuge at 25,000 *g*, 15 min, 4°C.
5. Remove the supernatant and let dry.

6. Add 30 µl of hybridization mix and dissolve the pellet thoroughly. Store at -20°C.

Hybridization mix

- 2 g Dextran sulfate
- 7 ml deionized Formamide
- 1 ml 20xSSC
- 2 ml dd H_2O
- pH 7.0
- Store in 500 µl aliquots at -20°C

Fluorescence in situ hybridization (FISH)

Microdissection of chromosomes and chromosomal regions can be carried out for two major purposes.

1. It provides the most elegant way for the characterization of chromosomal aberrations, e.g. the composition of marker and derivative chromosomes as well as the exact breakpoint of deletions and translocations. In these cases, the chromosomes in question are isolated, or special regions of interest are excised. In a so called reverse FISH experiment, the resulting DNA libraries are hybridized back to normal metaphase spreads thus revealing their chromosomal origin.

2. By use of microdissection, specifically tailored partial painting probes can be generated and applied in a forward FISH experiment in order to characterize chromosomal aberrations.

Probe preparation

In general, 4 µl of the unpurified PCR product after labeling, is largely sufficient for one hybridization. This corresponds to about 200–300 ng of DNA. Following the above mentioned purification procedure this amount of labeled DNA is contained in approximately 6 µl of the finally obtained hybridization mixture. Small aliquots (20 µl) of Cot1-DNA, each sufficient for one single experiment, are lyophilized. The 6 µl of the hybridization mix with the labeled DNA are added to one lyophilized aliquot of Cot1-DNA and mixed thoroughly. If only one probe is hybridized, the total volume is increased by adding 6 µl of hybridization mix. In case a second probe should be hybridized simultaneously, 6 µl of another probe mix that was prepared accordingly can be added instead. The probe is denatured at 75°C for 5 min, chilled down rapidly (e.g. ice water) and incubated at 37°C for 30 min.

Slide preparation

Conventional chromosome slides can be used for the hybridization. However, slides which have been spread on the same day the hybridization is to be carried out, are not suitable. Chromosome morphology is better retained if slides are aged at least overnight. Pretreatment of chromosome slides prior to the hybridization is optional. However, RNase digestion might reduce the background staining and digestion with a proteolytic enzyme (Pepsin or Proteinase K) improves the probe penetration. Slides should not be older than 2 weeks.

Chromosome denaturation

Denaturation solution

- 0.7 ml Formamide
- 0.1 ml 20×SSC
- 0.2 ml H_2O
- pH 7.0

1. Prewarm a heating plate to 70°C.
2. Place slides on the heating plate.
3. Apply 100 µl denaturation solution to each slide.
4. Overlay with a 24×60 mm coverslip.
5. Incubate for 3 min at 70°C.
6. Flip off the coverslip and.
7. Immediately transfer to a Coplin jar with 70% ethanol for 1 min.
8. Subsequently, transfer to a Coplin jar with 90% and 100% ethanol, incubate for 1 min each at RT.
9. Air dry.

Hybridization

The denatured and preannealed probes are transferred onto the denatured chromosome slides and overlaid with a coverslip. The edges of the coverslips are sealed with rubber cement (e.g. Fixogum) and the slides are stored in a moist chamber for hybridization overnight at 37°C.

Post-hybridization washes

Solutions required:

- 0.4×SSC, pH 7.0
  - 2.0 ml 20×SSC, pH 7.0
  - 98 ml H_2O
  - Preheat to 69°C
- 4×SSCT, pH 7.0
  - 20 ml 20×SSC, pH 7.0
  - 80 ml H_2O
  - 0.05 ml Tween 20
  - Room temperature (RT)
- DAPI stock solution
  - 1 mg DAPI
  - 10 ml 4×SSCT
  - Dissolve thoroughly and store aliquots (0.5 ml) at -20°C
- DAPI working solution
  - 0.003 ml DAPI stock solution
  - 1.5 ml 4×SSCT

1. Carefully remove rubber cement and coverslips.
2. Place slides in prewarmed (69°C) 0.4×SSC, incubate for 2 min.
3. Transfer slides to 4×SSCT, incubate for 1 min, RT.
4. Apply 50 μl of the DAPI-solution, overlay with a 24×60 mm coverslip, incubate for 2 min.
5. Flip off the coverslip and wash in 1×PBS for 2 min, RT.
6. Drain off fluid and blow dry with a rubber ball.

7. Apply 20 µl of an antifade solution (e.g. Vectashield) and cover with a 24×60 mm coverslip.
   Slides are ready for microscopic evaluation.

Reverse FISH

In reverse FISH experiments the labeled DNA-library of an aberrant chromosome, or parts of it, is hybridized to normal metaphase spreads. In order to ensure a precise breakpoint definition, metaphase plates should be well spread and chromosomes sufficiently long. A well defined DAPI banding is obtained when the slides are aged for 5–7 days at room temperature prior to hybridization.

When clinical cases are analyzed, precautions should be taken to ensure maximum reliability of the results. Especially when aberrant chromo-

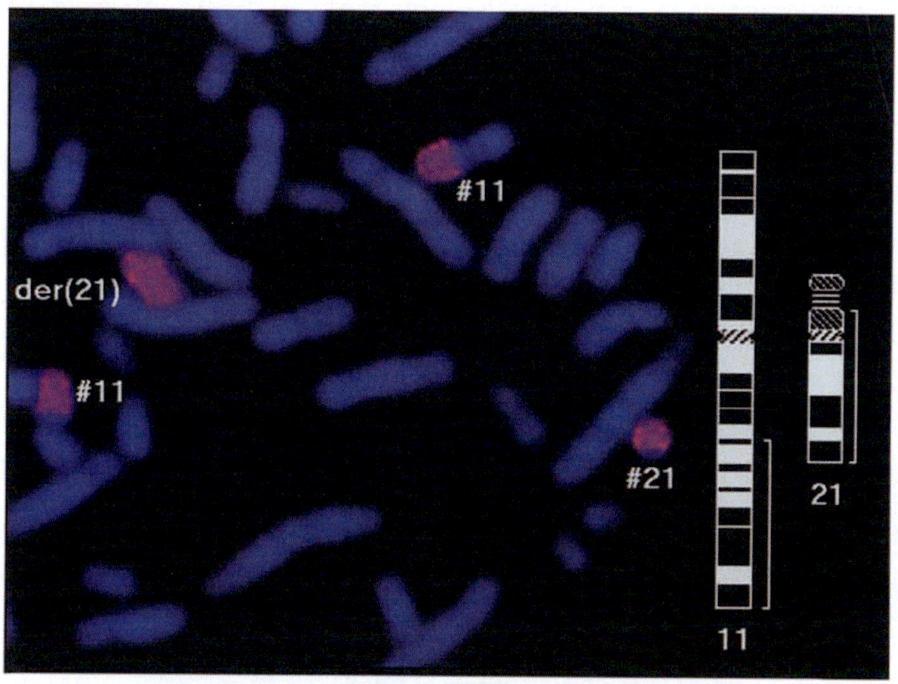

Fig. 2. Reverse FISH of a marker chromosome is shown. The DNA library is hybridized back to the patient's metaphase spreads. Interpretation of these results leads to the diagnosis of a der(21)(11qter11q14,2::21p11cenqter). Localization of the involved chromosomal regions into the formation of the derivative chromosome is indicated by *square brackets* next to the chromosomes ideograms

somes from prenatal cases are analyzed, time is the most limiting factor and repetition of microdissection experiments has to be avoided. We recommend collecting fragments into two different independent PCR tubes. In principle one single fragment or chromosome is enough to obtain a sufficiently strong fluorescence signal after hybridization. However, to be on the safe side, we recommend taking 3 fragments or chromosomes for each PCR tube.

After labeling of the PCR products, the probes should be hybridized to normal metaphase spreads as well as back to metaphases from the patient. This provides a good control for the correctness of the microdissection. However, reverse FISH results can be confusing. One has to be aware that the microdissection library hybridizes not only to the aberrant chromosome of the patient's metaphases, but to the corresponding normal chromosomes and chromosomal regions that are involved in the aberrant chromosome. An example is given in Fig. 2. A derivative chromosome 21 was found prenatally in the case of a girl with severe dysmorphic features and developmental delay (Novotna et al. 1998). Microdissection of the whole chromosome revealed the origin of the additional material as 11q14.2qter. Hybridized back to the patient's metaphases, fluorescent sig-

Fig. 3. Microdissection of a marker chromosome from a renal cell carcinoma revealed altogether six different distinct signals on chromosome 7. Localization of the signals is indicated next to the ideogram of chromosome 7

nals were observed not only on the whole der(21) but on the normal chromosome 21 and on both chromosomes 11 (11q14.2qter) as well.

A second example is presented in Fig. 3. Three copies of a small marker chromosome from a renal cell carcinoma (for details see Glukhova et al. 1998) were isolated and amplified. The labeled products were hybridized to normal metaphase spreads. Altogether, six distinct signals were observed, which were all localized on chromosome 7 (see Fig. 3). The question arising from these data was, whether all six chromosomal regions are contained in each of the marker chromosomes, or if a diversity of different marker chromosomes is present in the tumor, which contain different regions of chromosome 7. Hence, five single fragment microdissection experiments were carried out, where in five different amplification reactions one single fragment was amplified. Reverse FISH with each of the DNA libraries however, confirmed the initial results. All hybridizations exhibited the same fluorescent signals which were identical to the first result. These investigations made clear, that indeed all collected marker chromosomes were identical.

Generation of painting probes by microdissection

Microdissection is not only an elegant way to characterize undefined chromosomes or chromosomal regions by reverse FISH, but can also be applied to generate painting probes and partial painting probes for use in forward FISH experiments for characterizing chromosomal aberrations. In experienced hands, the quality of painting probes generated by microdissection and DOP-PCR is high and, they can even be used in mFISH experiments. The principle of mFISH is based on the combinations of different fluorochromes for the labeling of single chromosome specific libraries (Fig. 4; Schrööck et al. 1996, Speicher et al. 1996).When applying this labeling strategy to all 24 different chromosome-specific paints, a spectral signature for each chromosome is obtained. Five fluorochromes are required for unambiguous combinatorial labeling. With appropriate filter sets for the fluorochromes, digital images are captured which are further processed and evaluated by adapted imaging systems. One elementary function is to display the 24 different chromosome classes of the human genome in 24 different false colors according to the different combinations of fluorochromes (Fig. 5). In this manner, complex interchromosomal exchanges can easily be analyzed in one single FISH experiment. However, the mFISH technique fails to detect intrachromosomal aberrations such as small deletions, which inversion can. It has been shown in the literature that chromosome arm-

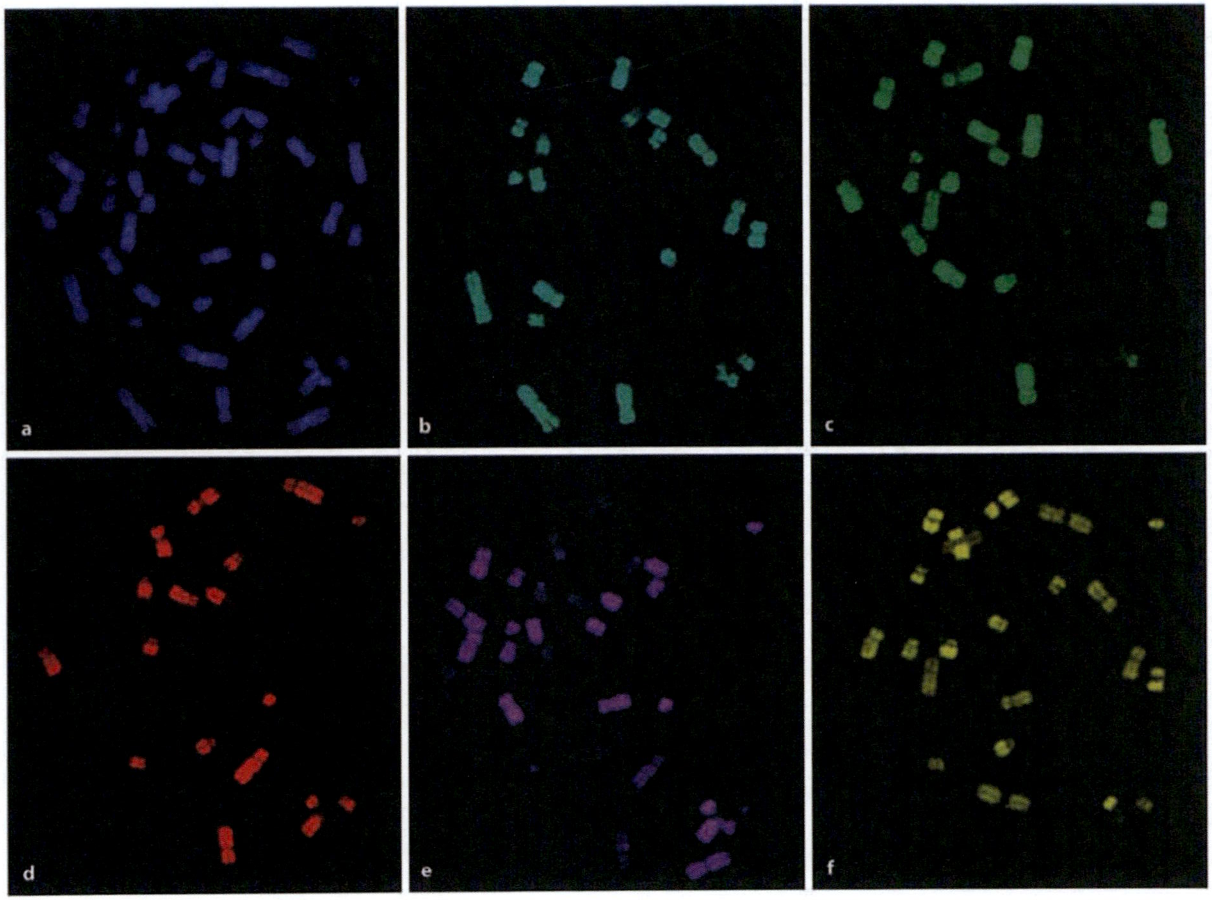

Fig. 4. A normal metaphase spread hybridized with a mix of painting probes for all human chromosomes (24*X*Cyte, MetaSystems) which are labeled in combination with altogether 5 different fluorochromes (DEAC, R110, Spectrum Orange, Texas Red, Cy 5). Images obtained with respective filter sets specific for the fluorochromes are shown (**b–f**) together with the image of the counterstain (**a**). In each color channel a different sub-set of the chromosomes is labeled

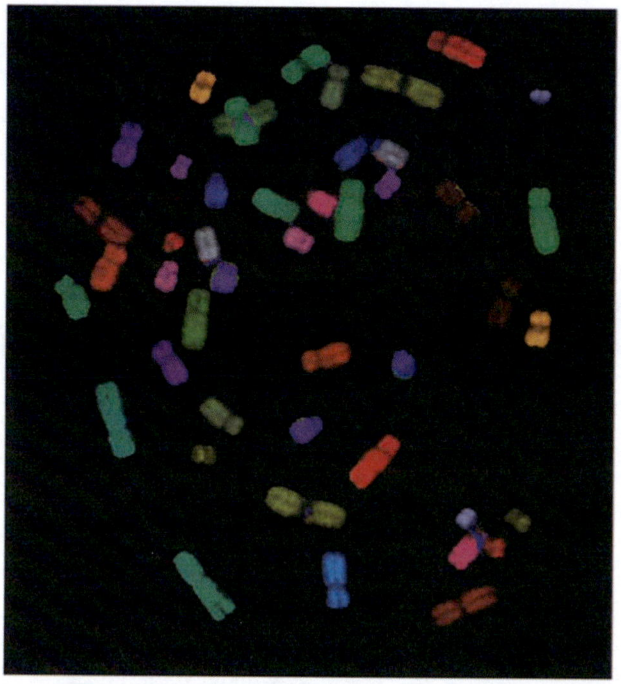

Fig. 5. The same metaphase spread as in Fig. 4 is displayed in false colors. A specific false color is assigned to each of the unique combinations of the fluorochromes resulting in 24 different false colors, each specific for one of the different chromosome classes of the human genome

specific partial paints improve cytogenetic analyses (Guan et al. 1996). Differently labeled p- and q-arm specific partial paints result in a modified painting signal when they are applied to pericentric inversions (Chudoba et al. 1996). The normal chromosome is displayed in two colors with one color on each chromosome arm, respectively, whereas the inverted chromosome is presented in an alternating pattern with regions labeled with both colors on both arms. Whole chromosome paints would not be informative in this kind of chromosomal aberrations.

Microdissected region-specific partial chromosome paints are the basic prerequisite for a chromosome specific multicolor banding (mBAND) which has been developed by our working group (Chudoba et al. 1999). Initially, this mBAND technique was realized for chromosome 5. Altogether, seven chromosomal regions of chromosome 5 were microdissected and amplified as described above. A specific feature of this probe set is that each partial paint covers a region in common with the adjacent partial paint. The partial paints are labeled with different fluorochromes and hy-

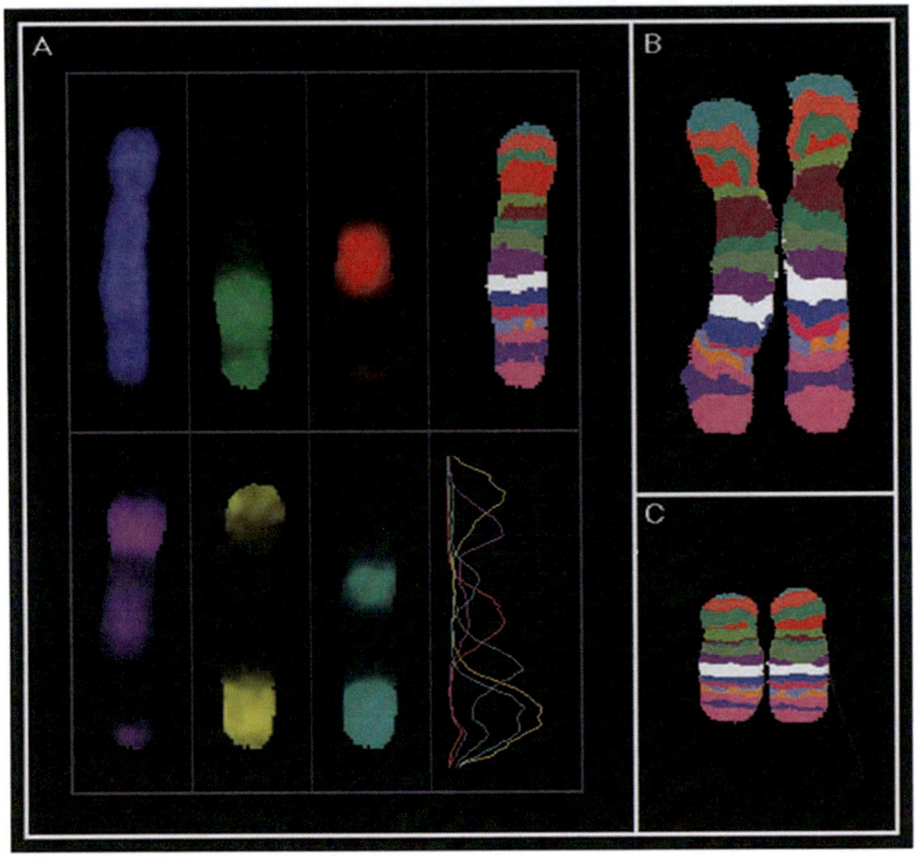

Fig. 6. A Gallery of the six color channels captured with a single band pass filter each together with the resulting mBAND pattern. The profiles represent the fluorescent intensities along the chromosome. B, C Examples for chromosomes 5 at different stages of condensation which show the same mBAND pattern resolution

bridized together. Due to the overlapping of probes next to each other, a multitude of different fluorescence intensity ratios is obtained (Fig. 6). A specific feature of the ISIS/mFISH software from MetaSystems (Germany), calculates the fluorescence intensity ratios within defined regions and assigns a false color. These calculations lead to the characteristic color banding (Fig. 6). Repeated hybridizations proved that this procedure is highly reproducible. Moreover, comparing investigations revealed that the banding pattern is largely independent of chromosome condensation and the same banding pattern can be obtained on chromosomes of different lengths (Fig. 6).

References

Chudoba I; Rubtsov N; Senger G; Junker K; Bleck C; Claussen U (1996) Improved detection of chromosome 16 rearrangements in acute myeloid leukemias using 16p and 16q specific microdissection libraries. Oncol Rep 3:829-832.

Chudoba I; Plesch A; LörchT; Lemke J; Claussen U; Senger G (1999) High resolution multicolor-banding: a new technique for refined FISH analysis of human chromosomes. Cytogenet Cell Genet 84:156-160.

Glukhova L; Goguel A-F; Chudoba I; Angevin E; Pavon C; Terrier-Lacombe M-J; Meddeb M; Escudier B; Bernheim A (1998) Overrepresentation of 7q31 and 17q in Renal Cell Carcinomas. Genes Chromosomes and Cancer 22:171-178.

Guan XY; Zhang H; Bittner M; Jiang Y; Meltzer P; Trent J (1996) Chromosome arm painting probes. Nat Genet 12:10-11.

Meltzer PS; Guan XY; Burgess A; Trent JM (1992) Rapid generation of region specific probes by chromosome microdissection and their application. Nat Genet 1:24-28.

Meltzer PS; Guan X-Y; Zhang J; Trent J M (1995) Microdissection and direct PCR amplification of human chromosomes . In: Verma R S; Babu A (eds) Human chromosomes - principles and techniques. McGraw-Hill, Inc, pp 271-279

Müller-Navia J; Nebel A; Schleiermacher E (1995) Complete and precise characterization of marker chromosomes by application of microdissection in prenatal diagnosis. Hum Genet 96:661-667.

Novotna D; Bartsch O; Chudoba I; Zumrova A; Palanova M; Kofer J; Chudoba D; Macek Sr M (1998) Prenatal diagnosis of a fetus with the rare syndrome of partial trisomy of 11q. Eur J Hum Genet 6: Supplement 1

Rubtsov N; Senger G; Kuzcera H; Neumann A; Kelbova C; Junker K; Beensen V; Claussen U; (1996) Interstitial deletion of chromosome 6q: precise definition of the breakpoints by microdissection, DNA amplification, and reverse painting. Hum Genet 97:705-709.

Scalenghe F; Turco E; Edströöm JE; Pirrotta V; Melli M (1981) Microdissection and cloning of DNA from a specific region of Drosophila melanogaster polytene chromosomes. Chromosoma 82:205-216.

Schröck E; du Manoir S; Veldman T; Schoell B; Wienberg J; Ferguson-Smith MA; Ning Y; Ledbetter DH; Bar-Am I; Soenksen D; Garini Y; Ried T (1996) Multicolor spectral karyotyping of human chromosomes. Science 273:494-497.

Senger G; Lüdecke H-J; Horsthemke B; Claussen U (1990) Microdissection of banded human chromosomes. Hum. Genet. 84, 507-511.

Speicher MR; Ballard SG; Ward DS (1996). Karyotyping human chromosomes by combinatorial multi-fluor FISH. Nature Genet 12:368-375

Telenius H; Carter NP; Bebb CE; Nordenskjold M; Ponder BA; Tunnacliffe A (1992) Degenerate oligonucleotide-primed PCR: General amplification of target DNA by a single degenerate primer. Genomics 13:718-725.

Viersbach R; Schwanitz G; Nothen MM (1994) Delineation of marker chromosomes by reverse chromosome painting using only a small number of DOP-PCR amplified microdissected chromosomes. Hum Genet 93:663-667.

Zhang J; Trent JM; Meltzer P (1993) Rapid isolation and characterization of amplified DNA by chromosome microdissection: identification of IGF1R amplification in malignant melanoma. Oncogene 8:2827-2831

Multicolor FISH Techniques for the Detection of Inter- and Intrachromosomal Rearrangements

Horst Zitzelsberger, Benjamin O'Brien, and Heinz-Ulrich Weier

Introduction

Fluorescence in situ hybridization (FISH) technology has evolved constantly during the past decade. Nowadays, a large variety of probes detected with novel fluorescent dyes, allows the simultaneous detection of several different nucleic acid targets. A simultaneous hybridization of different probes has led to "multicolor FISH" experiments in which more than two fluorescent dyes are used for probe detection. Probes of greatly differing complexity can be combined for multicolor FISH procedures, including whole or partial chromosome painting probes, locus-specific probes (YAC, BAC, P1, cosmids) and repetitive sequences. Since only a limited number of distinct fluorescent dyes can be resolved with the standard microscope setup, combinatorial labeling schemes have been developed. In most circumstances, this wide variety of mixed colors can only be distinguished by utilisation of digital image acquisition and analysis systems. Multicolor FISH procedures are also demonstrated elsewhere in this manual describing for instance the simultaneous staining of each chromosome homologue in the human or murine chromosome complement in a different color (Chapter 22). In this chapter, protocols are presented describing the application of multicolor FISH in radiation

✉ Horst Zitzelsberger, GSF-Forschungszentrum, Institut für Strahlenbiologie, Ingolstädter Landstr. 1, Oberschleißheim, 85764, Germany (*phone* +49-89-31873421; *fax* +49-89-31872873; *e-mail* Zitzelsberger@gsf.de), Ludwig-Maximilians-Universität, Institute of Radiation Biology, München, 80336, Germany
Benjamin O'Brien, University of California, Department of Subcellular Structures, Life Sciences Division, E. O. Lawrence Berkeley National Laboratory, Berkeley, CA, 94720, USA
Heinz-Ulrich Weier, University of California, Department of Subcellular Structures, Life Sciences Division, E. O. Lawrence Berkeley National Laboratory, Berkeley, CA, 94720, USA

and tumor cytogenetics for a quantification of radiation-induced chromosome aberrations and the detection of chromosomal rearrangements.

Proximity effects for the formation of chromosome aberrations are an interesting topic in radiation cytogenetics. It has been shown for asymmetrical and, thus, unstable aberrations that the yield ratios for radiation-induced intra- and interchromosomal exchanges are linear energy transfer (LET)-dependent. High LET radiation causes higher yields for intra-arm intrachanges than low LET radiation possibly due to different proximity mechanisms in the formation of aberration. To investigate potential bio-indicators for high LET radiation also for stable aberration types, symmetrical translocations as well as paracentric inversions must be quantified by FISH-painting approaches. A reliable and quick approach for scoring inversions is the application of Rx-FISH, a multicolor technique using gibbon DNA that hybridizes to noncontiguous discrete sections on human chromosomes and results in a distinct color banding pattern. Inversions on human chromosomes are visible as irregular banding patterns.

For a more precise delineation of chromosomal alterations at the gene level, the more sophisticated 'chromosomal rainbow (CR)' technique may be used. This is of particular interest in the field of tumor cytogenetics where small gene rearrangements might provide tumor-specific markers enabling the rapid detection of tumor cells. The method of CR FISH, developed at the Lawrence Berkeley National Laboratory, is based on the simultaneous hybridization of probes that bind on average every 8–10 megabase pairs (Mbp). Probe generation was facilitated by resources generated in the course of the International Human Genome Project such as large insert DNA libraries and detailed physical maps. CR probes sets are prepared and applied for one chromosome type at a time. Labeled with distinct alternating reporter molecules, the probes are distributed more or less evenly all along the target chromosome type to allow the detection of intra- and interchromosomal rearrangements with a resolution of a few Mbp. A rearrangement is indicated when the normal pattern of 'rainbow' colors along the hybridized chromosome appears changed. CRs are also a powerful tool to be used in combination with Comparative Genomic Hybridization (CGH) and Spectral Karyotyping (SKY). A single CR hybridization experiment can provide detailed information about the nature of a copy number change detected by CGH or the approximate position of chromosomal breakpoints.

Outline

- Optional: Irradiate peripheral lymphocytes in vitro or use peripheral blood of an exposed subject or cultivated tumor cells
- Treat cells with hypotonic solution and fix
- Prepare metaphase spreads
- Choose DNA probes (Rx or CR probes)
- Pretreat slides (optional for older preparations)
- Denature and hybridize probes with metaphase preparations
- Remove unbound probes
- Score slide and select rearranged metaphases
- Acquire images with an image analysis system
- Describe inversions by comparing FISH patterns and DAPI banding

Note: Caution: Most subprotocols should be seen as suggestions. Typically, previously established procedures should be integrated (image analysis system, labeling systems, etc.).

Materials

Equipment

- Centrifuge (Heraeus, Biofuge fresco)
- CO_2 incubator
- Dry bath (heat block) (Liebisch)
- Fluorescence microscope equipped with 40× and 63× oil immersion, lenses and motorized filter changer (Zeiss, Axioplan 2)
- Fluorimeter (Hoefer, DyNA Quant 200)
- Incubator oven (Memmert)
- Phase contrast microscope (Zeiss, Axioskop)
- Thermal cycler for in vitro DNA amplification (model 4800, Perkin Elmer)
- Water bath (GFL)

Chemicals and Supplies

- Citifluor AF 1 or Vectashield mounting media
- COT1 DNA solution, 1 mg/ml (Life Technologies)
- DAPI (4,6-diamidino-2-phenylindole) (Sigma)
- Distilled and sterile water
- Digoxigenin-11-dUTP, 1 mM solution (Roche Molecular Biochemicals)
- Ethanol, p.a. (Merck)
- Fluorescein-12-dUTP, 1 mM solution (Roche Molecular Biochemicals)
- Formamide, molecular biology grade (Sigma)
- Glacial acetic acid (Merck)
- Glass Coplin jars
- Glass coverslips (22×22 mm)
- HCl (2M)
- Methanol, p.a. (Merck)
- Microscope slides
- NaCl
- Nonidet P40
- Parafilm
- Phenol:chloroform:isoamyl alcohol (25:24:1) (Sigma)
- Rabbit anti-FITC antibodies (DAKO)
- Rubber cement
- Salmon sperm DNA solution, 10 mg/ml (Life Technologies)
- Sodium citrate
- Tween 20

Buffers and other solutions

- Antifading solution: 0.1% p-phenylenediamine dihydrochloride (Sigma), 0.1×PBS (Gibco), 45 mM $NaHCO_3$, 82% glycerol, pH 8.0
- Avidin-AMCA (Vector), 5 mg/ml in sterile distilled water, dilute 1:200 in PNM buffer.

- Avidin-Cy5 (BDS), 1 mg/ml in sterile distilled water, dilute 1.5:500 in PNM buffer.
- Biotinylated goat anti-avidin (Vector), 500 μg/ml in sterile distilled water, dilute 1:100 in PNM buffer.
- DAPI stock solution (10 μg/ml): add 1 mg DAPI to 100 ml 2×SSC and store protected from the light at 4°C for at least 3 months.
- DAPI counterstain solution (0.1 μg/ml): add 50 ml 2×SSC to 500 μl DAPI stock solution. Prepare fresh daily.
- Denaturation buffer: add 70 ml formamide to a solution of 10 ml 20×SSC and 20 ml distilled water and mix thoroughly. Adjust pH to 7.5 with 2 N HCl and store at 4°C for 1 week.
- Deoxythymidintriphosphate (dTTP) (Sigma), 100 mM solution pH 7.0, dilute with sterile distilled water to a final concentration of 1 mM.
- 70% ethanol: Mix 350 ml ethanol and 150 ml distilled water. Store at room temperature.
- 85% ethanol: Mix 425 ml ethanol and 75 ml distilled water. Store at room temperature.
- 90% ethanol: Mix 450 ml ethanol and 50 ml distilled water. Store at room temperature.
- F1 detection reagent: For each slide, add 199 μl 4×T solution to 1 μl rabbit anti-FITC antibody, mix gently and incubate at room temperature protected from the light for 10 min.
- F2 detection reagent: For each slide, add 198 μl 4xT solution to FITC-labeled goat anti-rabbit antibody, mix gently and incubate at room temperature protected from the light for 10 min.
- FITC labeled goat anti-rabbit IgG antibodies (Vector)
- FITC-labeled goat anti-mouse (Vector), 1.5 mg/ml in sterile distilled water, dilute 1:300 in PNM buffer.
- Formaldehyde solution: add 2 ml formaldehyde (37%, Sigma) to 72 ml 1×PBS/$MgCl_2$. Prepare fresh daily.
- 50% formamide/0.5×SSC: add 50 ml formamide to a solution of 25 ml 2×SSC and 25 ml distilled water and mix thoroughly. Adjust pH to 7.5 with 2 N HCl and store at 4°C for 1 week.

- Hybridization Master Mix 2.1: 14.3% w/v dextran sulfate, 78.6% formamide, 2.9×SSC, pH 7.0. Add 1.45 ml 20×SSC, 0.7 ml sterile distilled water and 1.43 g dextran sulfate to 7.86 ml formamide. Heat to 70°C to dissolve dextran sulfate. Aliquot in 1.5 ml tubes and store at -20°C.
- Mouse anti-FITC (DAKO), 430 µg/ml in sterile distilled water, dilute 1:200 in PNM buffer.
- 1×PBS/50 mM $MgCl_2$: add 50 ml 10×PBS and 25 ml 1 M $MgCl_2$ to 425 ml sterile distilled water.
- 10×PBS (Dulbecco's PBS): dissolve 95.5 g PBS Dulbecco in 1000 ml distilled water and sterilize the solution. Store at room temperature.
- Pepsin solution: prepare pepsin stock solution of 10 mg/ml in sterile distilled water. Add 10-50 µl stock solution to 100 ml 0.01 M HCl. Prewarm at 37°C in water bath, prepare fresh daily.
- PN buffer: (Na_2HPO_4-7 hydrate/NaH_2PO_4-1 hydrate, pH 8, 0.05% nonidet P-40)
- PNM buffer: 5% nonfat dry milk, 0.1% NaN_3 in PN buffer
- Proteinase K solution: prepare proteinase K stock solution of 10 mg/ml in sterile distilled water. Add 3 µl stock solution to a buffer containing 2 ml 1 M Tris/HCl pH 7.5 , 0.2 ml 1 M $CaCl_2$ and 97.8 ml distilled water. Prewarm at 37°C in water bath, prepare fresh daily.
- Rhodamine-labeled sheep anti-digoxigenin (Roche Molecular Biochemicals), 200 µg/ml in sterile distilled water, dilute 1:100 in PNM buffer.
- 20×SSC solution: dissolve 219.1 g sodium chloride and 110.3 g sodium citrate in 1000 ml distilled water. Verify that the pH is between 6.5 and 7.5. Autoclave the solution for extended storage at room temperature.
- 2×SSC solution: add 50 ml 20×SSC to 450 ml distilled sterile water and adjust pH to 7.0 with 2 M HCl. Store at room temperature until solution becomes cloudy.
- 2×SSC/0.1% Tween 20 buffer: add 0.1 % Tween 20 to 2×SSC
- 1×TE buffer: 10 mmol Tris HCl, 1 mmol EDTA, pH 8.0
- 4×T solution: add 25 ml Tween 20 to a solution of 100 ml 20×SSC and 400 ml distilled sterile water and adjust pH to 7.5 with 2 N HCl. Store at room temperature until the solution becomes cloudy.

Additional material needed for Rx-FISH

- RxFISH Color Chromosome Analysis FISH Kit (Applied Imaging)
- Image recording device (e.g. ISIS, MetaSystems)
- single excitation filters for DAPI, FITC, TRITC and Cy5 (Zeiss)

Additional material needed for CR-FISH

- Random Priming kit (e.g. BioPrime, Gibco/LTI, Gaithersburg, MD)
- Image recording device (e.g. QUIPS + SciIImage software)
- various filter sets (e.g. triple-and quadruple-color sets, Chroma Technology, Brattleborrow, VT)

Procedure

Slide preparation

Slide preparation should be performed as described in Chapter 3, FISH on Chromosome Preparations of Peripheral Blood by Thomas Liehr and Uwe Claussen.

Pretreatment of slides, hybridization and detection of bound probes

Depending on the age and nature of chromosome preparations a pretreatment of slides with pepsin or proteinase K solution might improve the results. This step is especially recommended, if preparations were not stored at -20°C or if cytoplasmic residue is visible around the metaphase spreads under the phase contrast microscope. It is also recommended to reduce background, which might occur due to proteins around the metaphases. A pepsin treatment is performed before the denaturation step, a proteinase K treatment after denaturation of preparations.

Pepsin treatment: Slides are incubated in 2×SSC for 1 h at 37°C and are transferred to prewarmed (37°C) pepsin solution. Incubation may vary from 1-3 min and digestion of cytoplasm should be checked under phase contrast microscope (briefly wash in 1×PBS). Slides are washed in 1×PBS twice, transferred in PBS/$MgCl_2$ solution for 5 min and fixed in formaldehyde solution for 5 min at room temperature. Slides are washed in 1×PBS and dehydrated in an ethanol series (70%, 90%, 100%).

Proteinase K treatment: Slides are denatured and dehydrated in an ethanol series (see above) and transferred to prewarmed (37°C) proteinase K solution. They are incubated for 7 min and dehydrated again in an ethanol series.

Hybridization for RX-FISH

The RxFISH probes are supplied as a ready-to-use hybridization mixture in a buffer containing formamide. The tube containing RxFISH probes is warmed in a 37°C water bath for 5 min. An aliquot of 10 µl sufficient for a single hybridization is placed in a fresh tube and denatured by incubating at 65°C for 10 min. The remaining RxFISH probes are kept at -20°C. The metaphase preparations are denatured at 70-75°C in 70% FA/2×SSC, pH 7.0 for 1.5-2 min. This step may require modification according to the type and age of samples. Older slides often require longer denaturation times. The denatured slides are immediately placed in ice-cold 70% ethanol for 2 min. Then, dehydrate the slides in the following ethanol series for 2 min each at room temperature: 70%, 2× 90%, 100% and air dry. The 10 µl denatured probe mixture is applied to the slide, which is covered by a 22×22 mm coverslip and sealed with rubber cement. The hybridization proceeds for 12-72 h in a moist chamber.

After hybridization, the slide is washed in 2×SSC, twice in 50% formamide/0.5×SSC, twice in 2×SSC and in 4×T solution at 45°C for 5 min each. The last washing step is performed for 10 min. For signal amplification of the FITC-labeled probe, the antibody solutions F1 and F2 (provided with the Kit) are centrifuged at 12,000 rpm for 10 min at room temperature and 200 µl of F1 is pipetted onto the slide, covered with a Parafilm coverslip and incubated at 37°C for 30 min in a moist chamber. After six washes in 4×T solution at 42°C this step is repeated with F2. After the last wash in 4×T solution, the slide is counterstained in DAPI counterstain solution for at least 2 min and mounted with 25 µl mounting media Citifluor AF1.

Hybridization for CR-FISH

The P1 DNAs are extracted from overnight cultures following an alkaline lysis protocol (Weier et al, 1995b; Chapter 18, Quantitative DNA Fiber Mapping by Heinz-Ulrich Weier). The YAC DNAs are isolated from 48-h liquid cultures following a standard protocol using yeast lytic enzyme (ICN, Aurora, OH) (Weier et al 1995b; Chapter 18, Quantitative DNA Fiber Mapping by Heinz-Ulrich Weier). The DNAs are then precipitated with isopropanol, washed once in 70% ice cold ethanol and resuspended in 1× TE buffer. To further improve purity, the DNA is extracted with phenol/chloroform mixture, again precipitated in isopropanol, resuspended in 1×xTE buffer and quantitated by Hoechst fluorometry using a Hoefer TK 100 instrument. Probes used in the CR 10 rainbow set are listed in Table 1. The probe labeling and detection scheme is summarized in Table 2.

Table 1. Composition of the probe set for chromosome 10 rainbow hybridization (CR10)

Probe Name	ProbeType	Cytogenetic Location	Gene or Marker	Fractional Length	Probe Label	Fluorochrome Emission
78 E 1	P1	10p15	D10S249	0.017	Dig	Red
91 A 2	P1	10p14	D10S547	0.097	Bio	Infrared
81 G 1	P1	10p13	D10S603	0.154	FITC	Green
87 G 10	P1	10p12	D10S588	0.214	Dig+FITC	Red and green
98 D 7	P1	10p11.2	D10S578	0.298	Dig	Red
58 C 11	P1	10q11.2	Ret	0.401	Bio	Infrared
101 C 7	P1	10q21	D10S539	0.443	FITC	Green
29 F 6	P1	10q21	D10S170	0.478	Dig+FITC	Red and green
106 H 5	P1	10q22	D10S537	0.541	Dig	Red
51 H 2	P1	10q22-23	D10S607	0.601	Bio	Infrared
60 D 11	P1	10q23	D10S541	0.671	FITC	Green
66 E 12	P1	10q24.1	NFkB-2	0.742	Dig+FITC	Red and green
911 D 9	YAC	10q24.3	D10S597	0.812	Dig	Red
858 H 7	YAC	10q25.2	D10S533	0.871	Bio	Infrared
937 A 6	YAC	10q25-26	D10S209	0.923	FITC	Green
932 F 11	YAC	10q26	D10S217	0.981	Dig+FITC	Red and green

Table 2. Probe detection scheme

Label	Layer 1	Layer 2	Fluorescence color
Biotin	-	Avidin-Cy-5	infrared
Digoxigenin	-	Rhodamine-labeled sheep-anti-digoxigenin	red
Fluorescein	Mouse-anti-FITC	FITC-labeled goat-anti-mouse	green

All probe DNAs are labeled by random priming using the BioPrime kit with slight modifications to the protocol (Weier et al 1995a,b; Chapter 18, Quantitative DNA Fiber Mapping by Heinz-Ulrich Weier).

Probes to be visualized in red are labeled by incorporation of digoxigenin-11-dUTP (B.M.) at a digoxigenin-11-dUTP:dTTP ratio of 35:65. Therefore, 1 mM digoxigenin-11-dUTP and 1 mM dTTP solution are simultaneously added in the random priming reaction.

Biotin-labeled probe DNA is prepared using a nucleotide mixture provided with the BioPrime kit, while fluorescein-12-dUTP (fluorescein-12-dUTP:dTTP ratio of 35:65) is used in the random priming mixture for probes to be visualized in green.

Biotinylated probes are detected with avidin-Cy5, which fluoresces in the infrared spectrum when excited around 640 nm.

Yellow visualization is obtained using probes which have been random primed in the presence of 1 mM digoxigenin-11-dUTP and 1 mM fluorescein-12-dUTP at a ratio of 50:50.

The hybridization mixture contains 70 μl of human COT-1 DNA, 17 μl of salmon sperm DNA and between 60-120 ng of each probe, depending on hybridization quality and signal intensity of individual probes. The DNA is precipitated in ethanol and resuspended in 3 μl water, before 7 μl of Master Mix 2.1 (Weier et al 1995b) is added so that the probe is dissolved in 55% formamide, 10% dextran sulfate and 2×SSC, pH 7.0. The probe mixture is then denatured by heating to 75°C for 7 min and placed at 37°C for 30 min. Slides are denatured at 75°C for 3.5 min in denaturing solution and dehydrated in a 70%,85%,100% ethanol series. About 10 μl of hybridization mixture are placed on top of the cells, covered with a 22×22 mm coverslip, sealed with rubber cement and incubated for 44-48 h at 37°C in a humidified chamber.

Posthybridization washes are performed as described previously (Weier et al., 1995a, 1995b), including three washes in PN buffer. Unspecific binding sites are blocked with PNM solution (Pinkel et al., 1986). The different layers of fluorochrome labeled avidin or antibodies (AB) are applied in the concentrations listed in Table 3. The yellow color is generated by overlapping green and red signals, i.e., by hybridization of probes labeled simultaneously with digoxigenin and FITC. After each AB layer, slides undergo three 10-min washes in PN buffer with agitation and are again blocked with PNM. After the final wash in PN buffer, slides are washed 10 min in 2×SSC/0.1% Tween 20 buffer, mounted with DAPI (0.1 μg/ml) in antifading solution and coverslipped.

Image acquisition

Images of the in situ hybridization results are recorded, for example on the Quantitative Image Microscope System (QUIPS) using the Scilimage software (Weier et al, 1995b) or on the MetaSystems image analysis system (see Chapter 2, Microscopy and Imaging Systems by Andreas Plesch). The image analysis systems are attached to a fluorescence microscope e.g., a Zeiss Axioplan 2 fluorescence microscope equipped with single-band excitation filters for DAPI, FITC, TRITC, Cy5, a motorized filter changer and oil immersion objectives (40×, 63×). For visual evaluation, triple and quadruple-color filter sets are recommended.

Results

RxFISH

Figure 1A shows a normal Rx-FISH pattern on human metaphase chromosomes. After in vitro irradiation of peripheral lymphocytes with 3 Gy X-rays and 1.2 Gy fission neutrons, para- and pericentric inversions became apparent due to atypical banding patterns (Fig. 1B, 1C). Breakpoints were determined by use of DAPI-banded images. In this way, inversions could be reliably quantified either for the whole chromosome complement or selected chromosomes. For an estimation of the LET dependent ratio between inversion and symmetrical translocations, the latter must be scored either in a separate three-color FISH painting experiment or by spectral karyotyping for the whole chromosome complement.

CR

On a normal metaphase chromosome 10, our scheme begins (from pter to qter) with a probe visualized in red at the distal short arm of chromosome 10, followed by a probe detected in infrared, next a probe seen in green, and then one in yellow. The next probe, detected in red, started a new cycle of colors (Fig. 2) creating a pattern of a chromosome-specific 'rainbow'. Probes derived from YAC clones gave somewhat stronger signals than probes prepared from P1 clones (Table 1), which provided further clues for the identification of marker chromosomes. A cytogenetic alteration is suspected, if the normal color sequence and signal pattern is altered. As an example, we applied a CR 10 set to metaphase spreads of the cell line TPC-

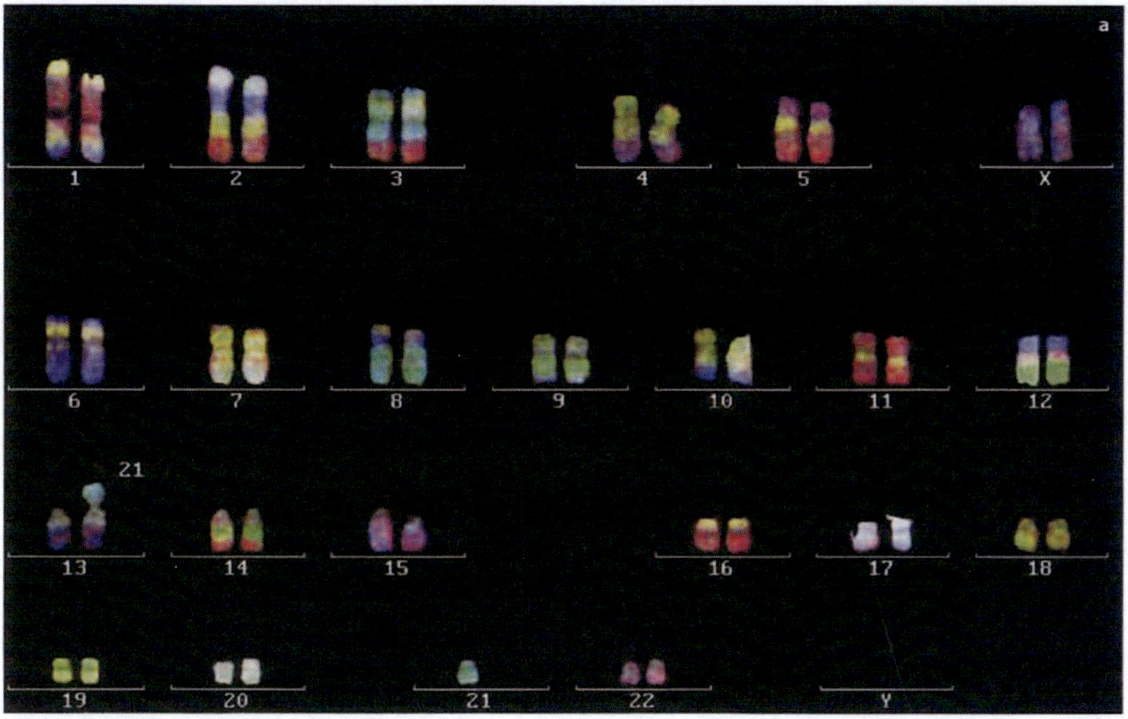

Fig. 1. RxFISH on human metaphase chromosomes. Labeled gibbon probe DNA was hybridized to a normal human metaphase (**1a**) and metaphases from 3 Gy X-rays irradiated (**b**) and 1.2 Gy fission neutrons (**c**) irradiated blood samples. In the irradiated metaphases a pericentric inversion on chromosome 1 (**b**) and a paracentric inversion on chromosome 2 (**c**) became apparent (Figs. b and c see page 420 and 421)

1 established from a papillary variant of thyroid carcinoma. As described earlier (Jossart et al., 1995, 1996), this cell line carries a complex t(1;10;21) translocation and a deletion involving the locus D10S170. The breakpoints along chromosome 10, the paracentric inversion and the orientation of the rearranged parts of chromosome 10 can be described easily based on the sequence of colored hybridization signals. The entire p-arm, the centromere and band 10q11 with the breakpoint distal to probe 58C11, is found reannealed to chromosome 1 fusing to the breakpoint on 1q with its p-terminus (Jossart et al., 1995; data not shown). The remainder of chromosome 10, i.e., part of the long arm, shows a paracentric inversion which is detected by the inverted location of probe 58C11 (purple) and probe 29F6 (yellow) (Fig. 3, *arrows*).

One homologue of chromosome 10 shows a deletion (indicated by *) of locus D10S170 (probe 29F6) (Fig. 4).

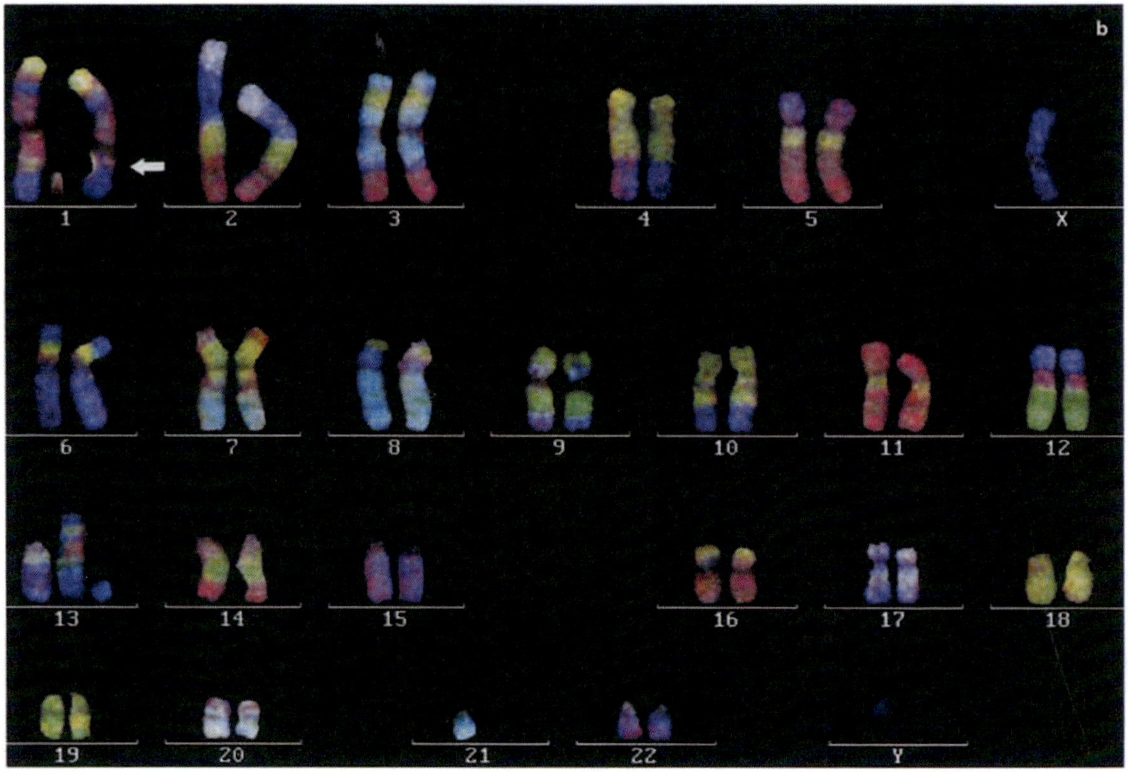

Fig. 1b.

Troubleshooting

- Once the probe panel is established and works, make up larger quantities of the probe mix as stock for convenience and consistency.
- If hybridization signals are weak or absent, increase denaturation times of slides and apply pretreatments like pepsin or proteinase K digestion; immediately after pretreatment check the slide under the phase contrast microscope for the presence of any cytoplasm around the metaphases.
- If swollen chromosomes or disrupted metaphases are present on the slide, reduce the denaturing time and increase postfixation after pretreatment procedures. In many cases, aging of slides, e.g. at 37°C for 1 week or at 60°C overnight may also help.

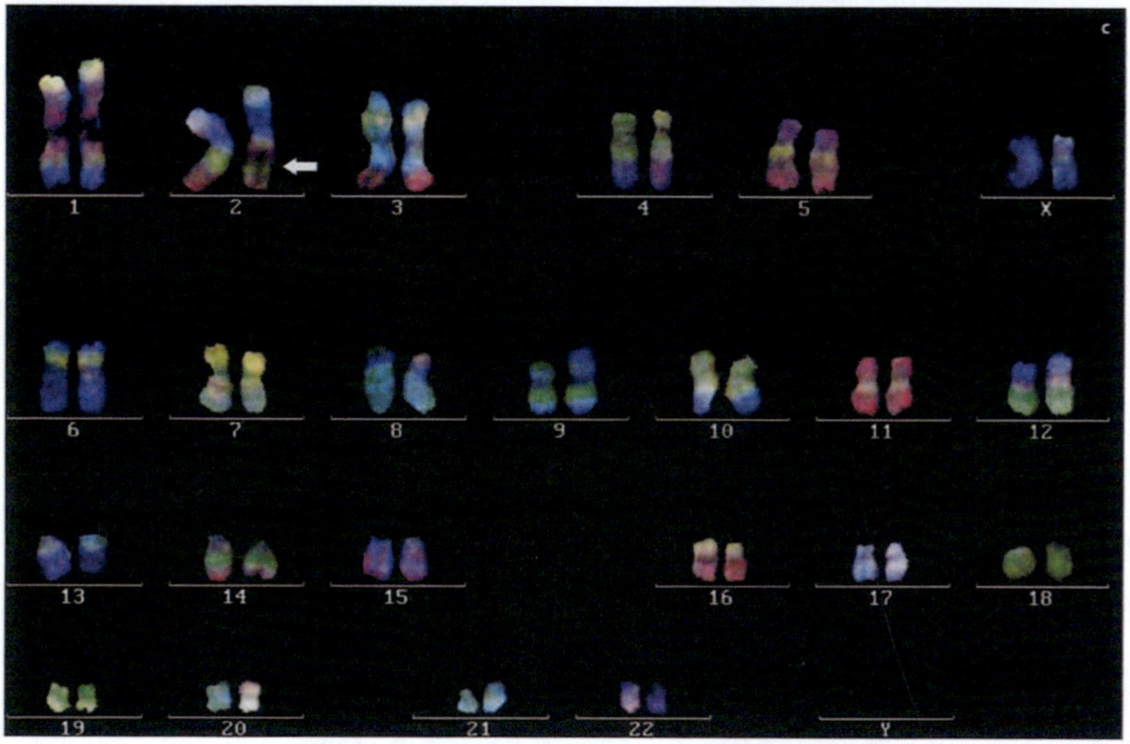

Fig. 1c.

- If background is a problem on the slides, apply an RNase digestion also in combination with other proteolytic pretreatment procedures.
- A high background signal can also be caused by large amounts of DNA probes in the hybridization mixture; high stringency posthybridization washes at higher temperatures or lower salt concentrations may also help.
- If probe specificity is poor and cross-hybridizations on other chromosomes occur, an increasing amount of COT1 DNA in the hybridization mixture may improve your results.
- For each labeling batch of DNA probes, check different probe and COT1 concentrations in the hybridization mix. This is most important, when using new DNA probes which have never been checked before.

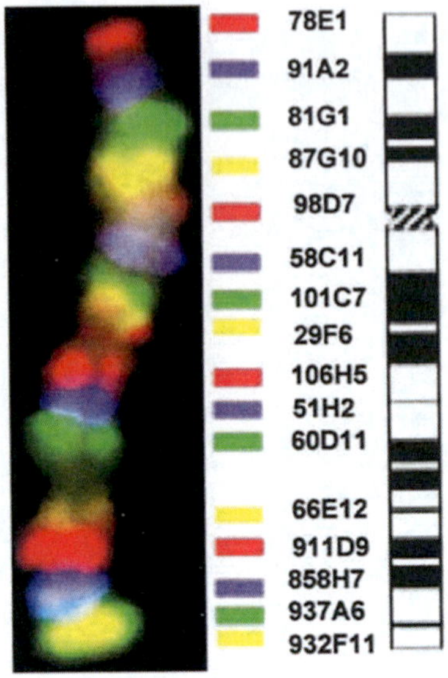

Fig. 2. Hybridization of a chromosomal rainbow set comprised of 16 locus-specific probes that map along chromosome 10. Normal copy of chromosome 10 in metaphase spreads prepared from short-term lymphocyte cultures of a healthy donor shows all 16 signals in the expected order

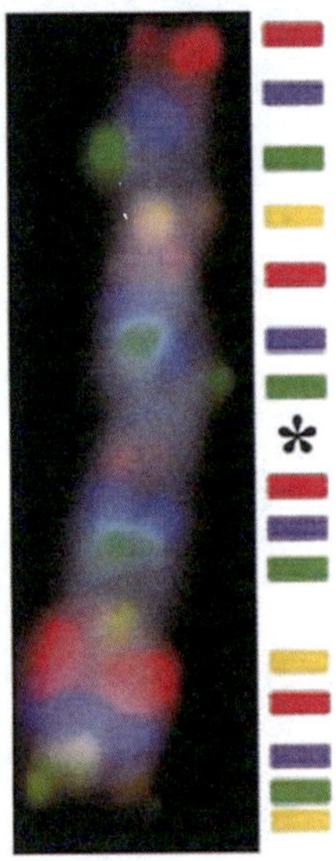

Fig. 3. The non-rearranged homologue of chromosome 10 in TPC-1 demonstrates deletion of locus D10S170 (probe 29F6, yellow, indicated by the *asterisk*)

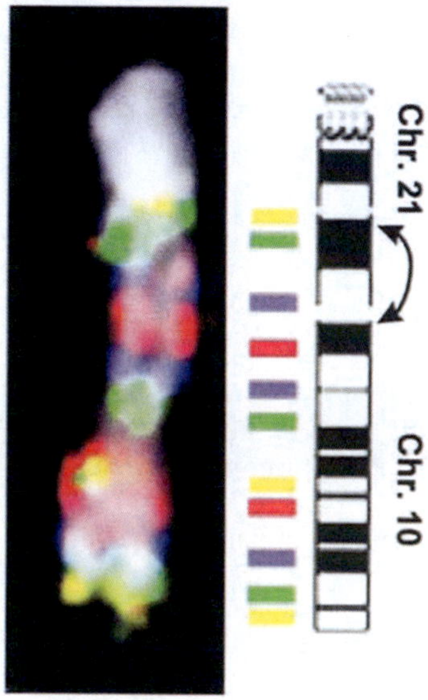

Fig. 4. The der(21) in TPC-1 carries material from the long arm of chromosome 10 from ret (probe 58C11, purple) to qter. The characteristic ret-H4/D10S170 inversion is indicated by the *arrows* (C)

- If the hybridization signal is weak in CR FISH, put slide under light protected conditions in the refrigerator for one day after detection and check it again under the microscope.
- Random priming reactions reveal better incorporation for CR FISH applications of labeled nucleotides than nick translation; if you decide to nick translate your probes, check the fragment sizes after nick translation by agarose gel electrophoresis. If an optimum range of fragment distribution between 300-1000 bp is not present, digest additionally with DNase I (time and concentration of incubation must be figured out experimentally). An improved incorporation of labeled nucleotides can also be obtained by using two different labeled nucleotides (e.g., biotinylated dUTP and dCTP) in the reaction.

Acknowledgements. This work was supported in part by the German Bundesamt für Wehrtechnik und Beschaffung, contract number E/B41G/V0132/T5922. Further support was provided by the US Department of Energy under contract DE-AC-03-76SF00098 and grants from the National Institute of Health (1R21 CA88258, 1RO1 CA80792).

References

Müller, S., O'Brien, P.M.C., Ferguson-Smith, M.A., Wienberg, J. A novel source of highly specific chromosome painting probes for human karyotype analysis derived from primate homologues. Hum. Genet., 101:149-153, 1997.

Müller, S., Rocchi, M., Ferguson-Smith, M.A., Wienberg, J. Towards a multicolor chromosome bar code for the entire human karyotype by fluorescence in situ hybridization. Hum. Genet., 100:271-278, 1997.

Teixeira, M.R., Micci, F., Dietrich, C.U., Heim, S. Cross-species color banding characterization of chromosomal rearrangements in leukemias with incomplete G-band karyotypes. Genes Chrom. Cancer, 26:13-19, 1999.

Micci, F., Teixeira, M.R., Dietrich, C.U., Saeter, G., Bjerkehagen, B., Heim, S. Combined RxFISH/G-banding allows refined karyotyping of solid tumors. Hum. Genet., 104:370-375, 1999.

Bauchinger, M. Retrospective dose reconstruction of human radiation exposure by FISH/chromosome painting. Mutat. Res., 404: 89-96, 1998.

Bauchinger, M., Schmid, E. LET dependence of yield ratios of radiation-induced intra- and interchromosomal aberrations in human lymphocytes. Int. J. Rad. Biol., 74:17-25, 1998.

Knehr, S, Zitzelsberger, H., Bauchinger, M. FISH-based analysis of radiation-induced chromosomal aberrations using different nomenclature systems. Int. J. Rad. Biol., 73:135-141, 1998.

Jossart, G.H., O'Brien, B., Cheng, J.F., Tong, Q., Jhiang, S.M., Duh, Q., Clark, O.H., Weier, H.U. A novel multicolor hybridization scheme applied to localisation of a transcribed sequence (D10S170/H4) and deletion mapping in the thyroid cancer cell line TPC-1. Cytogenet. Cell Genet., 75:254-257, 1996.

Weier H-UG, George CX, Greulich KM, Samuel CE: The interferon-inducible, double-stranded RNA-specific adenosine deaminase gene (DSRAD) maps to human chromosome 1q21.1-21.2. Genomics 30:372-375 (1995a).

Weier H-UG, Rhein AP, Shadravan F, Collins C, Polikoff D: Rapid physical mapping of the human trk proto-oncogene (NTRK1) to human chromosome 1q21-22 by P1 clone selection, fluorescence in situ hybridization (FISH) and computer-assisted microscopy. Genomics 26:390-393 (1995b).

Pinkel D, Straume T, Gray J: Cytogenetic analysis using quantitative, high sensitivity fluorescence in situ hybridization. Proc Natl Acad Sci USA 83:2934-2938 (1986)

Jossart GH, Greulich KM, Siperstein AE, Duh Q, Clark OH, Weier H-UG: Molecular and cytogenetic characterization of a t(1;10;21) translocation in the human papillary thyroid cancer cell line TPC-1 expressing the ret/H4 chimeric transcript. Surgery 118:1018-1023 (1995).

Jossart GH, O'Brien B, Cheng J-F, Tong Q, Jhiang SM, Duh Q, Clark OH, Weier H-UG: A novel multicolor hybridization scheme applied to localisation of a transcribed sequence (D10S170/H4) and deletion mapping in the thyroid cancer cell line TPC-1. Cytogenet Cell Genet 75:254-257 (1996).

Centromere-Specific Multicolor-FISH (cenM-FISH)

ANGELA NIETZEL, ANITA HELLER, HEIKE STARKE, UWE CLAUSEN, and THOMAS LIEHR

Introduction

24-color FISH using five different fluorochromes has recently been described as M-FISH or SKY, and is a useful technique for the characterization of complex chromosomal aberrations (Speicher et al., 1996; Schröck et al., 1996; for an overview of multicolor-FISH literature see http://mti-n.mti.uni-jena.de/~huwww/Mol_ZYTO/mFISHlit.htm). For technical reasons whole or partial chromosome libraries, each specific for an entire chromosome, exclude the centromeric regions of the color staining due to the blocking of repetitive sequences by COT1-DNA. Here, we describe a recently developed new multicolor FISH technique – called centromere-specific multicolor-FISH (cenM-FISH) (Nietzel et al. 2001). The technique is based on all available centromere-specific DNA probes, which are labeled in five different fluorochromes (i.e. TexasRed, SpectrumGreen, SpectrumOrange, Cy5 and DEAC) according to Fig. 1B and hybridized simultaneously. This approach allows the identification of all human centromeres by their individual coloring in one single step and therefore is a powerful tool in molecular cytogenetics. Centromere-specific multicolor FISH is a very helpful technique especially for the characterization of small

Angela Nietzel, Institute of Human Genetics and Anthropology, Kollegiengasse 10, Jena, 07743, Germany
Anita Heller, Institute of Human Genetics and Anthropology, Kollegiengasse 10, Jena, 07743, Germany
Heike Starke, Institute of Human Genetics and Anthropology, Kollegiengasse 10, Jena, 07743, Germany
Uwe Clausen, Institute of Human Genetics and Anthropology, Kollegiengasse 10, Jena, 07743, Germany
✉ Thomas Liehr, Institute of Human Genetics and Anthropology, Kollegiengasse 10, Jena, 07743, Germany (*phone* +49-3641-935533; *fax* +49-3641-935502; *e-mail* i8lith@mti-n.mti.uni-jena.de)

marker chromosomes with no – or nearly no – euchromatin and small amounts of available chromosome suspension (see Fig. 1A).

Outline

1. FISH is performed on metaphase spreads from a lymphocyte suspension, amniocytic suspension or bone marrow; slides are treated with RNase and pepsin prior to denaturation.
2. Denaturation of the slides is performed for about 2-3 min at 75°C. The probes are denatured for 5 min and hybridized on the slides for 1 night at 37°C in a humid chamber. After a postwashing series, detection of the biotinylated probe is performed with a Cy5-avidin system. After counterstaining with DAPI and addition of antifade solution, slides can be evaluated on a six-filter fluorescence microscope equipped with a CCD-camera and a computer with image-analyzing system.

Procedure

Slide pretreatment

1. Slides with the chromosome suspension are incubated in 2×SSC for 5 min at RT (in a 100-ml Coplin jar on a shaker).
2. Remove slides from the Coplin jar, add 100 μl of RNase solution/slide and cover with a 24×50 mm coverslip (RNase solution: per slide 100 μl 2×SSC plus 1 μl of RNase stock (=5 μg/μl)).
3. Incubate the slides in a humid chamber for 15 min at 37°C (see comment 1).
4. Put slides back into the Coplin jar with 100 ml 2×SSC (RT) and remove the coverslips with forceps. Leave slides in 2×SSC solution for 3 min with gentle agitation.
5. Discard 2×SSC and replace it with 100 ml 1×PBS (RT) for 5 min (shaker).
6. Replace 1×PBS with 100 ml prewarmed pepsin-buffer (37°C) and incubate the slides for 10 min at 37°C, without agitation (see comment 1). [Pepsin-buffer: Add 1 ml of 1 M HCl to 99 ml of distilled water and incubate at 37°C for about 20 min; then add 50 μl of the pepsin stock

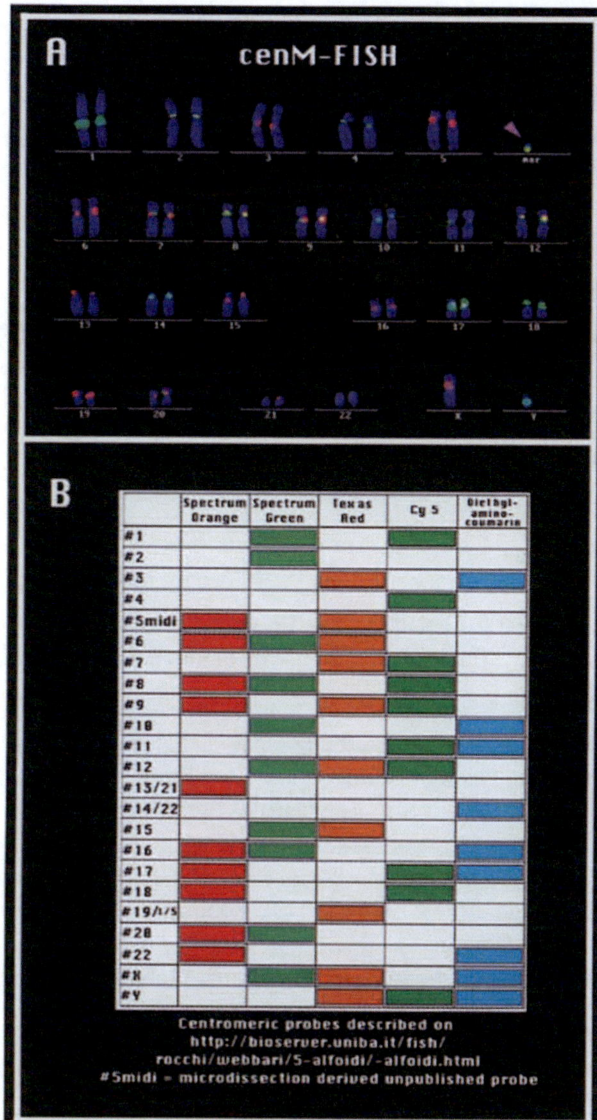

Fig. 1. cenM-FISH result of a prenatal case: the supernumerary marker chromosome (*arrowhead*) was determined as derived from chromosome 8. No specific signals could be observed using 24-color-FISH with whole chromosome painting probes (result not shown). Image was captured on a Zeiss Axioplan microscope using the isis digital FISH imaging system (MetaSystems, Altlussheim, Germany) using an XC77 CCD camera with on-chip integration (Sony); karyogram was created using the ikaros karyotyping system (MetaSystems, Altlussheim, Germany). **B** Labeling scheme of the centromere-specific probes used for cenM-FISH

solution 10% (w/v) (Serva 31855) and leave the Coplin jar at 37°C; make fresh as required].

7. Replace fluid with 100 ml $1\times PBS/MgCl_2$, incubate at RT for 5 min with gentle agitation. $MgCl_2$ will block the enzymatic activity of pepsin ($1\times PBS/MgCl_2$ = 5% (v/v) 1 M $MgCl_2$ in $1\times PBS$).

8. Postfix nuclei on the slide surfaces by replacing $1\times PBS/MgCl_2$ with 100 ml of formalin-buffer (3 ml acid free formaldehyde [37%; Roth 4979.1] in 100 ml 1xPBS) for 10 min (RT, with gentle agitation).

9. Formalin-buffer is replaced by 100 ml $1\times PBS$ for 2 min (RT, with gentle agitation).

10. Finally, slides are dehydrated by an ethanol series (70%, 90%, 100%, 3 min each) and air dried (see comment 2).

Preparation of probes

1. Centromere-specific probes can be obtained (for research use only) from Prof. Dr. Mariano Rocchi (internet: http://www.biologia.uniba.it/rmc).

2. Plasmid-DNA with centromere-specific DNA as inserts are labeled by nick translation using commercially available kits according to the manufacturers instructions (e.g. biotin labeling: Biotin-Nick Translation Kit, Boehringer, 1745824; labeling with DEAC-dUTP (NEN-Dupont; NEL455), Chromatide TexasRed-dUTP (Mol. Probes; C-7631), SpectrumOrange-dUTP (Vysis; 30-803000), SpectrumGreen-dUTP (Vysis; 30-803400) can be done by Nick-Translation Kit, Boehringer, 976776). Labeling is done according to the labeling scheme in Fig. 1B.

3. 1 μg of each probe can be labeled with one nick translation reaction; each labeling should be tested separately for its efficiency by hybridizing 10 ng of the probe on normal metaphase spreads.

4. All labeled probes are mixed and 50 aliquots are placed in 1.5-ml microtubes. Each aliquot is precipitated together with 10 μl of 1 μg/μl COT1-DNA (Gibco). The pellets are vacuum dried and stored until use at -20°C.

5. Prior to use the pellet is dissolved in 20 μl of the hybridization-buffer, which is sufficient for 1 slide covered with a 24×50 mm coverslip. (Hybridization buffer: Dissolve 2 g dextran sulfate in 10 ml 50% deionized

formamide/2×SSC/50 mM phosphate buffer for 3 h at 70°C. Aliquot and store at -20°C).

Fluorescence in situ hybridization (FISH)

1. Add 100 µl denaturation-buffer to the slides and cover with 24×50 mm coverslips. [Denaturation-buffer: 70% (v/v) deionized formamide, 10% (v/v) filtered double distilled water, 10% (v/v) 20×SSC, 10% (v/v) phosphate buffer; make fresh as required; phosphate buffer: prepare 0.5 M Na_2HPO_4 and 0.5 M NaH_2PO_4, mix these two solutions (1:1) to get pH 7.0, then aliquot and store at -20°C].
2. Incubate slides on a warming plate for 2-3 min at 75°C.
3. Remove the coverslips immediately with forceps and place slides in a Coplin jar filled with 70% ethanol (4°C) to conserve target DNA as single strands.
4. Dehydrate slides in ethanol (70%, 90%, 100%, 4°C, 3 min each) and air dry.
5. For each slide to be hybridized, use 20 µl of the labeled chromosome-specific satellite probe mix described before.
6. Denature probe-solution at 75°C for 5 min and prehybridize at 37°C for 5 min.
7. Add 20 µl of probe-solution onto each denatured slide, put 24×50 mm coverslips on the drops and seal with rubber cement (Fixogum; Marabu).
8. Incubate slides overnight at 37°C in a humid chamber (see comment 3).
9. Take the slides out of the 37°C chamber, remove rubber cement with forceps and the coverslips by letting them swim off in 4×SSC/ 0.2%Tween (RT, 100 ml Coplin jar) (see comment 3).
10. Postwash the slides 3×5 min in formamide-solution (45°C) followed by 3× 5 min in 2×SSC (37°C) in a 100 ml Coplin jar, with gentle agitation.
11. Put the slides in 4×SSC/0.2% Tween (100 ml, RT), for a few seconds.
12. Add 50 µl of solution I to each slide, cover with 24×50 mm coverslips and incubate at 37°C for 30 min in a humid chamber. [Solution I: Cy5-Amersham, PA45000)/4×SSC + 0.2% Tween + 5% BSA (1:50); make fresh as required].

13. Remove the coverslips and wash 3 x 3 min in 4×SSC/0.2% Tween (RT, with gentle agitation).
14. Add 50 µl of solution II to each slide, cover with 24×50 mm coverslips and incubate at 37°C for 75 min in a humid chamber. [Solution II: Biotinylated antiavidin (CAMON Vector Laboratories BA0300)/ 4×SSC + 0.2% Tween + 5% BSA (1:100); make fresh as required].
15. Repeat step 13.
16. Repeat steps 12 and 13.
17. Counterstain the slides with DAPI solution (100 ml in a Coplin jar, RT) for 8 min. [DAPI-solution: Dissolve 1 µl of DAPI (4,6-diamidino-2-phenylindol.2HCl stock-solution; Serva 18860) in 100 ml 4×SSC/ 0.2% Tween; make fresh as required].
18. Wash slides several times in water for a few seconds and air dry.
19. Add 15 µl of antifade Vectashield (CAMON Vector Laboratories H1000), cover with coverslips and look at the results under a fluorescence microscope equipped with a CCD-camera and an image analyzing system; the latter is necessary, as Cy5 is a dye invisible to the human eye. Moreover, false colors have to be created by the image analyzing system for a reliable evaluation.

Comments

1. RNase and pepsin pretreatment conditions should be tested in each laboratory on a single slide first. Both RNase and pepsin concentrations can be too stringent, resulting in clean slides without any remaining metaphase spreads or nuclei.
2. The pretreated slides can be hybridized immediately or stored at RT for up to 3 weeks. If longer storage is necessary slides are stable at -20°C for several months.
3. During the washing steps it is important to prevent the slide surfaces drying out, otherwise background problems may arise.

Acknowledgements. This work was supported by the Herbert Quandt Stiftung der VARTA AG and the Madeleine Schickedanz-Kinderkrebs-Stiftung, the Wilhelm Sander-Stiftung (99.105.1) and the EU (ICA2-CT-2000-10012 and QLRT-1999-31590). The continuos support of the Carl Zeiss GmbH (Jena, Germany) is gratefully acknowledged.

References

Liehr T, Thoma K, Kammler K, Gehring C, Ekici A, Bathke KD, Grehl H, Rautenstrauss B (1995) Direct preparation of uncultured EDTA-treated or heparinized blood for interphase FISH analysis. Appl Cytogenet, 21:185-188

Nietzel A, Rocchi M, Heller A, Starke H, Wlodarska I, Beensen V, Claussen U, Liehr 7 (2001) A new multicolor-FISH approach for the characterization of marker chromosomes: Centromere-specific multicolor-FISH (cenM-FISH). Hum Genet, 108: 199-204

Schröck E, du Manoir S, Veldman T, Schoell B, Wienberg J, Fergueson-Smith M, Ning Y, Ledbetter D, Bar-Am I, Soenksen D, Garini Y, Ried T (1996) Multicolor spectral karyotyping of human chromosomes. Science 273:494-497

Speicher MR, BAMLard SG, Ward DS (1996) Karyotyping human chromosomes by combinatorial multi-fluor FISH. Nature Genet. 12:368-375.

Simultaneous Multicolor-FISH and Immunocytochemical Analysis of Fresh Tumor Material

HEIKE STEILEN-GIMBEL and STEFFI URBSCHAT

Introduction

A variety of immuncytochemical (IC) and in situ hybridization (ISH) methods give insight into the presence or absence of specific proteins and nucleic acid sequences, or establish kinetic parameters of cells. The combination of these techniques in the same biological specimen has proven advantageous in, e.g., simultaneously demonstrating the presence of mRNA and its protein product in a given cell (Dirks et al. 1993; Trembleau et al. 1993), immunophenotyping cells containing a specific chromosomal aberration (Gendelman et al. 1985; Porter et al. 1990; Van den Berg et al. 1991; Weber-Matthiesen et al. 1993), or characterization of cytokinetic parameters of tumor cell populations that are genetically or phenotypically aberrant (Balazs et al. 1991; Van Dekken et al. 1991).

For a successful combination of IC and non-radioactive ISH, several factors must be taken into account. Preservation of cell morphology and protein epitopes, accessibility of nucleic acid targets, cross-reaction between the different detection procedures, and stability of enzyme cytochemical precipitates and fluorochromes all determine the success rate and sensitivity of the procedure. In a combined approach, it is necessary to perform immunohistochemistry prior to ISH because several steps in the ISH technique, e.g. enzymatic digestion, postfixation, denaturation at high temperatures, and hybridization in formamide, can destroy antigenic determinants.

Heike Steilen-Gimbel, Saarland University, Department of Neurosurgery, Institute of Human Genetics, Bldg. 68, Homburg / Saar, 66421, Germany (*phone* +49-6841-1626605; *fax* +49-6841-1626600; *e-mail* nchstg@med-rz.uni-sb.de)
✉ Steffi Urbschat, Saarland University, Institute of Human Genetics, Bldg. 60, Homburg / Saar, 66421, Germany (*phone* +49-6841-1626635; *fax* +49-6841-1626600; *e-mail* hgsmur@uniklinik-saarland.de)

Several such combined IC/ISH methods have been described with either enzyme precipitation reactions (Porter et al. 1990; Losada et al. 1991), fluorochromes (Van den Berg et al. 1991; Weber-Matthiesen et al. 1993; Steilen et al. 1994, Ermis et al. 1995), or a combination of both (Strehl and Ambros 1993). The methods can be subdivided into two groups, those which use fluorochromes for IC, and those which use enzyme reactions. For the detection with fluorochromes FITC, TRITC or AMCA are mainly used. Enzyme precipitation reactions have also been used efficiently for combined IC/ISH staining of proteins in cell preparations and tissue sections. Frequently used enzyme precipitation products that withstand the proteolytic digestion and denaturation steps in the ISH procedure are:

- Stain precipitates (Fast Red, New Fuchsin and BCIP/NBT) formed by alkaline phosphatase
- Diaminobenzidine precipitate formed by horseradish peroxidase

Non-random chromosomal abnormalities have been observed in many types of solid human tumors (Sandberg, 1990). Most information on the genomic changes of solid tumors is acquired by karyotype analysis of cell cultures, but in recent years FISH procedure has been used for targeted analyses of numerical chromosome abnormalities within interphase nuclei of malignancies. Since tumor samples usually consist of both non-tumorous and tumor cells and may therefore contain resident cells with a normal chromosome complement, the interpretation of FISH results can be difficult. Moreover, tumor cells may lack chromosome alterations at the regions detected by hybridization and, for this reason, may be indistinguishable from normal cells. Therefore, techniques that enable characterization of the tumor tissue e.g. by determining cell-surface antigens or proliferation markers, in combination with interphase cytogenetics offer several advantages. Preselection of cells that are positive or negative for a certain antigen reduces the number of cells that need to be examined and avoids the potential for misinterpretation due to the analysis of normal cells. To combine the detection of antigen staining and visualization of DNA probes in the same cell makes this procedure highly suitable for detection of tumor cell heterogeneity, detection of rare phenotypes, and in cases where only a few cells can be obtained for analysis (e.g. in needle biopsy material).

We conclude that the combination of immunocytochemistry and ISH (FISH) is a valuable method of examining the correlation between genotype and phenotype characteristics.

Outline

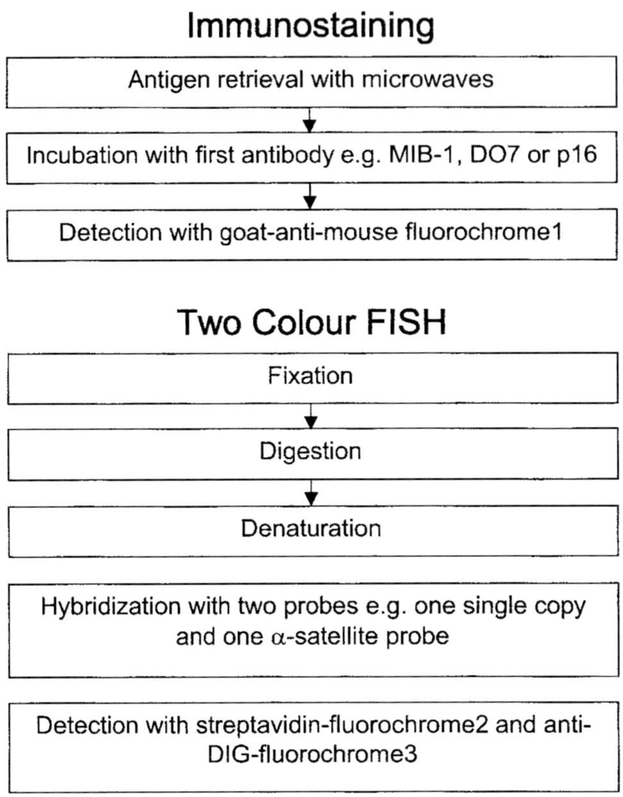

Materials

- Citrate buffer (0.01M citric acid-monohydrate, pH 6.0)
- 4% paraformaldehyde solution (diluted in PBS), freshly prepared
- Deionized formamide
- Hybridization mixture:
  - 50% deionized formamide
  - 2×SSC
  - 50 mM sodium phosphate buffer pH 7.0
  - 10% dextran sulfate
  - Store at -20°C

- Denaturation mixture:
  - 70% deionized formamide
  - 2×SSC
  - 50 mM sodium phosphate buffer pH 7.0
  - Store at -20°C
- 65% formamide/2×SSC, pH 7.0
- 50% formamdide/2×SSC, pH 7.0

Procedure

Immunostaining on smear preparations or cryosections

1. Wash the slides in PBS, pH 7.3 for 10 min.
2. Place the slides in a microwave-proof plastic tube containing citrate buffer solution and treat in a microwave oven for 5 min at 500 W.
3. Refill with citrate buffer to prevent the probes from drying.
4. Incubate the probes 1 min in the hot buffer.
5. Repeat the microwave treatment under the same conditions.
6. Keep the slides for 20 min in the hot citrate buffer.
7. Rinse the slides for a few seconds with distilled water and then incubate them in PBS/0.01%Triton X-100 for 10 min.
8. Apply 120 μl PBS/0.1% BSA to each slide (background blocking) and incubate 10 min at RT (room temperature) in a moist chamber.
9. Apply the first antibody diluted 1:50-1:100 in PBS/0.1%BSA to the slides e.g. MIB-1 (mouse antibody against Ki-67 antigen; DAKO) or Do-7 (mouse antibody against p53, DAKO) or mouse anti p16 (Santa Cruz Biotechnology).
10. Incubate for 1 h at 37°C in a moist chamber (p16 over night at 4°C).
11. Wash briefly 3×5 min with PBS/0.01%Triton X-100.
12. Apply 120 μl of the second antibody e.g. goat anti mouse fluorochromel (e.g. AMCA or FITC) diluted 1:50-1:100 in PBS/0.1%BSA to each slide.
13. Incubate for 1 h at 37°C in a moist chamber.

14. Wash 3×5 min with PBS/0.01%Triton X-100.
15. Fix the slides for 3-5 min in freshly prepared 4% paraformaldehyde solution (diluted in PBS) on ice.
16. Wash the slides with PBS/0.01%Triton X-100 for 10 min.

Dehydrate the slides in an ethanol series and air dry for the subsequent FISH procedure.

Multicolor-FISH with α-satellite and/or single copy probes

Probe preparation and hybridization with α-satellite probes:

1. Prepare 30 µl of the hybridization mixture containing 65% deionized formamide, 2 µl 20×SSC, 3 µl biotinylated DNA probe 1 (20-40 ng/µl), and 3 µl digoxigenin-labeled DNA probe 2 (20-40 ng/µl).
2. Apply the probe solution onto the immunostained region of the air-dried slides, cover with a 24×60 mm coverslip, and seal with rubber cement.
3. Denaturate probe and target DNA simultaneously on a heating plate for 12 min at 80°C.
4. Incubate the slides at 37°C overnight in a moist chamber.
5. Perform stringency washes in e.g. 65% formamide/2×SSC, pH 7.0 at 37°C for 3×5 min.
6. Incubate the slides in 4×SSC/0.5% blocking reagent (Roche Molecular Biochemicals) to adjust to the following signal detection.

Probe preparation and hybridization with single copy plus α-satellite probes:

Pretreatment of the slides

1. Wash the immunostained slides for 5 min with 2×SSC.
2. Place the slides for 10 min in a pepsin solution (7 mAnson U/100 ml 0.01 M HCl) at 37°C.
3. Wash 2×5 min in 2×SSC at RT.
4. Wash 1×5 min in 1×PBS at RT.

5. Place the slides for 10 min in a 1% acid-free formaldehyde/1×PBS/50 mM $MgCl_2$ solution at RT.
6. Rinse the slides 1 min in 1×PBS at RT.
7. Incubate the slides 10 min in 1×PBS at RT.
8. Dehydrate the slides through an ethanol series and air dry.
9. Apply 120 µl of denaturation mixture on each slide.
10. Cover the solution with a 24×60 mm coverslip.
11. Denature 2 min at 80°C on a heating plate.
12. Remove coverslips and place the slides immediately in 70% ethanol at -20°C for 5 min, then dehydrate in 80% and 95% ethanol at -20°C each for 5 min and air dry.
13. Pre-warm the slides on a heating plate at 37°C.

Probe preparation and denaturation

1. Take the required amount of single copy probe DNA (e.g. 150 ng).
2. Add 50× human Cot1 DNA.
3. Add 0.1× volume 3 M sodium acetate pH 5.5.
4. Add 2.5× volume 100% ethanol.
5. Mix, spin down and place it for 2 h at -20°C.
6. Spin down (30 min, 4°C, 14,000 rpm).
7. Remove supernatant and dry the pellet for 5 min in a speed-vac.
8. Resuspend the pellet in 20 µl hybridization mixture and dissolve for 1 h by gentle mixing.
9. Denature the DNA for 5 min at 80°C.
10. Place the DNA immediately on ice for 1 min and spin down.
11. Pre-anneal the DNA for 20 min at 37°C.

At the same time prepare the α-satellite DNA probe

1. Denature 10 µl of the a-DNA probe dissolved in hybridization mixture (6 ng/µl) for 10 min at 80°C, place it on ice for 1 min and warm it again at 37°C.
2. Add the α-satellite DNA probe to the pre-annealed single copy probe, mix well and spin down for a few seconds.

3. Apply the probe to the slide (target area) and cover with a 24×60 mm coverslip.
4. Seal with rubber cement.
5. Hybridize at 37°C in a moist chamber for 48 h.

If you use commercially available single copy DNA probes (e.g. ONCOR): First prepare the α-satellite DNA probe.

1. Denature 10 μl of the a-DNA probe dissolved in hybridization mixture (6 ng/μl) for 10 min at 80°C, place it on ice for 1 min and warm it again at 37°C.
2. Add the single copy probe to the α-satellite DNA probe, mix well and spin down for a few seconds.
3. Incubate the probes for 5 min at 37°C.
4. Apply the probe to the slide (target area) and cover with a 24×60 mm coverslip.
5. Seal with rubber cement.
6. Hybridize at 37°C in a moist chamber for 48 h.

Post hybridization washes

1. Perform stringency washes in e.g. 50% formamide/2×SSC, pH 7.0 and in e.g. 2×SSC at 37°C each for 3×5 min (formamide concentration in the stringency washes depends on the probes).
2. Incubate the slides in 4×SSC/0.05% Tween-20 to adjust to the following signal detection.

Signal detection

1. Apply 120 μl 4×SSC/0.5% blocking reagent to each slide to reduce background.
2. Incubate the slides for 30 min at 37°C in a moist chamber.
3. Apply 120 μl streptavidin-fluorochrome 2 (e.g. FITC or AMCA) diluted 1:100 in 4×SSC/0.5% blocking reagent to each slide.
4. Incubate for 1 h at 37°C in a moist chamber.
5. Wash briefly 3x5 min with 4×SSC/0.05% Tween-20.
6. Apply 120 μl of anti-streptavidin-biotin diluted 1:100 in 4×SSC/0.5% blocking reagent to each slide.

7. Incubate for 1 h at 37°C in a moist chamber.
8. Wash 3×5 min with 4×SSC/0.05% Tween-20.
9. Apply 120 µl streptavidin-fluorochrome 2 (e.g. FITC or AMCA) diluted 1:100 and anti-digoxigenin-fluorochrome 3 (e.g. rhodamine) diluted 1:25 in 4×SSC/0.5% blocking reagent to each slide.
10. Incubate for 1 h at 37°C in a moist chamber.
11. Wash briefly 3×5 min with 4×SSC/0.05% Tween-20.
12. Dehydrate the slides in an ethanol series and air dry.
13. Apply 25 µl embedding medium (e.g. Vectashield, Vector) and cover with a 24×60 coverslip.

Troubleshooting

Immunostaining

If no or weak signals are observed, then:

- Raise the concentration of the primary antibody.
- Raise the incubation time of the primary antibody (up to 16 h at 37°C, RT or 4°C).
- Modify the incubation time of the postfixation step after immunostaining.

Multicolor-FISH

If no signals are observed, then:

- Amplify the signals in an additional step.
- Check the labeling quality of the assigned in situ probes (dot blot).
- Dot blot protocol:
  - 1. Spot serial dilution e.g. 6 ng, 0.6 ng, 60 pg, 6 pg, 0.6 pg, and 0.06 pg of the labeled probe on a nitrocellulose or nylon membrane.
  - 2. Fix the probes to the membrane by baking for 60 min at 80°C on a heating plate or by cross-linking with UV light.
  - 3. Incubate the membrane in 6×SSC/1%BSA for 30 min.

- 4. Incubate the membrane in buffer 1 (0.05 M Tris, 0.075 M NaCl, pH 7.5) for 10 min
- 5. Dilute Streptavidin-alkaline phosphatase (2.25 U) or Anti-DIG-alkaline phosphatase (2.25 U) in 3 ml buffer 1.
- 6. Incubate the membrane in diluted antibody solution for 30 min.
- 7. Wash 2×5 min in buffer 1.
- 8. Wash 1×5 min in buffer 2 (0.05 M Tris, 0.05 M NaCl, 0.025 M $MgCl_2$, pH 9.5).
- 9. Mix 33 µl NBT (25 mg/ml, Roche Molecular Biochemicals) and 25 µl BCIP (50 mg/ml, Roche Molecular Biochemicals) with 3 ml buffer 2 (prepare this solution freshly!).
- 10. Incubate membrane in stain solution for 30-60 min.
- 11. Fix membrane in 70% ethanol.

- Check the fragment length of the labeled probe in agarose gel.
- Modify the denaturation step of the immunostained slides by varying temperature and denaturation time of the procedure.

If you have a high background fluorescence, then:

- Spin down the diluted fluorochrome-conjugated antibody (2 min, 14.000 rpm) and take only the supernatant (this avoids background induced by agglutinated fluorochromes).
- Reduce the concentration of primary and fluorochrome-conjugated antibodies.
- Make the wash steps between signal detection longer and at a higher temperature, e.g., 3×10 min at 37°C.
- Store the mounted slides in a dark moist chamber at 4°C for a few days and observe under the microscope again.

Acknowledgements. We thank K.-D. Zang and W. Henn for helpful discussions, and B. Mayer and B. Rohrschneider for technical assistance.

References

Balazs M, Mayall BH, Waldman FM (1991) Simultaneous analysis of chromosomal aneusomy and 5-bromodeoxyuridine incorporation in MCF-7 breast tumor cell line. Cancer Genet Cytogenet 57:93-102

Dirks RW, Van de Rijke FM, Fujishita S, Van der Ploeg M, Raap AK (1993) Methodologies for specific intron and exon RNA localization in cultured cells by haptenized and fluorochromized probes. J Cell Sci 104:1187-97

Ermis A, Henn W, Remberger K, Hopf C, Hopf T, Zang K.D. (1995) Proliferation enhancement by spontaneous multiplication of chromosome 7 in rheumatic synovial cells in vitro. Hum Genet 96:651-654

Gendelman HE, Moench TR, Narayan O, Griffin DE, Clements JE (1985) A double labeling technique for performing immunocytochemistry and in situ hybridization in virus infected cell cultures and tissue. J Virol Methods 11:119-30

Losada AP, Wessman M, Tiainen M, Hopman AHN, Willard HF, Sole F, Rosa Caballin M, Woessner S, Knuutila S (1991) Trisomy 12 in chronic lymphocytic leukemia: an interphase cytogenetic study. Blood 78:775-9

Porter HJ, Heryet A, Quantrill AM, Fleming KA (1990) Combined non-isotopic in situ hybridization and immunohistochemistry on routine paraffin wax embedded tissue: identification of cell type infected by human parovirus and demonstration of cytomegalovirus DNA and antigen in renal infection. J Clin Pathol 43:129-32

Sandberg AA (1990) The chromosomes in human cancers and leukemia 2nd ed., Elsevier, New York

Steilen H, Burk U, Zwergel T, Wullich B, Unteregger G (1994) Simultaneous characterization of in vitro growing cells by fluorescence in situ hybridization (FISH) and Ki-67 labeling. Analytical Cellular Pathol 6:83-87

Strehl S, Ambros PF (1993) Fluorescence in situ hybridization combined with immunohistochemistry for highly sensitive detection of chromosome 1 aberrations in neuroblastoma. Cytogenet Cell Genet 63:24-8

Trembleau A, Roche D, Calas A (1993) Combination of non-radioactive and radioactive in situ hybridization with immunohistochemistry: a new method allowing the simultaneous detection of two mRNAs and one antigen in the same brain tissue section. J Histochem Cytochem 41:489-98

Van Dekken H, Schervish EW, Pizzolo JG, Fair WR, Melamed MR (1991) Simultaneous detection of fluorescent in situ hybridization and in vivo incorporated BrdU in a human bladder tumour. J Pathol 164:17-22

Van den Berg H, Vossen JM, Van den Bergh RL, Bayer J, Van Tol MJD (1991) Detection of Y chromosome by in situ hybridization in combination with membrane antigens by two-color immunofluorescence. Lab Invest 64:623-8

Weber-Matthiesen K, Deerberg J, Müller-Hermelink A, Schlegelberger B, Grote W. (1993) Rapid immunophenotypic characterization of chromosomally aberrant cells by the new fiction method. Cytogenet Cell Genet 63:123-5

Simultaneous Fluorescence Immunophenotyping and FISH (FICTION)

BRIGITTE SCHLEGELBERGER, STEFAN GESK, YANMING ZHANG, and REINER SIEBERT

Introduction

In this chapter, the FICTION (**F**luorescence **I**mmunophenotyping and Interphase **C**ytogenetics as a **T**ool for **I**nvestigation **o**f **N**eoplasms) technique will be detailed. This method has been described in 1992 by Weber-Matthiesen et al. as the first technique to combine immunophenotyping and FISH. In contrast to previous methods like the morphology-antibody-chromosome (MAC) technique (Teerenhovi et al. 1984; Schlegelberger et al. 1992), the FICTION technique can be applied on interphase cells, which is of great advantage especially for tumor cytogenetics. To briefly introduce the technique: In general, conventional unstained peripheral blood or bone marrow smears, imprints and cytospin slides from fresh tissue and cryostat sections are applicable for FICTION. In order to preserve cellular antigens, preparing high quality slides is one of the critical steps for successfully performing FICTION. Freshly prepared slides should be air dried overnight at room temperature. Thereafter, slides can be processed immediately or stored airtight at -80°C for years. For the FICTION procedure, slides are fixed in acetone and then immunophenotyped by a direct or indirect fluorescence detection system. Immunophenotyping is followed by a fixation step to preserve the fluorescent immunostaining during the harsh hybridization procedure. After fixation, standard direct

✉ Brigitte Schlegelberger, Medizinische Hochschule Hannover, Institut für Zell- und Molekularpathologie, Carl-Neuberg-Str. 1, Hannover, 30625, Germany (*phone* +49-511-532-4522; *fax* +49-511-532-4521; *e-mail* Schlegelberger.Brigitte@mh-hannover.de)
Stefan Gesk, Universität Kiel, Institut für Humangenetik, Schwanenweg 24, Kiel, 24105, Germany
Yanming Zhang, Universität Kiel, Institut für Humangenetik, Schwanenweg 24, Kiel, 24105, Germany, Present address: University of Chicago, Medical Center, Section of Hematology/Oncology, 5841 S. Maryland Ave. MC2115, Chicago, IL, 60637-1470, USA
Reiner Siebert, Universität Kiel, Institut für Humangenetik, Schwanenweg 24, Kiel, 24105, Germany

or indirect fluorescence in situ hybridization (FISH) is performed using appropriately modified DNA probes. Using appropriate filter sets, the immunophenotype and the hybridization signals can be evaluated simultaneously under a fluorescence microscope (for review see Siebert and Weber-Matthiesen 1997; Siebert and Schlegelberger 1997).

For example, the following objectives can be studied using the FICTION technique:

- The immunophenotypic characterization of tumor cell clones with defined chromosome aberrations (Haferlach et al. 1996, 1997; Zhang et al. 1999, 2000);
- The definition of the differentiation state present in the tumor cell clone (Weber-Matthiesen et al. 1995a);
- The identification of the most immature precursor cells within the tumor clone representing the cell population from which the tumor clone may have originated (Zhang et al. 1996);
- The immunophenotypic characterization of bystander cells with a normal chromosome constitution (Weber-Matthiesen et al. 1995b);
- The chromosomal characterization of small tumor cell populations with a selective immunophenotype within a background of normal cells (Weber-Matthiesen et al. 1995c, 1996).

Materials

Reagents

- AMCA-conjugated avidin (Jackson 003-150-083)
- Biotin-labeled goat anti-avidin antibody (Vector BA-0300)
- Class-matched mouse antibodies IgG1, IgG2a, IgG2b, IgM (Immunotech 0571, 0572, 1266, 1268)
- Human Cot-1 DNA (Gibco BRL/Life Technologies 15279-011)
- Cy3-conjugated avidin (Jackson 003-160-083)
- Cy3-conjugated goat anti-mouse antibody (Jackson 115-165-068)
- Cy3-conjugated rabbit anti-goat antibody (Jackson 305-165-045)
- Cy3-conjugated donkey anti-rabbit antibody (Jackson 711-165-152)
- Cy3-conjugated rabbit anti-mouse antibody (Jackson 315-165-003)

- DABCO (1,4-Diazabicyclo-(2,2,2)-octane) (Sigma D2522)
- DAPI (4,6-Diamidino-2-phenylindole) (Sigma D9542)
- Deionized formamide (Gibco BRL/Life Technologies 15515-018)
- Dextran sulfate 50% (Q-BIOgene DEXT 5001)
- Digoxigenin-conjugated sheep anti-mouse antibody (Chemicon International AQ 300D)
- FITC-conjugated donkey anti-mouse antibody (Jackson 715-095-151)
- FITC-conjugated sheep anti-digoxigenin antibody (Roche 1207741)
- Mouse anti-FITC antibody (monoclonal) (Dakopatts M878)
- Mouse anti-digoxigenin antibody (monoclonal) (Roche 1333062)
- Normal mouse serum (Jackson 015-000-001)
- Rubber cement: Fixogum (Marabu 2901 10 000)

Buffers and solutions

- Antifade solution: To 90 ml glycerol add 10 ml PN buffer containing 23 mg/ml DABCO
- DAPI stock solution: 0.2 mg/ml DAPI in A. bidest
- Paraformaldehyde solution (1%): Dissolve 1 g paraformaldehyde powder (harmful!) in 60 ml A. bidest.; add 5 drops 10 M NaOH. Heat to 70-100°C until the solution is clear when viewed from above the bench. Cool down and add 10 ml 100 mM $MgCl_2$. Adjust pH to 7-7.5 and fill up to 100 ml with distilled water. Filter the solution. Store at 4°C.
- PN buffer: 0.1 M NaH_2PO_4, 0.1 M Na_2HPO_4, pH 8.0
- PNM buffer: Add 5% dry milk powder to PN buffer; heat to 50°C overnight while stirring; add 0.03% NaN_3 (poisonous); PNM buffer is stable at 4°C for at least 6 months.
- 20×SSC: ×3 M NaCl, 0.3 M sodium citrate; pH 7.0
- Mastermix for indirect centromeric probes:
  Mix 6 ml deionized formamide, 2 ml 50% dextran sulfate and 0.5 ml 20×SSC. Vortex thoroughly, adjust to pH 7.5 and make up to a final volume of 9 ml with A.bidest.

- Mastermix for single copy probes:
  Mix 5 ml deionized formamide, 2 ml 50% dextran sulfate and 1 ml 20×SSC. Vortex thoroughly, adjust to pH 7.5, fill up to 9 ml with A. bidest.

- Mastermix for indirect single copy probes plus Cot-1 DNA:
  Add 250 µl 100% ethanol to 100 µl Cot-1 DNA (1 µg/1µl) in a 1.5 ml Eppendorf tube and spin 30 min at 14,000 rpm in a lab centrifuge. Remove supernatant, add 500 µl 70% ethanol and vortex. Spin again 15 min at 14,000 rpm. Remove supernatant, dry pellet in a speedvac or allow to air dry. Add 9 µl prewarmed (37°C) mastermix for single copy probes and shake for at least 30 min.

  Note: All mastermixes can be aliquoted and stored at -20°C.

Probes and probe preparation for FISH

For a series of questions to be addressed in a tumor cytogenetic lab by FISH or FICTION, probes directly labeled with a fluorescent dye or indirectly labeled with biotin or digoxigenin are commercially available (Q-BIOgene, Vysis, etc.). For the detection of numerical aberrations, satellite probes specific for centromeric regions of distinct chromosomes should be applied. Chromosome painting probes can be used for the identification of marker chromosomes and structural chromosome aberrations in metaphase spreads but are generally not recommended for use in interphase nuclei. For some recurrent translocations, e.g. for the t(9;22), the t(15;17), or for some deletions, e.g. of the RB gene or the p53 gene, probes are also commercially available. For other structural aberrations probe sets have been published which can be obtained from the scientific community. The latter can be prepared according to routine protocols and labeled by random primed labeling or nick translation using commercially available kits according to the manufacturers' instructions (Life Technologies, Eggenstein, FRG; Roche, Mannheim, FRG). For hybridization, an appropriate amount of labeled probe DNA should be diluted in the mastermixes described above or in hybridization buffers supplied by the manufacturers (Schlegelberger et al. 1999).

In this paragraph a series of protocols developed for easy and universal application of commercial and non-commercial probes are described. For other commercial probes, follow the recommendations of the supplier.

Hybridization mixture for indirectly labeled centromeric probes (Oncor):

Add 1 µl of biotin labeled probe (Oncor Cat. No. P5000-BIO -P5090-BIO) or 1 µl of digoxigenin labeled probe (Oncor Cat. No. P5000-DG-P5090-DG) or 0.5 µl of biotin labeled probe (Oncor Cat. No. P5000-BIO -P5090-BIO) and 0.5 µl of digoxigenin labeled probe (Oncor Cat. No. P5000-DG-P5090-DG) as supplied by the manufacturer to 9 µl of "mastermix for centromeric probes". Non-commercial DNA probes should be adjusted to 10 ng/µl in TE buffer (1 mM EDTA, 10 mM Tris-HCl, pH 8) before being added to the mastermix.

Hybridization mixture for indirectly labeled commercial single copy probes (Oncor):

Indirectly labeled single copy probes (Oncor), e.g. probes for translocations or microdeletions, can be applied without prior modifications. For the combination of single copy probes with centromeric probes a 10:1 mixture, for the combination of two single copy probes a 1:1 mixture is recommended. If hybridization with the directly applied single copy probe or a mixture of two probes produces a high background noise a 1:2 or 1:4 (up to 1:10) dilution of the probe/probe mixture in "mastermix for single copy probes plus Cot-1 DNA" may enhance the hybridization quality.

Hybridization mixture for indirectly labeled non-commercial single copy probes with or without combination with indirectly labeled centromeric probes (Oncor):

For non-commercial single-copy probes the mixture of 1 µl labeled single-copy probe (40-100 ng/µl) and 9 µl "mastermix for single copy probes plus Cot-1 DNA" is recommended. If the probe produces a high background noise or results in a low signal intensity 1-5 µl labeled single-copy probe can be coprecipitated with 100 µl Cot-1 DNA using conventional ethanol precipitation (as described for "mastermix for single copy probes plus Cot-1 DNA") and resuspended in 10 µl "mastermix for single copy probes". For combining a single-copy probe with a centromeric probe (Oncor) add 1 µl of the latter to the respective probe mixture.

Hybridization mixture for indirectly labeled painting probes (Oncor, AGS):

Painting probes can be applied directly or can be handled as described for commercial single-copy probes. In general, predigestion of the slides is recommended.

Hybridization mixture for directly labeled probes (Vysis):

Directly labeled probes should be premixed in the supplied buffers according to the manufacturers' instructions. Alternatively, or for the combination with indirectly labeled probes, dilution in the mastermixes described for indirectly labeled centromeric or single copy probes is possible.

Procedure

Slide preparation for FISH

Preparing high quality slides is one of the most critical steps for successfully performing FISH. In general, cytogenetic preparations as described above, conventional unstained peripheral blood or bone marrow smears, touch preparations and cell suspensions from fresh and frozen tissue are applicable for FISH. The use of formalin-fixed paraffin-embedded tissue, tissue sections and stained blood or bone marrow smears is not recommended for routine use and should be restricted to labs experienced with the technical considerations of these techniques. In order to facilitate the correlation of the hybridization signals to individual nuclei, cells should not be packed too densely. The number of cells per cytospin slide should not exceed 3,000 cells (diameter of the area containing cells = 0.5 cm). In imprints and smears cells often overlap, but there are usually areas in the periphery where the cells lie separated from each other. Only such areas should be evaluated. The erythrocytes should lie flat on the slide and not be standing on end. Sometimes, this is only possible at the end of the smears. Prior to hybridization the quality of slides should be checked by phase contrast microscopy. Areas qualifying for hybridization are characterized by a high density of clearly separated grayish nuclei lacking any surrounding cytoplasm and should be marked on the back of the slide with a diamond pencil. In general, all specimens can be hybridized without any pretreatment. Nevertheless, in order to enhance the hybridization efficiency, predigestion of the slides according to the following or similar protocols is recommended for most tumor specimens.

Pepsin digestion of FISH slides

1. Incubate slide in a Coplin jar for 5 min at 37°C with freshly prepared (not prewarmed!) digestion solution.
2. Wash slide once in A. bidest.
3. Fix slide for 10 min in paraformaldehyde solution (1%).
4. Wash slide once in A. bidest.
5. Dehydrate in 70%, 85% and 100% ethanol 1 min for each step.
6. Air dry slide.

For different samples, optimal pepsin concentration and incubation duration may vary slightly.

Slide preparation for cytogenetic samples, imprints and blood/bone marrow smears: Slides from cytogenetic samples are prepared as described above. Blood and bone marrow smears as well as imprints are prepared as for routine morphologic evaluation.

Cytospin preparations

Cytospin preparations of cytogenetic specimens: The hybridization efficiency for interphase FISH from cytogenetic specimens can be enhanced by preparing cytospin preparations. This procedure is especially recommended for samples containing few cells or displaying no signals or a high background noise after conventional preparation.

1. Determine the cell number.
2. Add 1×PBS ad 10 ml to 1-3 ml cells containing Carnoy's fixative in a 15-ml tube.
3. Spin 10 min at 1800 rpm in a lab centrifuge.
4. Discard supernatant.
5. Resuspend pellet in 1×PBS to a cell count of 1×10^4 cells/ml.
6. Spin 200 µl of this solution (2000 cells/slide) in a cytocentrifuge for 5 min at 800 rpm.
7. Air dry slide.
8. Proceed with pepsin digestion of FISH slides.

Cytospin preparations of cell suspensions from solid tumors: This protocol is intended for FISH analyses of frozen specimens from lymph nodes and solid tumors, in which neither cytogenetic preparations nor freshly prepared imprints are available.

1. In a cooled Petri dish cut a piece of tissue (10-40 mg) from a tumor block stored in liquid nitrogen or at -80°C. Refreeze remaining tumor material immediately!
2. Per 10 mg tumor tissue add 60-120 µl 1×PBS (tissue-dependent).
3. Thaw tissue in 1×PBS for 15-30 min at RT.
4. Disaggregate tissue mechanically by cutting and scraping with a scalpel.
5. Collect 1×PBS solution containing the cells from the dish and transfer it to a 1.5-ml Eppendorf tube.
6. Disaggregate cells in solution by pipetting several times through a 10-100 µl pipette.
7. Count the number of cells in solution.
8. Transfer 3000 cells/slide into a new Eppendorf tube.
9. Add 200 µl digestion solution. (Do not prewarm!)
10. Incubate at 37°C for 10 min.
11. Immediately transfer tube onto ice; cytocentrifuge the solution for 5 min at 800 rpm.
12. Air dry slides for 10 min.
13. Fix slides for 10 min in paraformaldehyde solution (1%).
14. Wash slides once in A. bidest.
15. Dehydrate in 70%, 85% and 100% ethanol for 1 min each.
16. Air dry slides.

    Note: The remaining cell solution can be stored for several months. Add 2.3 vol 100% ethanol and freeze at -20°C. For preparing slides from this stored cell solution, transfer required amount of cells into an Eppendorf tube, pellet cells by short spinning in a dish centrifuge, discard supernatant, add 40 µl 1×PBS per 1×10^5 cells and incubate on ice for at least 15 min. Proceed with step 9.

Hybridization and post-hybridization washes

Hybridization

1. Apply 1.5 μl of the respective hybridization solution to the area of the slide containing the cells.
2. Cover with a round 10-mm coverslip.
3. Seal with rubber cement.
4. Place the slides at the bottom of a metal box on a wet paper towel, close the box and denature for 7 min in a water bath at 75°C.
5. Immediately transfer the hot metal box into a 37°C incubator. According to the probe, hybridize from 1 h to 3 days.

Post-hybridization washes

1. Prewarm three Coplin jars containing 0.1×SSC at 60°C in a water bath.
2. Remove rubber cement from the slides with a needle.
3. Shake slides in the first Coplin jar at 60°C in 0.1×SSC until all coverslips are removed.
4. Wash slides by shaking for 5 min at 60°C in each of the three Coplin jars containing 0.1×SSC.
5. Equilibrate slides in PN buffer for 2 min at RT.

Simultaneous detection of biotin- and digoxigenin-labeled probes with AMCA (blue) and FITC (green)

The following protocol allows the simultaneous detection of biotinylated probes with the blue fluorochrome AMCA and digoxigenin-labeled probes with the green fluorochrome FITC. In analogy, a series of alternative cascades allowing the detection of indirectly labeled probes by other colors or fluorescent dyes can be designed. The respective solutions can be incubated under coverslips followed by three washings for 2 min each in PN buffer in a Coplin jar. Nevertheless, the use of Shandon holders not only facilitates manual detection but also provides the prerequisite for automatization with pipette robots (e.g. Tecan SLT).

Preparation of slides

1. Incubate slides with a mixture of Cy3-conjugated-avidin (1:400 in PNM buffer) and mouse anti-digoxigenin antibody (1:100 in PNM buffer) at RT for 30 min.

2. Wash once with PN buffer.
3. Incubate with a mixture of biotinylated goat anti-avidin antibody (1:200 in PNM buffer) and digoxigenin-conjugated sheep-anti mouse antibody at RT for 30 min (1:100 in PNM buffer).
4. Wash once with PN buffer.
5. Incubate slides with a mixture of Cy3-conjugated avidin (1:400 in PNM buffer) and FITC-conjugated sheep anti-digoxigenin antibody (1:100 in PNM buffer) at RT for 30 min.
6. Wash once in PN buffer and for 5-10 min in 2xSSC.
7. For counterstaining blue incubate the slide for 5-10 min in DAPI-Solution (6 µl DAPI stock solution in 60 ml 2×SSC) (DAPI-concentration and/or incubation time can be adjusted according to preferred staining intensity).
   For counterstaining red – if the hybridization signals are visualized by AMCA and FITC- mount the slides with antifade solution containing 500 ng/ml propidium iodide (red).
8. Wash for 5-10 min in 2×SSC.
9. Mount the slide with antifade solution.

   Note: If necessary, the fluorescence intensity can be amplified by multiple repetitions of steps 3 to 5.

Slide preparation for FICTION

The quality of slides is even more important for successfully performing the FICTION technique than for FISH. For FICTION, cytospin slides, smears, imprints and cryostat sections may be used. Freshly prepared slides should be air-dried overnight at room temperature. Thereafter, slides can be processed immediately or stored at -80°C. Slides preserved in this way can be used for years; when single slides are taken out of the freezer, it is very important to make sure that the remaining slides in the freezer do not thaw. This is possible by taking the slides out of the container inside the freezer. If slides are stored at -20°C, the quality of the immunophenotyping may be diminished after a few months. However, the quality of the hybridization is not affected. Before starting the FICTION procedure, slides should be examined by phase contrast microscopy for the morphological quality of the cells. If many slides are available, only those with good cell morphology should be processed.

FICTION protocol for combined immunophenotyping and two-color FISH

Indirect immunophenotyping

1. Fix slides in fresh acetone for 10 min at room temperature and air dry for 10 min.

   Note: Cryopreserved slides should be air-dried prior to the fixation for at least 1 h after being taken out of the freezer.

2. Apply 100 µl of monoclonal mouse antibody (against the antigen that is to be stained) diluted in PNM buffer to the area of the slide containing the cells or the section. Incubate for 30 min at RT.
3. Wash once in PN buffer.
4. Incubate with Cy3-conjugated goat anti-mouse antibody (1:200 in PNM buffer) for 30 min.
5. Wash once in PN buffer.
6. Incubate with Cy3-conjugated rabbit anti-goat antibody (1:200 in PNM buffer) for 30 min.
7. Wash once in PN buffer.
8. Incubate with Cy3-conjugated donkey anti-rabbit antibody (1:100 in PNM buffer) for 30 min.
9. Wash once in PN buffer.

This four-step immunostaining results in very strong fluorescent staining. If strongly expressed antigens, such as CD3, have to be stained, it is possible to omit steps 8 and 9 or even steps 6 to 9.

Monitoring of immunophenotyping

At this point the success of the immunostaining can be monitored. Mount the slides with PN buffer (not with glycerol!) and examine them under the fluorescence microscope.

Note: The fluorescence intensity is relatively low if the slides are mounted in PN buffer. However, mounting in glycerol prior to in situ hybridization may cause hybridization artifacts.
Monitoring of immunophenotyping is always recommended if the processed slides have been stored for a long period or if the cells showed poor morphology prior to the staining procedure.

Hybridization

10. Put the slides into freshly prepared, ice-cold Carnoy's fixative for 10 min.

11. Wash in A. bidest for 2 min at RT.
12. Place the slides in 1% paraformaldehyde for 1 min at 4°C.
13. Wash in A. bidest for 2 min at RT.
14. Dehydrate the slides in a series of 70%, 85% and 100% ethanol at RT, for 2 min each. Air dry the slides for 10 min at RT.

Hybridization

15. Hybridize with indirectly labeled probes (as described above).

Detection

Simultaneous detection of biotin- and digoxigenin-labeled probes by AMCA (blue) and FITC (green):

1. Incubate slides with a mixture of AMCA-conjugated avidin and FITC-conjugated monoclonal mouse anti-digoxigenin antibody at RT for 30 min [both 1:200 in PNM buffer].
2. Wash once in PN buffer.
3. Incubate with a mixture of biotinylated goat anti-avidin antibody and mouse anti-FITC antibody at RT for 30 min [both 1:200 in PNM buffer].
4. Wash once in PN buffer.
5. Incubate with a mixture of AMCA-conjugated avidin and FITC-conjugated donkey anti-mouse antibody at RT for 30 min [both 1:200 in PNM buffer].
6. Wash once in PN buffer.

   Note: The fluorescence intensity can be amplified by multiple repetitions of steps 18 to 20.

7. Mount the slides with antifade solution.

Counterstaining

Directly labeled probes can now be counterstained blue by incubating for 5–10 min in DAPI solution (e.g. 6 <m›l DAPI stock solution in 60 ml 2×SSC) followed by washing for 5–10 min in 2×SSC and mounting in antifade solution. Alternatively, if green (e.g. FITC) and/or blue (e.g. AMCA) fluorochromes have been applied, the slides can be mounted with antifade solution containing 500 ng/ml propidium iodide (red).

If counterstaining is required, the following steps can be included before mounting. Nevertheless, to interfere with the immunophenotype counterstaining is not generally recommended.

1. Wash once in PN buffer and for 5-10 min in 2×SSC.
2. For counterstaining blue incubate the slide for 5-10 min in DAPI solution (6 µl DAPI stock solution in 60 ml 2×SSC) (DAPI concentration and/or incubation time can be adjusted according to preferred staining intensity).
   For counterstaining red – if the hybridization signals are visualized by AMCA and FITC- mount the slides with antifade solution containing 500 ng/ml propidium iodide (red).
3. Wash for 5-10 min in 2×SSC.
4. Mount the slide with antifade solution.

   Note: If necessary, the fluorescence intensity can be amplified by multiple repetitions of steps 3 to 5.

FICTION protocol for combined double-immunophenotyping plus single-color FISH

Immunophenotyping

Immunophenotyping against antigen A by Cy3 (red):

1. Incubate with monoclonal mouse antibody against antigen A (e.g. CD5) for 30 min at RT.
2. Wash once in PN buffer.
3. Incubate with Cy3-conjugated rabbit anti-mouse antibody (1:200 in PNM buffer) for 30 min at RT.
4. Wash once in PN buffer.
5. Incubate with Cy3-conjugated donkey anti-rabbit antibody (1:200 in PNM buffer) for 30 min at RT.
6. Wash once in PN buffer.
7. Incubate with 20% normal mouse serum diluted in PNM buffer for 15 min at RT. By incubating with normal mouse serum containing high concentrations of mouse immunoglobulins, the free binding sites of the Cy3-conjugated rabbit anti-mouse antibody (step 2) are blocked. This prevents the monoclonal mouse antibody against antigen B, which is applied in step 5, from being caught by free binding sites of the rabbit anti-mouse antibody.

   Note: Attention: Do not wash the slides after this step.

Immunophenotyping against antigen B by AMCA (blue):

8. Incubate with biotinylated monoclonal antibody against antigen B (e.g. CD19) in PNM buffer containing 20% normal mouse serum for 30 min at RT.
9. Wash in PN buffer.

   Note: The detection system is very similar to that used for the detection of biotinylated DNA probes.

10. Incubate with AMCA-conjugated avidin (1:200 in PNM buffer) for 30 min at RT.
11. Wash in PN buffer.
12. Incubate with biotinylated goat anti-avidin antibody (1:200 in PNM buffer) for 30 min at RT.
13. Wash in PN buffer.

    Note: Multiple amplifications of the AMCA staining are possible by repeating steps 10 to 13.

Hybridization

14. Fix in 1% paraformaldehyde for 1 min at 4°C.
15. Wash in A. bidest for 2 min at RT.
16. Put the slides into fresh ice-cold Carnoy's fixative for 10 min.
17. Wash in A. bidest for 2 min at RT.
18. Dehydrate the slides in a series of 70%, 85% and 100% ethanol at RT for 2 min each.
19. Air dry the slides for 10 min at RT.
20. Apply hybridization mixture containing digoxigenin labeled probe to the slide.

    Note: Attention: In this protocol a digoxigenin labeled probe is applied. Do not use biotin-labeled probes, because they bind nonspecifically to the antibody cascade of antigen B.

Detection

Detection of the digoxigenin labeled probe by FITC (green):

21. Incubate with an FITC-conjugated monoclonal mouse anti-digoxigenin antibody at RT for 30 min (1:200 in PN buffer).

22. Wash in PN buffer.
23. Incubate with FITC-conjugated donkey anti-mouse antibody at RT for 30 min (1:200 in PNM buffer).
24. Wash in PN buffer.
25. Incubate with monoclonal mouse anti-FITC antibody for 30 min at RT (1:200 in PNM buffer).
26. Wash in PN buffer.
27. Incubate with FITC-conjugated donkey anti-mouse antibody for 30 min at RT (1:200 in PNM buffer).
28. Wash in PN buffer.

    Note: If required, the fluorescence intensity can be amplified by repeating steps 25 to 28.

29. Mount the slides with antifade solution.

Storage of stained slides Mounted slides can be stored for several weeks at 4°C and for several months or up to years at -20°C. Always keep slides in the dark.

Other FICTION protocols

The FICTION technique can generally be performed with any kind of FISH procedure (Fig. 1). If immunophenotyping and, most important, paraformaldehyde fixation after immunophenotyping is done according to the FICTION protocol, then it is possible to proceed with individual FISH protocols as established in other laboratories. This way, all types of probes for detecting numerical or structural aberrations can be employed. The only restriction is that proteolytic treatment must be avoided because it would considerably impair the quality of immunophenotyping.

Results

Evaluation of FICTION analyses

Technical prerequisites: In general, the different fluorescent colors of the slides can be evaluated separately under each fluorescence microscope equipped with specific filter sets for AMCA, FITC and Cy3. Alternatively, if triple-dye fluorescence filter sets are used, all three fluorescent colors

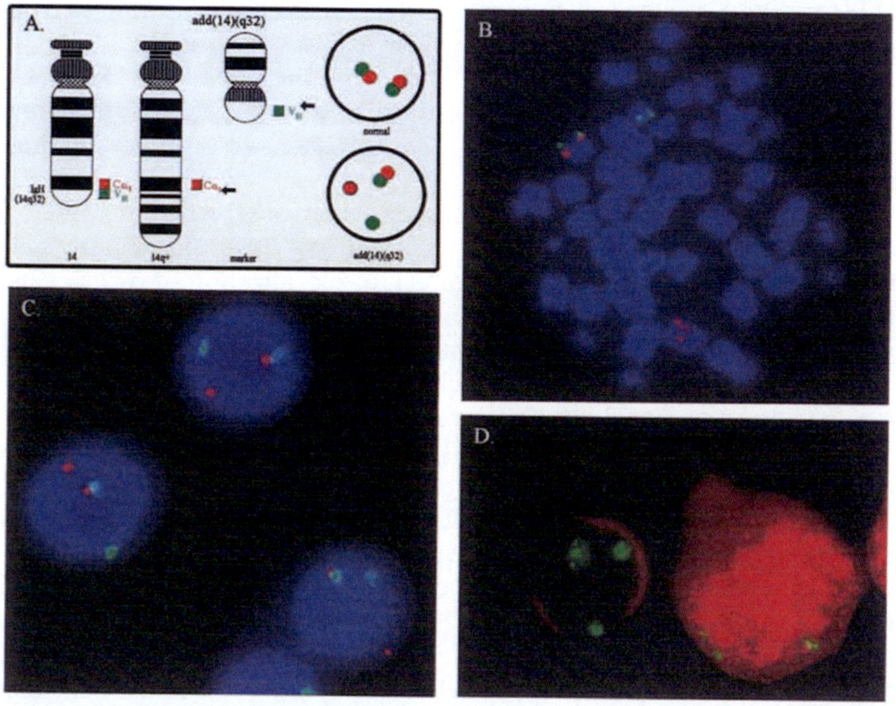

Fig. 1. FICTION study on breast carcinoma cell line MCF-7. Combined immunophenotyping with monoclonal mouse anti-human estrogen receptor antibody visualized by Cy3 (red) and FISH with the YAC 19I11F containing the ESR gene visualized by FITC (green). One positive cell showed strong nuclear staining (red) while the adjacent cell was negative. Both cells showed three hybridization signals (green), indicating both ESR positive and negative cells to contain three copies of the ESR gene

can be evaluated simultaneously. Appropriate filter sets are available from microscope manufacturers and from filter producers. Documentation is possible by conventional photography using a 400ASA film and exposure times up to 20 s. Digital imaging using commercially available software like the ISIS system supplied by MetaSystems Hard & Software GmbH is an easy alternative.

Evaluation of FICTION analyses: For controlling the specificity of the immunophenotyping process in FICTION, class-matched mouse antibodies, non-reactive with human tissue, should be applied on a control slide. A major problem of evaluating FICTION slides is autofluorescence, which must not be confused with non-specific antibody binding. Typically, autofluorescence can be observed with all different filter sets appearing dull red (Cy3 filter), yellow (FITC filter) or white (AMCA filter), whereas true Cy3 fluorescence is not visible if the AMCA filter set is used.

For the evaluation of FICTION slides the same rules have to be observed as for conventional FISH. Controls have to be performed for each immunofluorescence as this may interfere with probe binding and signal evaluation.

Evaluation criteria and cut-off levels: For inter-observer comparability the evaluation criteria have to be defined exactly. This is particularly necessary for translocation probes, in which a colocalization of two signals indicates a translocation. As to the organization of the nucleus in interphase cells, a spatial separation of the two colocalized signals is possible. Thus, it has to be defined up to which distance between the signals, (calculated in relative signal size, i.e. once or twice the signal diameter) a real colocalization is assumed.

In each laboratory, diagnostic thresholds have to be defined for each probe. This is done by analyzing at least five control cases with a minimum of 200 interphase nuclei each. For each slide preparation process, these controls have to be performed independently as there are significant differences in the false positive and negative rates according to the protocols used. The cut-off levels are usually defined as mean percentage of false positives in these controls plus three standard deviations. Positive controls on tumor specimens with a known karyotype have to be performed to verify the specificity of each assay.

Evaluation of tumor specimens: For the evaluation of tumor specimens the same rules have to be applied as for the controls. At least 100 interphase nuclei have to be evaluated by the observer. Only good quality slides and clearly analyzable cells can be counted. In cases with low percentages of aberrant cells (in the range of the thresholds) at least 200 nuclei have to been counted. Each result should be confirmed independently by a second observer.

Troubleshooting

In situ hybridization

Too much background

- Unspecific cross-hybridization: Add human Cot-1 DNA to probe and/ or coprecipitate.
- Too low hybridization stringency: Increase hybridization and/or wash temperature. Increase formamide concentration. Decrease salt concentration. Prepare new wash solutions.

- Signal amplification too strong: Omit some amplification steps.

No hybridization signals

- Defective DNA probe: Test probe on freshly prepared slides. If unsuccessful, try a new probe.
- Paraformaldehyde solution too old (pH!): Replace.
- Defective detection reagents: Use another detection cascade as control.

AMCA fluorescence intensity consistently too low

- AMCA unstable: Buy new AMCA reagents. Store AMCA reagents in 50% glycerol at -20°C.

Immunophenotyping

No staining/high background staining

- Inappropriate slide preservation: Use freshly prepared slides. Preserve slides correctly.
- Defective monoclonal primary antibody: Test antibody on different freshly prepared slides.
- Defective secondary antibodies: If unsatisfactory immunostaining occurs on different freshly prepared slides, all antibodies have to be tested separately. Exchange all antibodies for new ones (only one per test).

Immunostaining is sufficient prior to FISH but less intensive after FISH

- Paraformaldehyde solution is too old. Replace.

References

Haferlach T, Winkemann M, Löffler H, et al. 1996 Blood, 87:2459-2463
Haferlach T, Winkemann M, Nickenig C, et al. 1997 Br J Haematol, 97:99-106
Schlegelberger B, Weber-Matthiesen K, Eumann S, et al 1992 Cytogenet Cell Genet, 59:10-11
Siebert R, and Weber-Matthiesen K 1997 Histochem Cell Biol, 108:391-402
Siebert R, and Schlegelberger B 1997 Rev Clin Exp Hematol, 4:41-55
Teerenhovi L, Knnutila S, Ekblom M et al. 1984 Blood, 64:1116-1122
Weber-Matthiesen K, Winkemann M, Müller-Hermelink A, Schlegelberger B, and Grote W 1992 J Histochem Cytochem, 40:171-175

Weber-Matthiesen K, Winkemann M, Grote W, et al. 1995a Genes Chrom Cancer:14, 22-27
Weber-Matthiesen K, Deerberg J, Poetsch M, et al. 1995b Cytogenet Cell Genet, 70:243-245
Weber-Matthiesen K, Deerberg J, Poetsch M, et al. 1995c Blood, 86:1464-1468
Weber-Matthiesen K, Deerberg-Wittram J, Rosenwald A, et al. 1996 Am J Pathol, 149:463-468
Zhang Y, Poetsch M, Weber-Matthiesen K, et al. 1996 Br J Haematol, 92:673-680
Zhang Y, Wong KF, Siebert R, Matthiesen P, Harder S, Eimermacher H, Feller AC, Schlegelberger B 1999 Br J Haematol, 105:737-742
Zhang Y, Siebert R, Matthiesen P, Harder S, Theile M, Scherneck S, Schlegelberger B 2000 Virchows Arch, 436:271-275

Part G

Not FISH but Almost

PRINS: Hybridization and Primed in Situ Labeling in One Step

JORN KOCH

Introduction

PRINS (**PR**imed **IN** Situ labeling) is a technique which is primarily used for the detection of tandemly repeated target sequences in chromosomes and nuclei ("in situ") by oligonucleotide probes. Such sequences can be detected down to a target size of a few hundred bases with a simple short procedure rendering virtually 100% of potential targets labeled.

The PRINS concept was introduced by the author of this text (Koch et al. 1989) for the study of minute details in centromere DNA in situ; details so small that they would be difficult to study by in situ hybridization (in casu sequence variation among 16-bp target sequences within an array of repeated sequences). The extent to which the technique can discriminate among closely related sequences has more recently also been illustrated by Pellestor and coworkers (1994,1995), and by us (Koch et al. 1995), both discriminating between the alpha satellite DNA of chromosome 13, and that of chromosome 21, on the basis of a single base difference. A further advantage of the PRINS technique is that the probes used are very small, and thus penetrate easily to the target, even if this is in a very tight formation. This feature of the technique is being exploited by the group of Franck Pellestor to enable efficient staining of target sequences in sperm cells (e.g. see Pellestor et al. 1997).

More recently, a new version of the PRINS technique, the "dideoxy-PRINS", has been developed to enable highly sensitive, quantitative staining of simple repeat DNA, including telomeric repeats and trinucleotide repeats. The dideoxy-PRINS can be understood from the basic PRINS principle outlined below (Fig. 1). In PRINS, the probe is unlabeled, and thus does not itself label the target binding it. Labeling is only ob-

Jorn Koch, Cancercytogenetics Laboratory, Department of Hematology, Tage Hansens Gade 2, Aarhus C, 8000, Denmark (*phone* +45-8949-7398; *fax* +45-8949-7399; *e-mail* Jorn.Koch@aas.auh.dk or jornkoch@hotmail.com)

tained when the probe is turned into a primer by a DNA polymerase, and elongated with the DNA flanking the hybridization target as template for the in situ synthesis of DNA. Feeding the DNA synthesis labeled nucleotides results in site-specific labeling where the probe was elongated by the polymerase. In the dideoxy-PRINS, one or more of the normal nucleotides fed to the polymerase has been replaced by dideoxynucleotides. Consequently, only oligonucleotides hybridizing at sites providing a template for DNA synthesis that can be copied in the presence of this (these) dideoxynucleotide(s) will be elongated and induce labeling at the binding site. Thus, if ddGTP e.g. is included in the reaction mixture, only probes hybridizing next to DNA lacking C-residues can be copied. In general this added selectivity increases the signal to noise ratio of the PRINS reaction by about an order of magnitude.

The advantages of the dideoxy-PRINS approach is first of all that even very small target sequences can be detected in situ (Koch, 2000). Further-

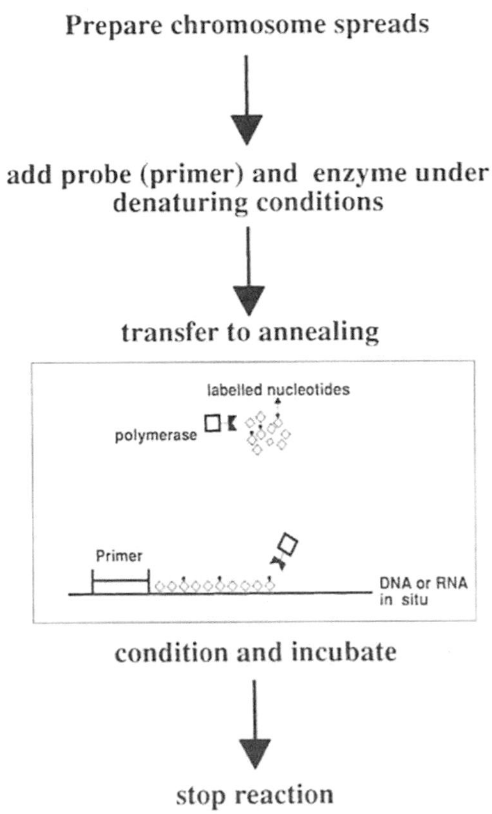

Fig. 1. Reaction format of PRINS and flowchart of PRINS on spreads of metaphase chromosomes

more, the blocking of non-specific priming also allows for prolonged incubations with high concentrations of reagents, leading to saturation of the hybridization targets, and a quantitative detection reaction (Krejci and Koch 1998, Serakinci and Koch 1999).

Quantitative PRINS reactions can be evaluated in either of two ways. The first is simply to measure the staining frequency (how many out of 100 potential targets are actually stained). The higher the staining frequency, the larger the target (Therkelsen et al. 1995). With the dideoxy-PRINS, this often becomes impractical as the staining efficiency in many cases is close to 100%. In normal blood lymphocytes the main telomeric repeat is e.g.

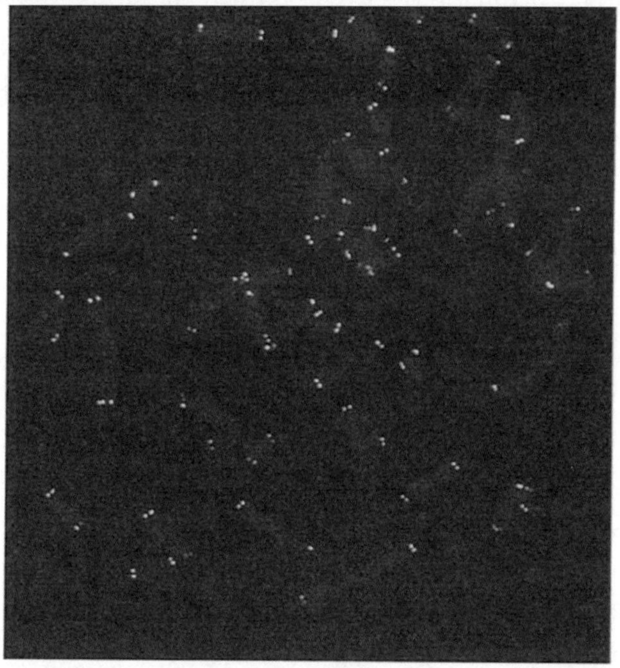

Fig. 2. Dideoxy-PRINS of telomeric repeats, using the $(CCCTAA)_7$ probe as PRINS primer (Therkelsen et al. 1995). The reaction was done as described in Materials and Methods, except that an equimolar amount of ddGTP was used instead of dGTP. The image represents the "raw" recording of the green fluorescence used for the detection of the PRINS (no counterstaining added, no image processing done). Virtually all chromosome ends are labeled

stained with an efficiency of 99.8% (Fig. 2). In such cases, it is preferable to measure the light intensities of the signals instead. We have described a procedure for this, employing the preexisting image analysis program QUIPS from VYSIS for the purpose (Krejci and Koch 1998). Currently, another analysis program is becoming available from DAKO. This program seems equally suited for signal quantification. Expanded trinucleotide repeats (Fig. 3) are by contrast stained with a low enough efficiency to be evaluated for relative size by determining the percentage of cells labeled.

Staining of very small target sequences has been employed for the study of variant telomeric repeats at individual chromosome ends in a variety of species (Fig. 4). The combination of this high sensitivity and the ability to discriminate single base variations enables individual repeat variants to be selectively studied (Krejci and Koch 1999, Serakinci et al. 1999). Furthermore, since one strand of the target is flanked at its 5-end by

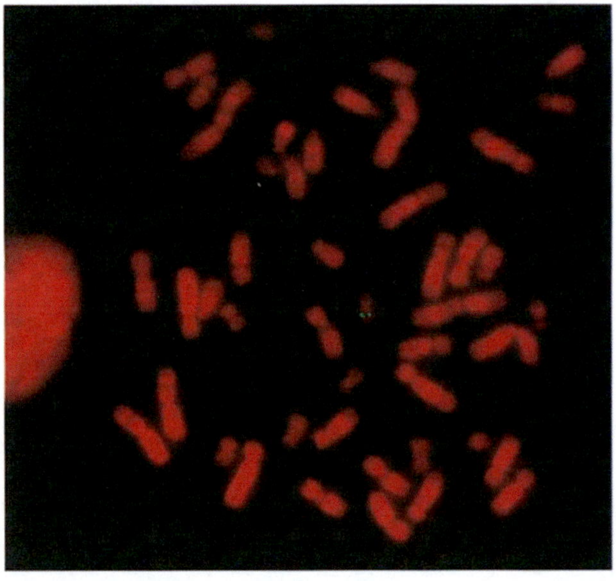

Fig. 3. Staining of the CTG-repeat expanded in myotonic dystrophy. As (this strand of) the repeat is void of Adenosine residues, the reaction was incubated with ddATP instead of dATP. Apart from this, and the $(CTG)_7CT$-primer used, the reaction was done as in Fig. 2. The image is produced by combining the PRINS-signal, in green, with counterstained chromosomes, in red. The myotonic dystrophy locus is stained on only one of the two chromosomes 19, as the normal allele on the other chromosome 19 is too small to be detected. The staining frequency of the abnormal allele was higher in this boy than in his mother, who had a less severe disease with a later onset. (Courtesy of K. Krejci)

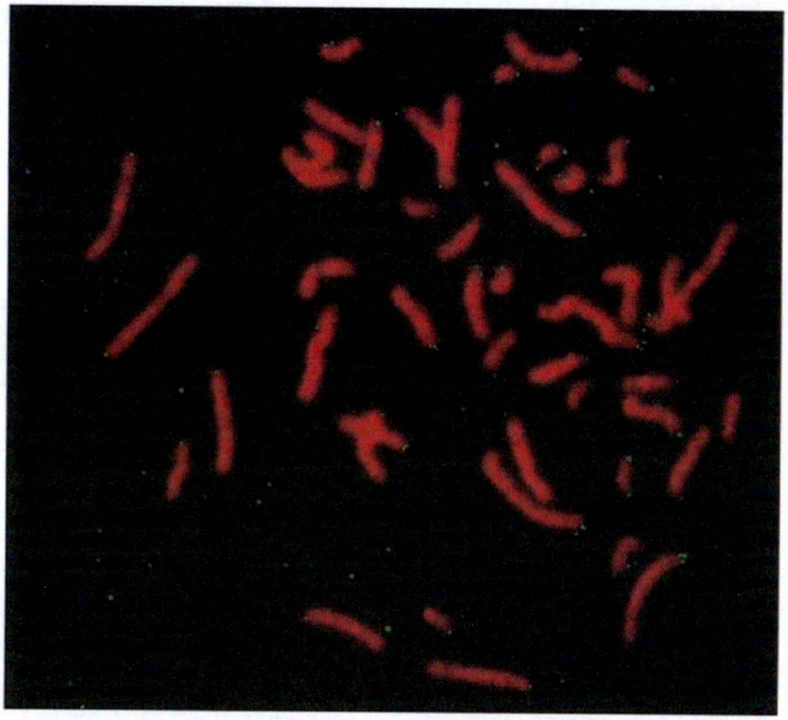

Fig. 4. Dideoxy-PRINS with the oligonucleotide $(GGTGAG)_4GGTG$ (Telo7, Krejci and Koch 1999) staining a less abundant telomeric repeat located next to the repeat detected in Fig. 2. In this case ddCTP was used instead of dCTP. The image is produced by combining telomeric signals, in green, with counterstained chromosomes, in red. The target stained by Telo7 does not light up on all chromosome ends, but when it does light up, it often presents itself with three or four signals on some chromosome ends, rather than just two

one sequence, and the other strand by another, it is possible to also determine the orientation of the target. It has e.g. been possible to determine the relative orientation of various telomeric repeat variants with respect to each other by this method (Krejci and Koch, 1999; Serakinci et al., 1999).

Materials

Oligonucleotide probes

- 10×dNTP: 1.0 mM each of dATP, dCTP, and dGTP and 0.1 mM of labeled dUTP in 50% glycerol, prepared from normal nucleotides (Lithium salt, Roche), digoxigenin-dUTP or fluorescein-dUTP (Roche). For dideoxy-PRINS, the nucleotide not needed for the specific chain elongation is simply replaced with a similar molar amount of the corresponding dideoxy-analogue (Roche).

- Tth DNA polymerase and 10×polymerase buffer: enzyme and buffer are supplied together by Roche
- Stop buffer: 50 mM NaCl, 50 mM EDTA, pH 8.0.
- Washing buffer: 4×SSC, pH 7.0, 0.05% Tween 20. (1×SSC: 150 mM NaCl, 15 mM sodium citrate)
- Blocking solution: 5% non-fat dry milk dissolved in washing buffer. Centrifuge for 2 min in an Eppendorf centrifuge and use supernatant.
- Antifade solution: Vectashield from Vector Laboratories
- Counterstain: either propidium iodide (Sigma) or DAPI (Sigma)
- Either, two **thermoblocks** with a metal surface to ensure good heat transfer, one with a surface temperature of 94°C and the other with a surface temperature of 55-65°C. The temperature of the 94°C incubator should be scrupulously measured, since it must not deviate more than 1-2°C from the desired value. To ensure this, it is important that the slide and the hot plate is covered with an insulating lid.
- **Or** a special **PRINS/in situ PCR machine** (from Hybaid, MJ-Research or Techne) preset (simulated slide function) to incubate the slide at 91-94°C for 2-4 min and at 55-65°C for 5-60 min (depending on the oligonucleotide used). This is the more convenient set-up and ensures absolute reproducibility of the reaction conditions.
  Both types of incubators should be covered with a lid (if a lid is not built in), to optimize temperature control.
- For a multicolor PRINS an ice-cold (-20°C) **ethanol series** (70%, 90%, 99%) is also needed.

Procedure

Standard PRINS

1. To cover a standard slide completely, prepare 50 μl reaction mixture containing: 38 μl ddH_2O, 5 μl 10 x polymerase buffer, 5 μl 10×dNTP, 1 μl DNA probe (from 500 ng and up) and 1 μl DNA polymerase (1 unit) and mix gently by tapping a finger on the Eppendorf tube.

To cover a smaller area, prepare less reaction mixture (and use a smaller coverslip). The amount (in μl) of mixture needed is roughly proportional to the length of the coverslip (in mm).

2. Denaturation of the chromosomal DNA: As soon as the polymerase has been added to the reaction mixture, it should be applied to a freshly prepared slide and spread with a coverslip of a suitable size (25×50 mm to cover a full slide). Denature the chromosomal DNA by placing the slide for 4 min on a 94°C hot plate covered by a lid.
3. Probe annealing and chain elongation: Transfer the slide quickly to 55-65°C and incubate for 5-60 min (depending on the probe). If a specialized incubator is used, the machine "handles" steps 2 and 3, so just spread the reaction mixture with the coverslip, put the slide in the machine, press the button to start the reaction and transfer the slide to stop buffer when the program is completed.
4. Place the slide in stop buffer preheated to the temperature used in step 3, for 1 min to terminate PRINS labeling.
5. Transfer the slide to 50 ml of washing buffer and wash for 3 min at room temperature.

6a. If fluorochrome labeled nucleotides have been used, the slide can now be counterstained, mounted and evaluated under the microscope.

6b. If digoxigenin labeled nucleotides have been used, they need to be visualized with anti-digoxigenin antibody as described below.

Dideoxy-PRINS

Perform the PRINS procedure exactly as above, but replace the nucleotide(s) not needed for DNA synthesis from correctly hybridized probes with the corresponding dideoxynucleotides (e.g. replace dGTP when synthesizing the CCCTAA element of the human telomere).

Multicolor PRINS and PRINS-FISH

1. Perform the PRINS procedure from above using, for instance, digoxigenin-11-dUTP (Roche) as the labeled nucleotide.

2. After incubation in stop buffer (step 4), **immediately** dehydrate the slide in a -20°C ethanol series (70%, 90% and 99%, 3 min each).
3. Remove the slide from the 99% ethanol, drain and air-dry.
4. Prepare and mix a new reaction mixture as before, but using a different type of labeled dNTP (in this case, e.g. fluorescein-dUTP (Roche)).
5. Annealing and chain elongation: The slide should NOT be denatured prior to the second PRINS, but is preheated to 55-65°C for 1 min. Then the likewise preheated reaction mixture is put on the slide and spread with a coverslip. Incubation is as above.
6. Place the slide in preheated stop buffer for 1 min at the temperature used in step 5.
7. Wash the slide for 3 min at room temperature in washing buffer.
8. The slide is now ready for antibody staining and counterstaining/mounting.
9. If a third PRINS reaction is desired, the wash (step 7) is omitted. The slide is instead immediately dehydrated in ethanol once more to repeat steps 2-8.

PRINS-FISH is done by adding the denatured FISH probe to the slide instead of the second PRINS probe (step 5). The usual FISH procedures can be employed except that the slide should not be (re)denatured after the PRINS, as this may cause the PRINS signal to disappear. Also, it is generally not necessary to do any proteinase treatments prior to PRINS-FISH. We use PRINS-FISH in PRINS-painting (Hindkjær et al. 1995) and PRINS-CGH formats to combine painting data with locus specific data (unpublished).

Visualization of digoxigenin-labeled DNA in situ

1. Apply 50 µl of fluorochrome conjugated anti-digoxigenin in blocking solution (2 ng/µl) to the slide. Incubate under a coverslip for 30-60 min in the dark.
2. Wash 2×5 min in 50 ml of washing buffer. The slide is now ready for counterstaining and mounting.

Counterstaining and mounting

a) Blue counterstaining of DNA is obtained by including 0.4 μM DAPI in the antifade solution.

b) Red counterstaining of DNA is obtained by including propidium iodide in the antifade solution. Mount the slide in 20 μl of antifade solution containing 0.5 μg/ml propidium iodide.

Signal quantification with the CGH program QUIPS from VYSIS

This program compares the intensity of two samples, a "test" and a "reference". We assign the telomere signals as "test", whereas the DAPI counterstain is chosen as "reference". The advantage of including the DAPI counterstain in the quantification is that some of the factors causing variation in in situ detection results would be expected to have a somewhat similar influence on the DAPI staining. The error possibly induced by such factors as uneven illumination, variable penetrability of the specimen to reagents, or locally increased background fluorescence could thus be reduced as they would influence both test and reference values, either increasing or decreasing both together, leaving the ratio less affected. The disadvantage of the approach is that non-homologous chromosome ends can not be compared directly, as the DAPI staining is not equal on all chromosomes.

By this CGH-like approach, signal intensities are transformed into ratio profiles. Since the significance of peak heights in individual profiles would be difficult to evaluate, we always analyze two types of cells together, using one as a standard against which the other can be compared. For this comparison, we consider it important that the two types of cells are spread together on the same slide, and thus are treated identically all the way from spreading to analysis. In experiments where we compare different cells within a single sample, this is automatically achieved. In experiments where we compare cells between samples, we mix male and female cells, so that the two types of cells to be compared can be easily identified through the presence or absence of a Y-chromosome. Apart from these adaptations, the program is used as specified by the supplier for CGH.

Acknowledgements. This work has been supported by the Danish Cancer Society, by the Meyer Foundation, and by Boehringer Mannheim (now Roche).

References

Hindkjær J, Brandt CA, Koch J, Lund TB, Køølvraa S, Bolund L (1995) Simultaneous detection of centromere specific probes and chromosome painting libraries by a combination of PRimed IN Situ labeling and chromosome painting (PRINS-painting). Chromosome Research 3:41-44

Koch J, Kølvraa S, Gregersen N, Bolund L (1989) Oligonucleotide-priming methods for the chromosome-specific labelling of alpha satellite DNA in situ. Chromosoma 98:259-265

Koch J, Hindkjær J, Kølvraa S, Bolund L (1995) Construction of a panel of chromosome specific oligonucleotide probes (PRINS-primers) useful for the identification of individual human chromosomes *in situ*. Cytogenet. Cell Genet. 71:142-147

Koch J (2000) PRINS: PRimed *In Situ* labeling and hybridization in one step. In Kessler C. (ed.): Nonradioactive labelling and detection of biomolecules (2nd edition, Springer Verlag, 407-416).

Krejci K, Koch J (1998) Improved detection and comparative sizing of human chromosomal telomeres *in situ*. Chromosoma 107:198-203

Krejci K, Koch J (1999) An *in situ* study of variant telomeric repeats. Genomics 58:202-206

Pellestor F, Girardet A, Andréo B, Charlieu JP (1994) A polymorfic alpha satellite sequence specific for human chromosome 13 detected by oligonucleotide primed *in situ* labelling (PRINS). Hum. Genet. 94:346-348

Pellestor F, Girardet A, Lefort G, Andréo B, Charlieu JP (1995) Rapid *in situ* detection of chromosome 21 by PRINS technique. Am. J. Med. Genet. 56:1-8

Pellestor F, Girardet A, Andréo B, Lefort G, Charlieu JP (1997) Incidence of chromosome 1 disomy in human sperm estimated by the primed *in situ* (PRINS) labeling technique. Cytogenet. Cell Genet. 76:192-195

Serakinci N, Koch J (1999) Detection and sizing of human telomeric repeat DNA *in situ*. Nature Biotechnology 17:200-201

Serakinci N, Krejci K, Koch J (1999) Telomeric repeat organization – a comparative *in situ* study between man and rodent. Cytogenet. Cell Genet. 86:204-211.

Therkelsen AJ, Nielsen A, Koch J, Hindkjær J, Kølvraa S (1995) Staining of Human telomeres with primed *in situ* labeling (PRINS). Cytogenet. Cell Genet. 68:115-118

Microarrays and DNA-Chips in the Analysis of Genetic Aberrations

UWE R. MÜLLER, Y. PAUL BAO, GUANGSHAN LI, XINYING XIE, and NALINI RAGHAVACHARI

Introduction

Fluorescence in situ hybridization (FISH) is a precise technique to determine copy numbers of a particular sequence in individual cells with a resolution better than 30 Kb, but requires a priori knowledge of the target sequence to be analyzed. Comparative genomic hybridization (CGH) on the other hand allows a global analysis of the whole genome, but at significant loss in resolution (on the order of 10-15 Mb). DNA chip technology (so called genosensor-based or matrix-based CGH) combines the high resolution of FISH with the inherent multiplex capability of CGH, and adds simplicity to the analysis. This technology represents the latest addition to chip based genetic analyses.

Indeed, three classes of DNA chips have been described so far which vary significantly in their structure, composition, and intended functionality. The feature in common is the small size of their target array, typically on the order of a few square centimeters (Note: we use the term target for the DNA on the chip to be consistent with the CGH and gCGH literature, but realize that an alternative nomenclature has emerged that refers to the DNA on the chip as the probe). Several different manufacturing technol-

✉ Uwe R. Müller, Corning Incorporated, Biochemical Technologies, SP-FR-01, Corning, NY, 14831, USA (*phone* 607-974-9994; *fax* 607-974-5957; *e-mail* MullerU@Corning.com)
Y. Paul Bao, Corning Incorporated, Biochemical Technologies, SP-FR-01, Corning, NY, 14831, USA
Guangshan Li, Corning Incorporated, Biochemical Technologies, SP-FR-01, Corning, NY, 14831, USA
Xinying Xie, Corning Incorporated, Biochemical Technologies, SP-FR-01, Corning, NY, 14831, USA
Nalini Raghavachari, Corning Incorporated, Biochemical Technologies, SP-FR-01, Corning, NY, 14831, USA

ogies have evolved to print 3 to 6 thousand individual target spots per cm^2, if each spot has a diameter of 80-100 μm and spots are separated by 50-80 μm from each other. Using microfabrication technologies pioneered by the semiconductor industry, spot sizes of less than 10 μm are quite feasible, providing for over a quarter of a million different target sites in the same area.

The intended use of the DNA chip determines its design with regard to target DNA size and composition. Since for short oligonucleotides the specificity of the hybridization reaction at a given stringency is mainly a function of the length of the DNA, the conditions can be chosen such that a single point mutation can be detected in the sequence to be analyzed (herein referred to as probe). If multiple target sequences are used simultaneously, cross-hybridizations are inevitable and a different strategy has to be employed. Based on this principle, chips can be designed with as many as 65,536 (all possible 8-mer sequences) or even more target spots on a single chip to unambiguously deduce the sequence of a specific DNA probe. This process, called sequencing by hybridization (SbH) (Yershov et al.1996; Southern, 1996; Drmanac et al.1998; Chee et al.1996; Drmanac et al.1989), generates very complex hybridization patterns and requires sophisticated computer software to interpret the image. In addition, the sequence to be analyzed must be fairly short and must be isolated and amplified from the rest of the genome through PCR (polymerase chain reaction) before it is applied to the chip for sequence analysis. Nevertheless, SbH is a very powerful technique that lends itself to any application where the exact nucleotide sequence has to be determined for an appropriate diagnosis or therapy. It is less useful, however, for diagnoses of gross chromosomal changes or alterations in the level of gene expression.

Judged by the number of publications and commercial efforts, the emphasis in the past few years has clearly been on the development of arrays for expression profiling. Both arrays of oligonucleotides as well as cDNAs have been employed for that purpose (Zhu et al. 1998; Schena 1996; Noordewier and Warren 2001; Rew 2001), though they differ with respect to a variety of funtional parameters. Expecting higher specificity and increased power to discriminate between very similar genes and splice variants, the trend is now shifting toward oligomer targets for expression applications (Hughes et al. 2001).

Because of the demand for higher sensitivity and accuracy the development of arrays for genomic analyses started with chips containing BAC, PAC, or P1 clones, each containing an insert of human genomic DNA (Pinkel et al. 1998; Solinas-Toldo et al. 1997). It is the size of these inserts, typically ranging from 80 Kb to 150 Kb, and their spacing along the human genome, that determines the resolution and sensitivity of this technique. Since the genetic complexity of a target spot is, at least theoretically, directly proportional to the number of probe nucleotides that it can bind (as long as the target DNA concentration is not rate limiting), increased complexity results in higher signal strength with a concomitant loss in resolution. However, absolute signal strength is not the only factor that determines assay sensitivity. In practice, a host of parameters such as probe complexity and concentration, label type and density, hybridization time and temperature, and a variety of factors related to detection mechanism and instrumentation are of equal importance (Müller 2001).

For diagnostically relevant CGH applications, the hybridization probe mixture typically contains on the order of 200 – 400 ng of total human DNA from each of the test and reference tissue. Thus, the total number of available probes for each target spot is relatively low, placing much higher demands on the sensitivity of this system than what is needed for standard FISH. The GenoSensor™ System (Vysis, Inc.) was the first commercial system to meet these high demands on sensitivity with the development of a unique chromium chip surface, small arrays of PAC or P1 clones, and a novel Large Field Multicolor Imaging System (Che et al. 2001). While the sensitivity of this system was shown to be sufficient for the reliable detection of monosomic and trisomic regions in clinical material (Bao et al. 1999; Lese et al. 1999), the low resolution of these large genomic clones still limits the use of this technology. More recently this limitation was overcome by the use of cDNA arrays for CGH applications (Pollack et al. 1999; Monni et al. 2001). We describe here a protocol for array CGH that provides for improved sensitivity, allowing the use of cDNA arrays with probe concentrations of less than 80 ng sample DNA per µl of hybridization solution.

Outline

As in CGH, the DNA from a sample tissue is compared to normal human DNA in a dual color hybridization that includes the following major steps:

- Extract DNA from a sample tissue
- Label sample DNA with a fluorescent dye (e.g., Cy3-dCTP) and reference DNA with a different color fluorescent dye (e.g., Cy5-dCTP) by primer extension with DNA polymerase

- Mix labeled sample and reference DNA with Cot-1 DNA to suppress repeat sequences
- Hybridize to DNA chip and remove unbound probes through washing
- Image chip to determine the fluorescence intensities in each color for each target spot
- Statistical analysis of color ratios and normalization to determine relative copy numbers

Procedure

Array fabrication

The arrays employed in this study were fabricated by a novel, high-throughput contact printing technology developed at Corning Incorporated that allows for the simultaneous deposition/printing of ›1000 elements within 6 seconds. Briefly, four grids were printed on each GAPS coated slide, whereby each slide contains 3392 human genes (including house-keeping genes and positive controls) and several *Bacillus subtilus* genes (negative controls).

Genomic DNA labeling

Since DNA extraction is a common practice, it will not be described. Genomic DNA was labeled by primer extension as follows. A 28 µl solution containing 4 µg of human genomic DNA and 4 µg of random hexamers was denatured for 5 min at 95 °C, briefly chilled on ice, and then combined with 4 µl of 10 x EcoPol buffer, 2 µl of 0.1 M DTT, 3 µl of a dNTP mixture (10 mM each dGTP, dATP, dTTP, and 1 mM of dCTP), 2 µl of Cy3- or Cy5-dCTP (1 mM, PerkinElmer, Boston, MA), and 1 µl of Klenow fragment (New England Biolabs, Inc., Beverly, MA). This resulting 40 µl reaction mixture was incubated at 37 °C for two hours. Probes were purified using a QIAquick PCR purification kit according to the manufacturer's instructions (Qiagen, Inc., Valencia, CA). The DNA concentration and the amount of Cy3/Cy5 incorporation was measured on the Agilent 8453E UV-Vis spectrometer. It is important to note that signal intensities will vary even for target spots that have similar test/reference ratios. This is most likely due to differences in hybridization kinetics due to variations in the size and sequence (i.e. secondary structure) of both the probe and target. We strongly advise to check the sample and reference DNAs for purity before labeling, and afterwards for efficiency of label incorporation.

Pre-hybridization

Arrays were prehybridized and washed as follows:

1. 2 x SSC/0.05% SDS/0.2%BSA at 42 °C for 20 min.
2. 1 x SSC at room temperature for 5 min, repeat once.
3. 0.2 x SSC for 5 min, repeat once.
4. After the final wash, slides were spin dried in a centrifuge at 2000 rpm for 1 min.

Hybridization

Combine in a 1.5 ml microcentrifuge tube:

Hybridization solution

- 20 μl 100% formamide
- 5 μl 20 x SSC
- 1 μl 10 μg/μl poly A
- 2 μl 10 μg/μl Human Cot-1 DNA
- 1 μl 5% SDS
- 2-5 μl Cy3 test DNA (4 μg)
- 2-5 μl Cy5 reference DNA (4 μg)

add DI water to a final volume of 50 μl

1. Vortex the hybridization solution, then microcentrifuge at 12,000 rpm for 5 s.
2. Incubate in a 95 °C water bath for 5 min in order to denature all DNAs.
3. Microcentrifuge at 12,000 rpm for 1 min at room temperature.
4. Add 5 μl of 2% BSA to the hybridization solution and mix well.

5. Apply entire hybridization mixture (final volume 50 µl) onto the array and then spread over the entire surface using a 24 mm x 50 mm cover slip.
6. Incubate chip(s) at 42 °C in the dark for overnight (15-18 h).

Post-hybridization processing

7. Wash chip(s) in a series of 8 Coplin jars as follows:
   - 1 wash in 2 x SSC/0.05% SDS at 42 °C for 1 min to remove the cover slip
   - 2 washes in 2 x SSC/0.05% SDS at 42 °C for 5 min
   - 2 washes in 1 x SSC for 5 min each at RT
   - 3 washes in 0.2 x SSC for 2 min each at RT

8. Remove chips from last wash, and dry by centrifugation at 2000 rpm for 1 min.

   Note: Adequate washing is vital to assay performance. When transferring chips from one bath to the next agitate the biochip with an up and down and back and forth motion to ensure proper washing.

   Note: Do not allow chips to dry between washes!

   Note: Any fingerprints, dust or lint on the chip will affect the image. Thus, it is critical to remove any dust with a clean source of compressed air.

Image acquisition and analysis

The GenePix 4000A (or 4000B) Array Scanner was used to obtain the Cy3/Cy5 fluorescence images using a PMT setting of 750-950 volts. All images were analyzed using the GenePix Pro 3.0 analysis software (Axon Instruments, Inc., Foster City, CA). For determination of copy number changes we make the assumption that most genes are represented by 2 copies. The background corrected test/reference fluorescent intensity ratios are then devided by the median ratio for all spots to obtain the normalized ratio of 1. We typically use self/self hybridizations to test the performance of our assay system with model arrays and find that the normalized ratios have a CV of approximately 28%.

Results

The Corning cancer array contains approximately 1700 unique cDNAs in 2 replicates and is designed for measuring gene expression profiles and genomic amplifications in tumor DNA. We have tested the performance of these arrays in CGH applications using two well-studied tumor cell lines. Our hybridizations were carried out with a mixture of total human reference DNA (labeled with Cy3) and DNA extracted from the tumor cell line (labeled with Cy5).

The cell line Colo320 (ATCC No. CCL-220.1) has an average of 52 chromosomes per cell and was chosen because of the well known amplification of the cMYC oncogene (Bolzan et al. 2000; Che et al. 2001; Pollack et al. 1999), which can be readily seen in Figure 1. The background corrected normalized test/reference fluorescent intensity ratio for the cMYC target reveals a 32 fold amplification, in good agreement with the literature (23-35 fold). In addition the data in Figure 2 suggest previously unreported copy number increases for genes ABCD1 (7 fold), MCL1 (12 fold), and POU5F1 (6 fold) located on chromosomes Xq28, 1q21 and 6q21, respectively.

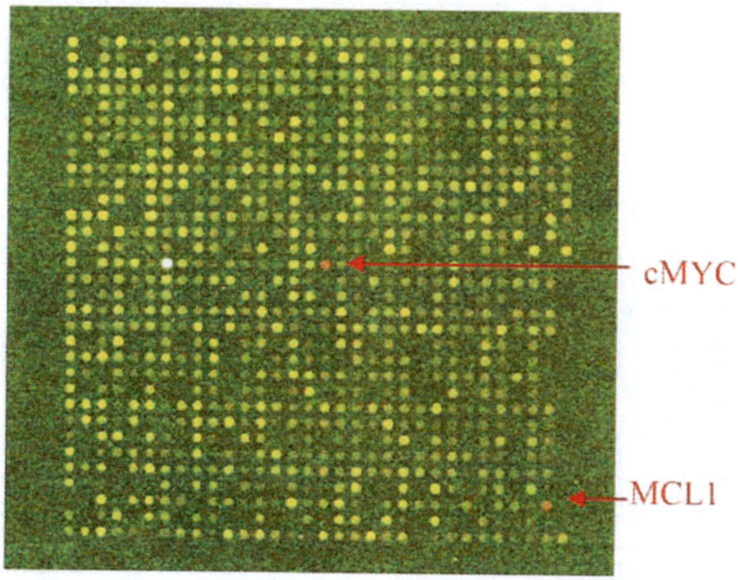

Fig. 1. Image obtained after hybridization of a Corning cancer array with a mixture of normal human reference DNA (labeled with green dye) and tumor cell Colo320 DNA (labeled with red dye). Only one of the 4 subgrids is shown.

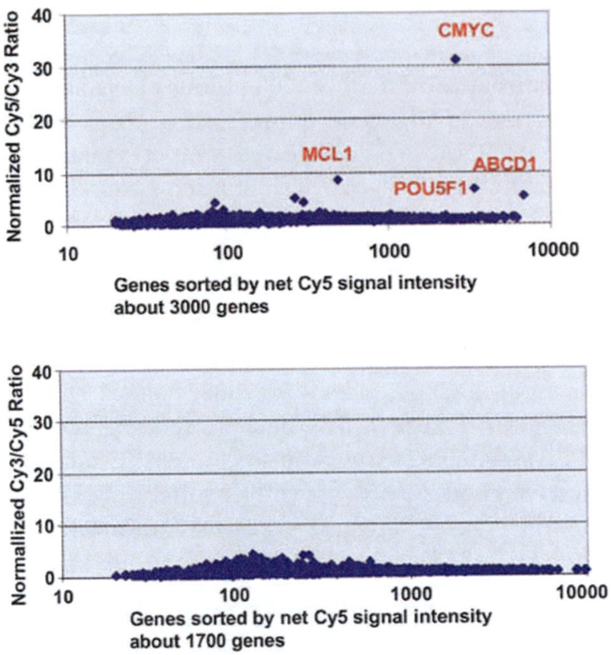

Fig. 2. Gene copy number analysis of tumor cell line Colo320 with cDNA arrays. The top panel shows all ~1700 genes on the cancer cDNA array sorted by CY5 signal intensity (reference DNA). The amplified onco-genes cMYC, ABCD1, MCL1, and POU5F1 are marked. The bottom panel shows a self/self hybridization with normal DNA for comparison.

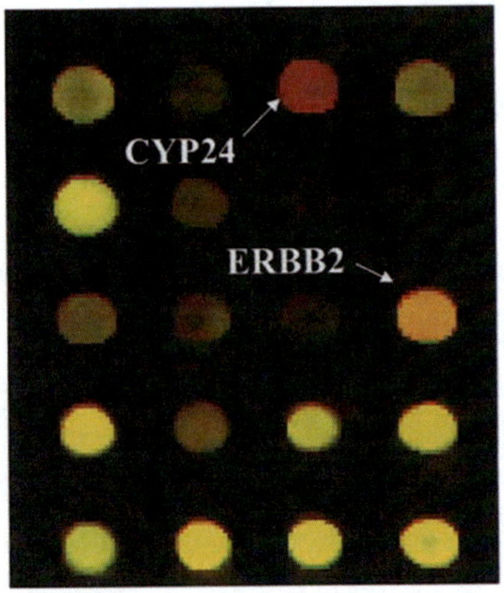

Fig. 3. Detection of DNA copy-number changes in tumor cell line BT474 with Corning cDNA arrays. Hybridization was carried out with a mixture of normal human reference DNA (labeled with green dye) and BT474 DNA (labeled with red dye). Only a portion of the array is shown.

Analysis of DNA extracted from the breast cancer cell line BT474 (ATCC No. HTB-20) with our cDNA arrays shows the known amplification of ERBB2 (6 fold) (Pinkel et al. 1998, Pollack et al. 1999) in Figure 3. Our finding of a 6 fold amplification for the gene CYP24 confirms a recent report (Albertson et al. 2000) that identifies CYP24 gene as a candidate oncogene located on an approximately 2-Mb amplified (8-12X) region on chromosome 20q13.2. Note that this cell line has an average number of 103 chromosomes/cell, meaning that because of our normalization algorythm copy numbers for "amplified" genes are likely to be underestimated.

Troubleshooting

As for any in situ hybridization protocol, failure of the assay translates into poor or no signal on all or part of the target spots. One of the major problem areas is the extraction and labeling of sample DNA, which can not be reviewed here. We have summarized below the major types of hybridization related problems and various potential causes.

Table 1.

Hybridization Related Problem	Potential Causes
No or low hybridization signal in either color	- no probe was added - hybridization temperature too high - washing temperature too high - SSC concentration too low - probe insufficiently denatured - probe mixture degraded
Green signal is unusually high; ratios appear incorrect	- auto-fluorescence in green chanel
Green and/or red signal is unusually high; ratios appear incorrect	- insufficient blocking of repeat sequences
High background due to non-specific binding of probe DNA	- incomplete washing - chip dried during hybridization or washing process
Hybridization signal in either color is unusually low for some target spots	- air bubbles in the hybridization reaction
Irregular and strongly fluorescent spots	- dust or other debris on the chip surface

Applications

Rapid and cost efficient testing for gene copy number changes in pathological tissues is likely to result in a considerable increase in our understanding of genetic disease. The ideal test would allow whole genome scanning at gene level resolution or better with a single chip and a minimum of sample. The sensitivity of array based CGH is already sufficient for detection of low-level amplifications and aneuploidies if the tumor tissue is not too diluted with normal cells and if most of the cells carry the mutation. In very homogeneous tissues, such as cell lines, certain leukemias, or tissues from patients with hereditary disease the detection of single copy deletions or additions has already been shown to be feasible with arrays of large genomic clones (Pinkel et al. 1998; Bao et al. 1999; Lese et al. 1999). Further improvements in sensitivity and robustness of this technology will soon make high resolution genome-wide scanning possible for cytogenetic applications with the use of cDNA arrays that cover the whole genome.

An important consideration is the amount of sample material required for this analysis, since tumor specimens are getting smaller as diagnoses are being made earlier. Especially for clinical applicability, this technology should work with samples obtained by biopsy. Until a robust and reliable total genome PCR amplification method becomes available, that does not distort the sequence ratios, this technique has to rely on non-amplified sample material. To that end we have achieved satisfactory results with as little as 3 μg of tumor DNA in overnight hybridizations without the aid of signal amplification or hybridization systems that are designed to overcome the barriers of diffusion. Preliminary results in our labs suggest that significantly greater sensitivity is possible. More demanding applications, such as analysis of individual clones in tumor tissue or the analysis of single fetal cells obtained from maternal blood, will clearly demand a reliable amplification technology for the sample DNA. Promising reports by other laboratories (Beltinger et al. 1997; Wells et al. 1999) suggest that such applications may not be very far off.

Acknowledgements. The cancer arrays used in our work were generated by the dedicated work of many of our colleagues in the pilot line and development teams at Corning Incorporated.

References

Albertson, D. G., Ylstra, B., Segraves, R., Collins, C., Dairkee, S. H., Kowbel, D., Kuo, W.-L., Gray, J. W., and Pinkel, D. (2000) Quantitative mapping of amplicon structure by array CGH identifies CYP24 as a candidate oncogene. Nature Genetics 25, 144-146

Bao, P., Che, D., Lermer, N., Shi, J., Prokhorova , A., Müller , U. (1999) Microarray based detection of single copy sequence changes in genomic DNA. American Journal of Human Genetics. 65, A157

Beltinger, C. P., Klimek, F., and Debatin, K. M.(1997) Whole genome amplification of single cells from clinical peripheral blood smears. Mol. Pathol. 50, 272-275

Bolzan, A. D. , Paez, G. L., Bianchi, M. S., and Bianchi, N. O. (2000) Analysis of telomeric repeats and telomerase activity in human colon carcinoma cells with gene amplification. Cancer Genet. Cytogenet. 15, 166-170

Che, D., Bao, Y. and Müller, U.R. (2001) A novel surface and multicolor charge coupled device-based fluorescent imaging system for DNA microarrays. J. Biomedical Optics 6, in press

Chee, M., Yang, R., Hubbell, E., Berno, A., Huang, X. C., Stern, D., Winkler, J., Lockhart, D. J., Morris, M. S., and Fodor, S. P. A. (1996) Accessing genetic information with high-density DNA arrays. Science 274, 610-614

Drmanac, R., Labat, I., Brukner, I., and Crkvenjakov, R. (1989) Sequencing of megabase plus DNA by hybridization: Theory of the Method. Genomics 4, 114-128

Drmanac, S., Kita, D., Labat, I., Hauser, B., Schmidt, C., Burczak, J. D., and Drmanac, R. (1998) Accurate sequencing by hybridization for DNA diagnostics and individual genomics. Nat. Biotechnol. 16, 54-58

Hughes, T.R., Mao, M. Jones A.R. et al. (2001) Expression profiling using microarrays fabricated by an ink-jet oligonucleotide synthesizer. Nature Biotechnol. 19, 342-347

Lese, C.M., Zhang ,X., Pinkel, D., Bao, P., Lermer, N., Che, D., Shi, J., Müller, U., Ledbetter, D.H. (1999) Comparative genomic hybridization arrays: Towards a "telomere chip". American Journal of Human Genetics. 65, A9

Monni, O., Bärlund, M., Mousses, S., Kononen, J., Sauter, G., Heiskanen, M., Paavola, P., Avela, K., Chen, Y., Bittner, M. L., Kallioniemi, A. (2001) Comprehensive copy number and gene expression profiling of the 17q23 amplicon in human breast cancer. PNAS 98, 5711-5716

Müller, U.R. (2001) Microarrays for quantitative analysis of sequence copy numbers. Clin. Applied Immunol. Rev. 1, 255-265

Noordewier, MO and Warren PV. (2001) Gene expression microarrays and the integration of biological knowledge. Trends Biotechnol. 19, 412-415

Pinkel, D., Seagraves, R., Sudar, D., Clark, S., Poole, I., Kowbel, D., Collins, C., Kuo, W. L., Chen, C., Zhai, Y., Dairkee, S. H., Ljung, B. M., Gray, J. W., and Albertson, D. G. (1998) High resolution analysis of DNA copy number variation using comparative genomic hybridization to microarrays. Nature Genet. 20, 207-211

Pollack, J. R., Perou, C. M., Alizadeh, A. A., Eisen, M. B., Pergamenschikov, A., Williams, C. F., Jeffrey, S. S., Bostein, D., Brown, P. O. (1999) Genome-wide analysis of DNA copy-number changes using cDNA microarrays. Nature Genetics. 23, 41-46

Rew, DA. (2001) DNA microarray technology in cancer research. Eur. J. Surg. Oncol. 27, 504-508

Schena, M. (1996) Genome analysis with gene expression microarrays. BioEssays 18, 427-431

Shalon, D., Smith, S. J., and Brown, P. O. (1996) A DNA microarray system for analyzing complex DNA samples using two-color fluorescent probe hybridization. Genome Res. 6, 639-645

Solinas-Toldo, S., Lampel, S., Stilgenbauer, S., Nickolenko, J., Benner, A., Döhner, H., Cremer, T., and Lichter, P. (1997) Matrix-Based Comparative Genomic Hybridization: Biochips to Screen for Genomic Imbalances. Genes Chromosomes Cancer 20, 399-407

Southern, E. M. (1996) DNA chips: Analysing sequence by hybridization to oligonucleotides on a large scale. Trends Genet. 12, 110-115

Wells, D., Sherlock, J. K., Handyside, A. H,, and Delhanty, J. D. A. (1999) Detailed chromosomal and molecular genetic analysis of single cells by whole genome amplification and comparative genomic hybridisation. Nucleic Acids Res. 27, 1214-1218

Yershov, G., Barsky, V., Belgovskiy, A., Kirillov, E., Kreindlin, E., Ivanov, I., Parinov, S., Guschin, D., Drobishev, A., Dubiley, S., and Mirzabekov, A. (1996) DNA analysis and diagnostics on oligonucleotide microchips. Proc. Natl. Acad. Sci. USA 93, 4913-4918

Zhu, H., Cong, J. P., Mamtora, G., Gingeras, T., and Shenk, T. (1998) Cellular gene expression altered by human cytomegalovirus: Global monitoring with oligonucleotide arrays. Proc. Natl. Acad. Sci. USA 95, 14470-14475

Suppliers

All arrays were printed on Corning GAPS™ slides which can be obtained at: http://www.corning.com/lifesciences/ Telephone (for US): 800-492-1110 / 978-635-2200 Fax: 001 - 978-635-2476 Email: actoncs@acton.corning.com

Abbreviations

ABCD1:	ATP-binding cassette, sub-family D (ALD), member 1
BAC:	bacterial artificial chromosome
CMYC:	v-myc avian myelocytomatosis viral oncogene homolog
CV:	standard deviation devided by average
CYP24:	cytochrome P450, subfamily XXIV (vitamin D 24-hydroxylase)
DTT:	Dithiothreitol
ERBB2	v-erb-b2 avian erythroblastic leukemia viral oncogene homolog 2 (neuro/glioblastoma derived oncogene homolog)
GAPS™	gamma amino propyl silane
MCL1	myeloid cell leukemia sequence 1 (BCL2-related)
PAC	phage artificial chromosome
P1	phage 1
PMT	photo multiplier tube
SSC	sodium chloride, sodium citrate
RT	room temperature

Internet sites Related to Cytogenetics and Molecular Cytogentics

Thomas Liehr

Probes for FISH and/or information about DNA-probes

Resources for molecular cytogenetics: *http://www.biologia.uniba.it/rmc/*

Deutsches Ressourcenzentrum: *http://www.rzpd.de/*

CEPH YAC library: *http://www.cephb.fr/bio/ceph_yac.html*

A new map of human genome: *http://www.ncbi.nlm.nih.gov/genmap98/*

IMAGE "world's largest public collection of genes": *http://image.llnl.gov/*

YACs mappes by FISH at the Molecular Cytogenetics Group, (Max Plank Inst., Berlin)
http://www.mpimg-berlin-dahlem.mpg.de/~cytogen/probedat.htm

FISH-pages

The multicolor-FISH database: *http://mti-n.mti.uni-jena.de/~huwww/MOL_ZYTO/mFISHlit.htm*

The Bio-Rad Fluorescence Database: *http://fluorescence.bio-rad.com/*

Tavi's FISH page: *http://info.med.yale.edu/genetics/ward/tavi/FISH.html*

CGH database (created 1997): *http://www.nhgri.nih.gov/DIR/LCG/CGH/*

Molecular Cytogenetics of Solid Tumors: *http://web.tiscali.it/costb19/*

Resources for molecular cytogenetics: *http://www.biologia.uniba.it/rmc/*

Cytogenetics

Atlas of Genetics and Cytogenetics in Oncology and Haematology:
http://www.infobiogen.fr/services/chromcancer

Breakpoint map of recurrent chromosome aberrations:
http://www.ncbi.nlm.nih.gov/CCAP/mitelsum.cgi

Geneatlas: *http://bisance.citi2.fr/GENATLAS/menu_an.html*

Collection of web sites for chromosome research:
http://infofarm.cc.affrc.go.jp/~shignak/chromosome.net/index2.html

Cytogenetic gallery: *http://www.pathology.washington.edu/Cytogallery/*

Cytogenetic association

European cytogenetic association (ECA): *http://www.biologia.uniba.it/eca/*

FISH-Courses

EGF courses: *http://www.eurogene.org*

Medical Literature on the Internet

PubMed including OMIM: *http://www.ncbi.nlm.nih.gov/PubMed/*

BioMedNet: *http://www.bmn.com/*

Genetics and Biology

hum-molgen: news in human genetics and molecular biology:
http://www.hum-molgen.de/NewsGen/index.php3

Resource for Cell and Molecular Biologists: *http://www.cellbio.com/*

Biology links: *http://mch.harvard.edu/BioLinks.html*

Image Library of Biological Macromolecules:
http://www.imb-jena.de/IMAGE.html

Subject Index

D

FSC
www.fsc.org
MIX
Papier | Fördert gute Waldnutzung
FSC® C083411

Kolibri 360 GmbH
Ferdinand-Jühlke-Straße 7
99095 Erfurt, Deutschland
produktsicherheit@kolibri360.de